Interfacial Photoprocesses:
Energy Conversion and Synthesis

Interfacial Photoprocesses: Energy Conversion and Synthesis

Mark S. Wrighton, EDITOR

Massachusetts Institute of Technology

Based on a symposium sponsored by the Division of Colloid and Surface Chemistry at the 176th Meeting of the American Chemical Society, Miami Beach, Florida, September 11–13, 1978.

ADVANCES IN CHEMISTRY SERIES **184**

AMERICAN CHEMICAL SOCIETY

WASHINGTON, D.C. 1980

Randall Library UNC-W

Library of Congress CIP Data

Interfacial photoprocesses.
(Advances in chemistry series; 184 ISSN 0065–2393)

Includes bibliographies and index.

1. Photochemistry—Congresses.
I. Wrighton, Mark S., 1949-   . II. American Chemical Society. Division of Colloid and Surface Chemistry. III. Series.

QD1.A355 no. 184 [QD701]   540'.8s   [541'.35]
                                          79–26245
ISBN 0-8412-0474-8   ADCSAJ 184 1-315 1980

Copyright © 1980

American Chemical Society

All Rights Reserved. The appearance of the code at the bottom of the first page of each article in this volume indicates the copyright owner's consent that reprographic copies of the article may be made for personal or internal use or for the personal or internal use of specific clients. This consent is given on the condition, however, that the copier pay the stated per copy fee through the Copyright Clearance Center, Inc. for copying beyond that permitted by Sections 107 or 108 of the U.S. Copyright Law. This consent does not extend to copying or transmission by any means—graphic or electronic—for any other purpose, such as for general distribution, for advertising or promotional purposes, for creating new collective works, for resale, or for information storage and retrieval systems.

The citation of trade names and/or names of manufacturers in this publication is not to be construed as an endorsement or as approval by ACS of the commercial products or services referenced herein; nor should the mere reference herein to any drawing, specification, chemical process, or other data be regarded as a license or as a conveyance of any right or permission, to the holder, reader, or any other person or corporation, to manufacture, reproduce, use, or sell any patented invention or copyrighted work that may in any way be related thereto.

PRINTED IN THE UNITED STATES OF AMERICA

Advances in Chemistry Series

M. Joan Comstock, *Series Editor*

Advisory Board

David L. Allara

Kenneth B. Bischoff

Donald G. Crosby

Donald D. Dollberg

Robert E. Feeney

Jack Halpern

Brian M. Harney

Robert A. Hofstader

W. Jeffrey Howe

James D. Idol, Jr.

James P. Lodge

Leon Petrakis

F. Sherwood Rowland

Alan C. Sartorelli

Raymond B. Seymour

Gunter Zweig

FOREWORD

ADVANCES IN CHEMISTRY SERIES was founded in 1949 by the American Chemical Society as an outlet for symposia and collections of data in special areas of topical interest that could not be accommodated in the Society's journals. It provides a medium for symposia that would otherwise be fragmented, their papers distributed among several journals or not published at all. Papers are reviewed critically according to ACS editorial standards and receive the careful attention and processing characteristic of ACS publications. Volumes in the ADVANCES IN CHEMISTRY SERIES maintain the integrity of the symposia on which they are based; however, verbatim reproductions of previously published papers are not accepted. Papers may include reports of research as well as reviews since symposia may embrace both types of presentation.

CONTENTS

Preface ix

1. Evaluation of Heterogeneous Photosensitizers for Use in
   Energy Storage 1
   R. R. Hautala and J. L. Little

2. Photoactivation of Polymer-Anchored Catalysts: Iron Carbonyl
   Catalyzed Reactions of Alkenes 13
   R. D. Sanner, R. G. Austin, M. S. Wrighton, W. D. Honnick,
   and C. U. Pittman, Jr.

3. Photoeffects on Reactions over Transition Metals 27
   B. E. Koel, J. M. White, J. L. Erskine, and P. R. Antoniewicz

4. Photochemistry of Surfactant and Hydrophobic Compounds in
   Organizates and at Interfaces: Effects of Environment and Solute
   Penetration on Excited State Quenching and Reactivity 47
   D. G. Whitten, J. A. Mercer–Smith, R. H. Schmehl,
   and P. R. Worsham

5. Characterization and Chemical Reactions of Surfactant
   Monolayer Films 69
   S. J. Valenty

6. Effect of the Condensed Phase, Homogeneous Solution, and
   Micellar Systems on Photoionization 97
   J. K. Thomas and P. Piciulo

7. Interfacial Photoprocesses in Molecular Thin Film Systems 113
   L. R. Faulkner, H. Tachikawa, F.-R. Fan, and S. G. Fischer

8. Photoelectrochemistry and Spectroscopy of Metal Phthalocyanine
   Films on a Transparent Semiconducting Electrode 139
   C. D. Langford, B. R. Hollebone, and T. Vandernoot

9. Charge Transfer at Illuminated Semiconductor–Electrolyte
   Interfaces 155
   A. J. Nozik, D. S. Boudreaux, R. R. Chance, and F. Williams

10. Short-Lived Radicals at Photoactive Surfaces: Spin Trapping
    and Mechanistic Consequences 173
    M. L. Hair and J. R. Harbour

11. Luminescent Properties of Semiconductor Photoelectrodes 185
    A. B. Ellis and B. R. Karas

12. Photoelectrochemical Solar Cells: Chemistry of the Semiconductor–
    Liquid Junction 215
    A. Heller and B. Miller

13. Photochemical Processes at the Solid–Gas Interface—The Adsorption and Reactions of Gaseous CO_2 and H_2O on Pt–$SrTiO_3$ Single-Crystal Sandwiches 233
    J. C. Hemminger, R. Carr, W. J. Lo, and G. A. Somorjai

14. Titanium Dioxide and Platinum/Platinum Oxide Chemically Modified Electrodes with Tailormade Surface States 253
    H. O. Finklea, H. Abruna, and R. W. Murray

15. Chemically Derivatized Semiconductor Photoelectrodes: A Technique for the Stabilization of n-Type Semiconductors 269
    M. S. Wrighton, A. B. Bocarsly, J. M. Bolts, M. G. Bradley, A. B. Fischer, N. S. Lewis, M. C. Palazzotto, and E. G. Walton

Index 295

PREFACE

Interfaces—solid–liquid, liquid–gas, and solid–solid—are key to important practical devices. Electronic materials, batteries, heterogeneous catalysts, and solar cells, to name a few examples, all depend on interfaces. It is therefore not too surprising that there is interest in developing an understanding of interfaces. This volume comprises the proceedings of a symposium which I organized for the Colloid and Surface Chemistry Division of the American Chemical Society. The symposium was held in Miami, Florida at the Fall, 1978 National Meeting of the American Chemical Society. From the title, this symposium was limited to interfacial photoprocesses. The aim was to bring together a diverse group of chemists interested in exploiting light-induced reactions occurring at interfaces. Accordingly, photochemists, surface scientists, organic, and inorganic chemists were invited and contributed to the program.

Interfacial systems, specifically solid semiconductor–liquid junctions, provide the best man-contrived chemical systems for solar energy conversion. This area is amply covered by the contributions from A. Heller and A. J. Nozik. In chapters by R. W. Murray and M. S. Wrighton syntheses of photosensitive interfaces by chemical functionalization are described offering new approaches to energy conversion materials important in electrochemistry. Preparing new bulk photosensitive materials from molecular solids is the approach described by L. R. Faulkner and C. H. Langford, III in their papers on photoinduced charge transfer employing thin molecular film electrodes. A. B. Ellis describes an in situ technique for characterizing near surface properties of semiconductor photoelectrodes used in liquid junction solar cells. Production and characterization of radicals by interfacial photochemistry is the subject of M. L. Hair's contribution. These contributions reflect a vigorous effort toward fundamental understanding of semiconductor–liquid junctions and the improvement of energy conversion devices based on them.

Solid–gas interfaces lend themselves to comprehensive and detailed studies aimed at elucidating just what surface reactions can be affected by light. G. A. Somorjai summarizes his recent studies of photoeffects at semiconductor–gas interfaces, bringing to bear the complete arsenal of surface sensitive measuring techniques to characterize important photosensitive surfaces. J. M. White's article provides the most definitive work to date on photoeffects of bulk metal exposed to carbon monoxide.

Unusual organic media—micelles, monolayers, and polymer-anchored catalysts and sensitizers—completes the coverage of the symposium. It is fair to say that photochemistry in such media has been one of the most exciting areas of research in recent years. D. G. Whitten, who can be credited with providing much of the stimulus for this area, describes his research directed towards manipulating interfacial charge transfer processes. Chapters by S. J. Valenty and J. K. Thomas also concern this important area. R. R. Hautala and C. U. Pittman, Jr. describe their work on polymer-anchored triplet sensitizers and photocatalysts, respectively. Photochemistry in unusual media already has wide application in photography. In the years ahead, energy conversion and chemical synthesis may become developed.

I wish to thank the patient contributors to this volume for devoting some of their most important work to this effort.

Massachusetts Institute of Technology   MARK S. WRIGHTON
Cambridge, Massachusetts 02139
July, 1979

Evaluation of Heterogeneous Photosensitizers for Use in Energy Storage

RICHARD R. HAUTALA and JAMES L. LITTLE

Department of Chemistry, University of Georgia, Athens, GA 30602

Two solid matrices, polystyrene and silica, have been functionalized with the dimethylaminobenzophenone chromophore, and the resulting immobilized photosensitizers have been quantitatively evaluated for their potential use in the norbornadiene–quadricyclene energy storage reaction. The polystyrene derivative exhibits a high limiting quantum efficiency (78%) for this sensitization process (ostensibly endothermic energy transfer), whereas the silica functionalized sensitizer is significantly less effective (24%). Homogeneous models for both systems are totally efficient (100%). The lower effectiveness of the silica derivative is attributed to a decrease in triplet yield and a low triplet energy, caused by the highly polar silica surface. Absorption spectra, emission spectra, and photostationary state isomer ratios for stilbene isomerization support this interpretation.

Interest in functionalized insoluble matrices for use as photosensitizers has steadily increased in recent years, paralleling a similar interest in immobilized reagents for chemical synthesis (1). The most obvious benefit of using insoluble sensitizers for solution photolyses is the ease with which the photoproduct can be isolated from the sensitizer. However, there are numerous other potential advantages. Sensitizers that tend to self-quench in solution could, in principle, be isolated on a matrix. The adverse influence of certain solvents on sensitizer properties could be minimized by using a matrix where the formation of a complete solvent cage would be impaired. An adsorptive affinity of the heterogeneous interface for specific substrates could result in more efficient and selective photochemistry. Other practical advantages pertinent to a solar energy storage system have been recently detailed (2).

In 1964, Moser and Cassidy (3) reported the first example of a polymeric photosensitizer used in solution. Soluble polyacrylophenone was found to sensitize the isomerization of 1,3-pentadiene in benzene. Leermakers and James (4) used the polymer in a solvent, isopentane, in which the polymer is totally insoluble. They successfully demonstrated the sensitized photochemistry of 1,3-pentadiene, norbornadiene, and myrcene with the system. A significant advance in this area came from the work of Blossey, Neckers, Thayer, and Schaap (5, 6). They covalently grafted rose bengal onto a preformed polystyrene bead and showed that this insoluble polymer was an effective sensitizer of singlet oxygen in solution. Blossey and Neckers (7) also grafted p-benzoylbenzoic acid onto a Merrifield polymer (chloromethylated styrene–divinylbenzene copolymer) and demonstrated the sensitized photodimerization of coumarin and indene and the sensitized cycloaddition of tetrachloroethylene to cyclopentadiene. Unfortunately this polymeric analog of benzophenone readily photodegrades, ostensibly through hydrogen abstraction from the polymer backbone. A more photostable fluorinated version of this polymer has recently been reported by Neckers (1).

At the outset of our work, no quantitative characterization of an immobilized photosensitizer has been made. Because of our interest in the application of immobilized sensitizers in an energy storage reaction (2), the factors of quantum efficiency and long-range stability were more crucial than in synthetic applications. This chapter is part of a comprehensive study to evaluate the immobilized dimethylaminobenzophenone chromophore as a sensitizer for the conversion of norbornadiene to quadricyclene, a potentially attractive system for the chemical storage of solar energy (8).

The Norbornadiene–Quadricyclene Interconversion

The features of the norbornadiene (**N**) to quadricyclene (**Q**) inter-

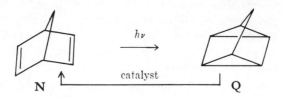

conversion that make it attractive as a potential system for the storage of solar energy have been described previously (2, 8, 9). The necessity of using a photosensitizer for this reaction derives from the complete lack of overlap between the absorption spectrum of **N** and the solar radiance spectrum. Numerous sensitizers effect this transformation, some of which absorb in or near the visible ($\lambda > 400$ nm) spectral region (2, 8–11). All effective sensitizers appear to operate by energy transfer to populate the triplet state of **N**, which then is transformed to **Q** (8, 10, 11). The triplet energy of **N**, about 70 kcal mol^{-1} (8, 11, 12), sets a limit, fortunately a very flexible one, on the longest wavelength of light that can be used in effecting the photochemical reaction. This wavelength limit is determined by the onset of the singlet absorption band of the sensitizer. For those sensitizers with triplet energies equal to or exceeding that of **N**, the wavelength limit would be near 380 nm (equivalent to ca. 75 kcal mol^{-1} with the excess energy of approximately 5 kcal · mol^{-1} necessary for the energy loss during relaxation of the sensitizer from the singlet to the triplet state). As indicated, this limit is not a rigorous one, and several very efficient sensitizers with triplet energies well below that of **N** are known. The energy deficiency is apparently made up through thermal activation or "thermal upconversion." A more detailed discussion of endothermic energy transfer, along with strong experimental evidence for it in analogous systems, has recently been made by Jones (8).

Dimethylaminobenzophenone as a Sensitizer

Several features make 4-(N,N-dimethylamino)benzophenone an attractive sensitizer for study with the norbornadiene system. Under certain conditions the quantum efficiency for sensitization of **Q** from **N** approaches 100%. The intense absorption band maximizing between 300 and 400 nm is sufficiently broad to tail into the visible spectral region. The high molar extinction coefficient ($\sim 25,000$) of this band allows for the use of very low sensitizer concentrations. The photophysical, photochemical, and sensitizing properties of this compound are very sensitive to medium polarity. For example, the absorption maximum varies from 332 nm in cyclohexane to 355 nm in ethanol. The intersystems crossing yield (triplet yield) falls from 100% in hexane to 6% in acetonitrile, paralleling similar behavior of the parent compound 4-aminobenzophenone (13). The triplet energy of 4-(N,N-dimethylamino)benzophenone ($E_T \leq 66$ kcal mol^{-1}) is also medium dependent, but in any case, triplet energy transfer to **N** is clearly endothermic. Finally, relatively straightforward synthetic techniques are available to allow grafting of the chromophore onto solid matrices for immobilization. The exhibited sensitivity to medium effects and the high extinction coefficient (allowing for relatively low loading) make this chromophore a particularly interesting probe for studying heterogeneous photosensitization. Structures and the corresponding acronyms for the sensitizers employed in this study are shown in Figure 1. Simple derivatives of dimethylaminobenzophenone behave similarly but not identically. Thus homogeneous models more

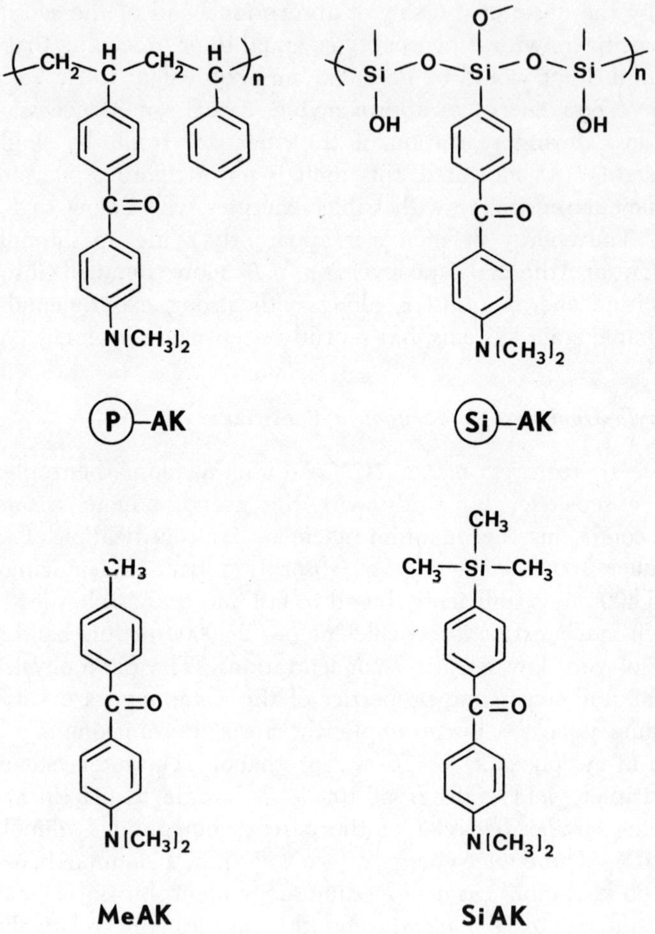

Figure 1. Structures of the sensitizers used in this study

closely resembling the immobilized sensitizers were prepared and are included for comparison. Preparation of the polystyrene-grafted sensitizer and general experimental procedures have been reported previously (2). Preparation of the silica-functionalized sensitizers will be reported elsewhere, however a reaction scheme is indicated below.

Scheme for Preparation of (Si)-AK

Quantum Yield Studies

The relationship between quantum efficiency for **Q** formation (ϕ) and the concentration of **N** is given by Equation 1 (2). Thus a reciprocal plot of quantum efficiency vs. **N** concentration should yield a straight line

$$\phi = \phi_T \phi_P \left(\frac{k_2[\mathbf{N}]}{k_2[\mathbf{N}] + k_1} \right) \quad (1)$$

where: k_2 = the effective rate constant for sensitization
k_1 = the composite rate constant for deactivation of the sensitizer (including a bimolecular term for quenching by oxygen)
ϕ_T = the triplet yield of the sensitizer
ϕ_P = the fraction of **N** triplets which give rise to **Q**

with an intercept of $1/\phi_T\phi_P$ and a slope of $k_1/k_2\phi_T\phi_P$. These parameters are quite useful in characterizing the performance of a sensitizer. The values for the slopes are dependent on the purity of the medium (due to quenching by impurities) and should not be regarded as values intrinsic to the sensitizer. Consequently, this discussion will be limited to comparisons of the intercepts or limiting quantum yields.

Data obtained for the polystyrene sensitizer and corresponding model are presented in Figure 2. Clearly the functionalized polymer, ⓟ-AK, is an effective and highly efficient sensitizer for this conversion. The small inherent inefficiency (limiting quantum yield of 78%) is likely due to light reflection from the polymer surface rather than to values of less than one for ϕ_T or ϕ_P. Simple visual inspection of the polymeric beads bathed in benzene reveals some discernible light reflection from the surface. In dramatic contrast to the monomeric model, ⓟ-AK is virtually unaffected by a change to polar solvents. For example, the quantum yield for a solution of **N** in acetonitrile (1M in **N**) using ⓟ-AK was found to be 0.4, significantly higher than the limiting quantum yields obtained in acetonitrile for any of the monomeric derivatives.

Among the reasons for examining the behavior of sensitizers grafted to silica or glass surfaces were expectations of reducing any reflective light losses and reducing the long-term deterioration (similar to, but far

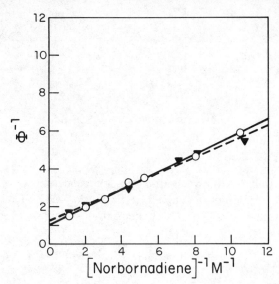

Figure 2. Reciprocal plot of the quantum yield for Q formation as a function of N concentration for the polystyrene immobilized sensitizer and homogeneous model. Solvent: benzene; irradiation wavelength: 366 nm; conversions: approximately 5%. Solutions were not deoxygenated. (—◯—) MeAK, (---▼---) ⓟ-AK.

less pronounced than that reported by Blossey and Neckers (7)) observed with the polystyrene derivatives. Both of these expectations have been met. However, as is apparent from the data plotted in Figure 3, (Si)–AK is inferior in comparison with either (P)–AK or the monomeric silicon model (limiting $\phi = 100\%$). We attribute the lower limiting efficiency (24%) to the same factor, inefficient intersystems crossing, that reduces the inherent efficiency of the parent chromophore, namely a polar environment.

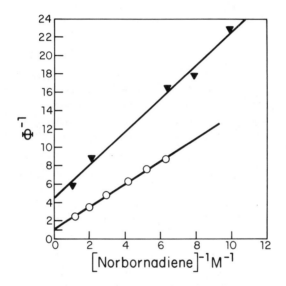

Figure 3. Reciprocal plot of the quantum yield for Q formation as a function of N concentration for the silica functionalized sensitizer and homogeneous model. Solvent: benzene; irradiation wavelength: 366 nm; conversions: approximately 5%. Solutions were not deoxygenated. (○) SiAK, (▼) (Si)–AK.

Spectral Studies

Absorption data (Table I) and emission spectra (Figure 4) of (P)–AK are rather unexceptional and closely resemble those of the model. However, the corresponding spectra for (Si)–AK are substantially red-shifted. The absorption maximum is identical to that observed for dimethylaminobenzophenone in ethanol. The phosphorescence spectrum (Figure 5) is not only red-shifted but substantial loss of structure is apparent, which is characteristic of the more polar environment. The triplet energy of (Si)–AK, available from the phosphorescence spectrum, is 57 kcal mol^{-1}.

Thus triplet energy transfer to **N** should be endothermic by some 13 kcal mol^{-1}, somewhat remarkable for a system yielding measured efficiencies as high as 15%.

Table I. Spectroscopic Properties and Sensitization Performance Indices of Heterogeneous Photosensitizers and Models

Sensitizer	Absorption^a (λ_{max}, nm)	Emission^b (λ_{max}, nm)	Triplet Energy^c (kcal mol^{-1})	Stilbene PSS^d ([cis]:[trans])	ϕ_{lim} NBD^e
ⓟ-AK	336	452	63.1	1.50	0.78
Me AK	337	442	64.7	1.78	1.0
ⓢⁱ-AK	355	500	57.0	2.33	0.24
Si AK	341	446	64.1	1.86	1.0

^{a} Benzene; benzene–silica slurry for immobilized sensitizers.
^{b} 0–0 band: methylcyclohexane glass at 77 K.
^{c} From emission spectra.
^{d} Stilbene isomerization: photostationary state ratio; benzene.
^{e} Limiting quantum yield for **N** to **Q** isomerization extrapolated to infinite **N** concentration; benzene.

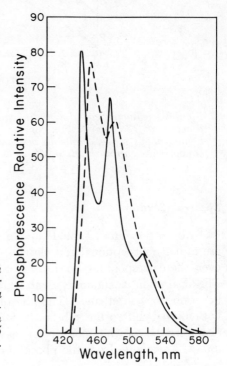

Figure 4. Phosphorescence spectra of the polystyrene-grafted sensitizer and homogeneous model. Spectra were taken at 77 K in a methylcyclohexane glass. The emission slit width was 1.0 nm for the polymer and 1.5 nm for the homogeneous sensitizer. (———) MeAK, (– – –) ⓟ-AK.

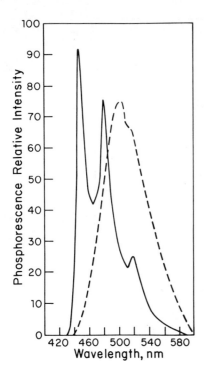

Figure 5. Phosphorescence spectra of the silica-functionalized sensitizer and homogeneous model. Spectra were taken at 77 K in a methylcyclohexane glass. The emission slit was 1.7 nm for the silica and 1.6 nm for the homogeneous sensitizer. (——) SiAK, (– – –) (Si)–AK.

Sensitization of Stilbene Isomerization

We were interested in using a second photochemical system to evaluate the heterogeneous photosensitizers. The trans–cis isomerization of stilbene was chosen because: it is a rather well-studied system (*14*); triplet energy transfer to *trans*-stilbene ($E_T \sim 50$ kcal mol^{-1}) would be exothermic; and the photostationary state ratio of *cis*- to *trans*-stilbene has been correlated with the triplet energies of numerous sensitizers (*15*). Minimally, the measurement of cis:trans photostationary state ratios provides a check on our measured triplet energies. Additionally, we anticipated the possibility that an unusually high preference for decay of the stilbene triplet to the sterically less-crowded *trans*-stilbene might result from sensitization at the crowded interface of the heterogeneous surfaces. Carefully measured values of these ratios complete the compilation of data presented in Table I. Comparison with other sensitizers is made in the Saltiel plot (Figure 6). It is clear that no dramatic preference for formation of *trans*-stilbene exists for the immobilized sensitizers. Furthermore, the values obtained are consistent with the measured triplet energies.

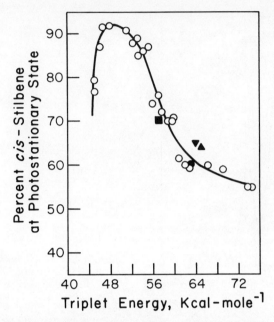

Figure 6. Saltiel plot of the photostationary state composition of stilbene isomerization sensitized by the immobilized sensitizers and homogeneous models. Open circles represent other data taken from Herkstroeter and Hammond (15). Photostationary state ratios were obtained using both cis- and trans-stilbene with agreement better than ±1%. Benzene was used as the solvent. (▲) *MeAK,* (○) (P)-*AK,* (▼) *SiAK,* (■) (Si)-*AK.*

Adsorptive Affinity of the Polymeric Support

During the process of analyzing solution compositions involving polymer systems with polar solvents, we noted that substrate concentrations in the supernatant liquid changed considerably following a brief (15 min) equilibration period with the polymer. The ability of the polystyrene (crosslinked with 20% divinylbenzene) to extract **N** and **Q** from a solvent such as methanol or acetonitrile can be appreciable, as indicated in Figure 7. Considerable caution must be taken in analyzing photolysis mixtures in such cases. No detectable partitioning was noted when benzene was used as the fluid medium. This phenomenon is of no particular value for the norbornadiene system because the photoproduct is adsorbed as well as, if not better than, the photoreactant, and because in a practical application, neat **N** would be used. However, one can easily envision other photochemical reactions where advantage of this effect could be taken. In systems where the photoproduct exhibits a low affinity for the solid support and a high affinity for the substrate, high local concentrations of the photoreactant in the vicinity of the sensitizer

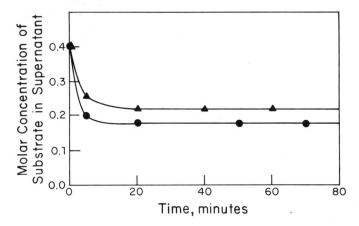

Figure 7. Adsorptive affinity of functionalized polystyrene for N and Q. The initial substrate concentration was 0.40M in methanol solution. In each case, 0.70 mL solution was exposed to 0.28 g polymer. (▲) N, (●) Q.

could be maintained during the entire course of the photolysis. Thus, not only would the quantum efficiency remain high, but undesirable secondary photoprocesses could be avoided entirely.

Other Heterogeneous Sensitizers for the Norbornadiene System

Several other common organic chromophores have been grafted onto solid supports such as polystyrene (2, 16). Preliminary evaluation of these immobilized sensitizers yielded performances clearly inferior to those discussed above. Another system under study by Kutal and Grutsch deserves mention, however. The complex $Ir(phen)_2Cl_2^+$ is totally ineffective ($\phi < 4 \times 10^{-3}$) at sensitizing the conversion of N to Q in solution (20% ethanol: 80% cyclohexane). When this cation is ionically attached to a sulfonated polystyrene, it sensitizes the conversion with a quantum efficiency of approximately 5% (0.5M N in cyclohexane) (17). The iridium complex cannot be studied under homogeneous conditions in nonpolar media such as cyclohexane due to total insolubility. The addition of alcohols, necessary for solubilization, is apparently deleterious to the photosensitizer. Thus even ionic species, which otherwise might be incompatible with organic systems, can be effective sensitizers when immobilized on solid supports.

Conclusion

The use of the dimethylaminobenzophenone chromophore has proved to be an incisive probe for heterogeneous photosensitization. Remarkably high energy transfer efficiencies can be achieved with this technique, and many advantages anticipated for immobilized sensitizers have been realized. In spite of the fact that grafting of the sensitizer chromophore occurs only at or near the surface of the solid support, the surface properties of the solid support are more influential in determining sensitizer behavior than are the properties of the surrounding fluid medium. The high observed efficiencies demand that the fluid medium be readily accessible to the attached sensitizer, yet the potentially adverse influence of a polar solvent is not felt by the sensitizer grafted to polystyrene. Likewise a favorable solvent, such as benzene, is unable to overcome the deleterious influence of polar silica surface. Although it was of no apparent use in the experiments described above, the potential for using a specific adsorptive affinity of the solid surface to achieve selective photochemistry is exciting.

Acknowledgment

Generous financial support from the Department of Energy is gratefully acknowledged. Syntheses of the silica compounds were carried out by Edward M. Sweet and Audrey W. Shields. Helpful discussions with R. Bruce King and Charles R. Kutal are acknowledged.

Literature Cited

1. Neckers, D. C. *CHEM TECH.* **1978**, *8*, 108.
2. Hautala, R. R.; Little, J.; Sweet, E. M. *Sol. Energy* **1977**, *19*, 503.
3. Moser, R. E.; Cassidy, H. G. *J. Polym. Sci., Polym. Lett. Ed.* **1964**, *2*, 545.
4. Leermakers, P. A.; James, F. C. *J. Org. Chem.* **1967**, *32*, 2898.
5. Blossey, E. C.; Neckers, D. C.; Thayer, A. L.; Schaap, A. P. *J. Am. Chem. Soc.* **1973**, *95*, 5820.
6. Schaap, A. P.; Thayer, A. L.; Blossey, E. C.; Neckers, D. C. *J. Am. Chem. Soc.* **1975**, *97*, 3741.
7. Blossey, E. C.; Necker, D. C. *Tetrahedron Lett.* **1974**, 323.
8. Jones, G., II. In "Chemical Conversion and Storage of Solar Energy"; Hautala, R. R., King, R. B., Kutal, C. R., Eds.; Humana: Clifton, NJ, in press.
9. Kutal, C.; Schwendiman, D. P.; Grutsch, P. A. *Sol. Energy* **1977**, *19*, 651.
10. Murov, S. Hammond, G. S. *J. Phys. Chem.* **1968**, *72*, 3797.
11. Barwise, A. J. G.; Garman, A. A.; Leyland, R. L.; Smith, P. G.; Rodgers, M. A. J. *J. Am. Chem. Soc.* **1978**, *100*, 1814.
12. Turro, N. J.; Cherry, W. R.; Mirbach, M. F.; Mirbach, M. J. *J. Am. Chem. Soc.* **1977**, *99*, 7388.
13. Cohen, S. G.; Saltzman, M. D.; Guttenplan, J. B. *Tetrahedron Lett.* **1969**, 4321.
14. Saltiel, J.; D'Agostino, J.; Megarity, E. D.; Metts, L.; Neuberger, K. R.; Wrighton, M.; Zafiriou, O. C. In "Organic Photochemistry"; Chapman, O. L., Ed.; Marcel Dekker: New York, 1973; Vol. 3, pp. 1–105.
15. Herkstroeter, W. G.; Hammond, G. S. *J. Am. Chem. Soc.* **1966**, *88*, 4769.
16. Little, J. L. M.S. Thesis, University of Georgia, 1978.
17. Kutal, C.; Grutsch, P. A., unpublished results.

RECEIVED October 2, 1978.

2

Photoactivation of Polymer-Anchored Catalysts

Iron Carbonyl Catalyzed Reactions of Alkenes

ROBERT D. SANNER, RICHARD G. AUSTIN, and MARK S. WRIGHTON[1]

Department of Chemistry, Massachusetts Institute of Technology, Cambridge, MA 02139

WILLIAM D. HONNICK and CHARLES U. PITTMAN, JR.[1]

Department of Chemistry, University of Alabama, University, AL 35486

Photoactivation of polymer-anchored iron carbonyl catalysts is reported. Prototypic reactions are 1-pentene isomerization and reaction of 1-pentene with $HSiEt_3$, which can be photocatalyzed at 25°C by near-UV irradiation of suspensions of the polymer-anchored catalyst systems. The basic polymer system is a styrene–1% divinylbenzene resin derivatized with either $-PPh_2$ or $-P(Ph)CH_2CH_2PPh_2$ anchoring sites for catalyst precursors. $Fe(CO)_n$'s (n = 3, 4) are attached to the phosphine sites by reaction with $Fe_3(CO)_{12}$. The polymer-anchored catalysts show turnover numbers exceeding 2×10^4 in some cases and observed quantum yields exceed unity, indicating the photogeneration of a thermal catalyst. From the data we conclude that the anchoring Fe–P bond is relatively photoinert and that the catalysis is initiated by the photoinduced dissociation of CO.

Coordinatively unsaturated transition metal organometallic complexes are believed to play a key role in homogeneous catalytic processes (1). Coordinative unsaturation can be generated photochemically by

---

[1] To whom correspondence should be addressed.

light-induced ligand dissociation and metal–metal bond cleavage (2,3). Such photochemistry has been exploited to initiate catalytic processes under relatively mild thermal conditions (4,5,6). Light-generated catalysts may be genuinely unique since the catalyst is the result of some excited-state decay process. Further, catalytic processes that are run at lower temperatures may yield greater product distribution selectivity when more than one product can be formed. Light also offers a degree of control over catalytic processes not necessarily attainable in conventional catalytic systems, since the product formation depends on two stimuli, light and heat. Examples of organometallic photocatalysis have demonstrated the concept of initiating catalysis under relatively mild conditions in the reactions of olefins with hydrogen and silicon hydrides and in unimolecular isomerization reactions of olefins (4,5,6).

In this chapter we summarize our recent findings (7) on the exploitation of photochemical catalyst generation in situations where the catalyst precursor is anchored to a polymer. Additionally, we present new data to further characterize such systems. A large number of examples of thermally activated, polymer-anchored catalyst systems have been studied (8,9,10). Anchoring a catalyst to a polymer takes advantage of the molecular specificity of organometallic catalysts while still being able to separate the products easily from the catalyst system. Viewing the polymer as a ligand, albeit an unusual one, one can also effect some control over the catalytic processes by the nature of the polymer-anchoring system. Photochemical activation of polymer-anchored catalyst systems allow the generation of unique catalyst systems. It has been demonstrated that extensive, multiple, coordinative unsaturation can be photogenerated by irradiating metal carbonyls in situations where the complexes are matrix isolated (11,12,13). The polymer-anchored complex can be viewed as a situation where the complex can be "matrix isolated," provided the rigidity of the polymer is great enough and the density of complexes on the polymer is low enough to permit isolation. Homogeneous organometallic catalytic species have previously been attached to polymer matrices and there is evidence for varying degrees of matrix isolation (14–19). However, this approach has never been applied to photogenerated catalytic species. Of course, the exploitation of any photoactivation of a polymer-anchored catalyst depends on the photoinertness of the anchor bond. Photoinduced selective loss of ligands other than the polymer ligand is the objective.

Detailed results have been obtained for $Fe(CO)_n$ ($n = 3, 4$) bound to a phosphinated styrene–divinylbenzene resin. Two types of anchoring sites have been used and are represented in **I** and **II**. Most of the work thus far has involved the phosphinated polymer **I**. This anchoring system approximates a triphenylphosphine ligand; consequently the $Fe(CO)_n$-

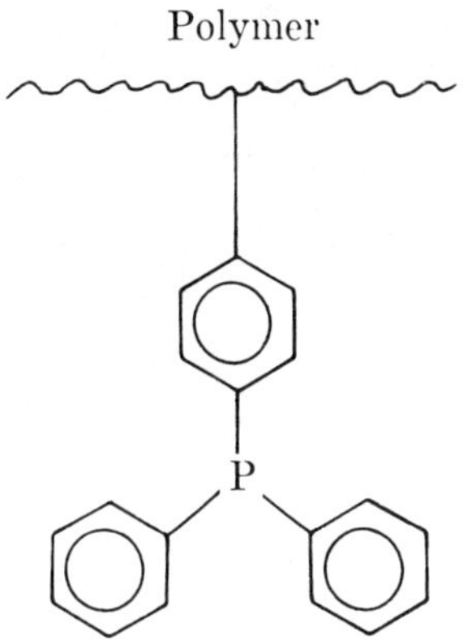

Polymer   Polymer

[Structure I: Polymer-phenyl-P(Ph)₂]

[Structure II: Polymer-phenyl-P(Ph)-CH₂-CH₂-P(Ph)₂]

      I 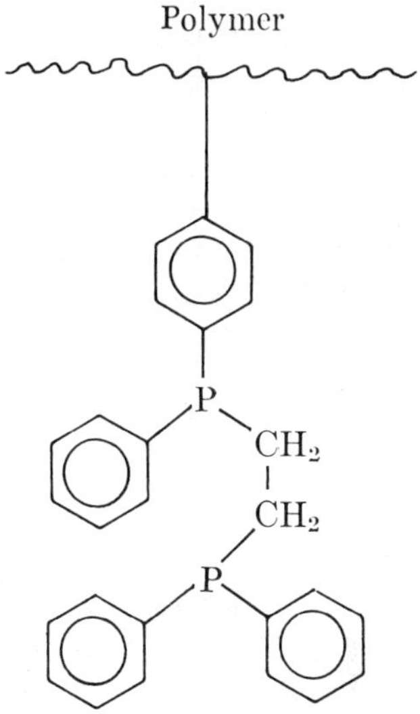   II

$(PPh_3)_{5-n}$ ($n = 3, 4$) complexes have been used as homogeneous models for the polymer-anchored system. A direct comparison of these phosphine derivatives with $Fe(CO)_5$ also has been made to assess the effect of having the phosphine ligand(s) in the coordination sphere. The probe catalytic chemistry is 1-pentene isomerization and reaction with $HSiEt_3$. Both reactions can be photocatalyzed using $Fe(CO)_5$ as the catalyst precursor (6).

Experimental

The general procedures and synthesis associated with Catalyst System I (see Table I) have been detailed elsewhere (7), including the synthesis and characterization of the derivatized polymer. The irradiation source

Table I. Analytical Data for Polymer Catalyst Systems

Anchoring[a] System	% of Phenyl Rings Substituted	Catalyst System	% P	% Fe	P:Fe
I	3.3	A	0.92	0.58	2.9
I	29	B	4.47	4.10	2.0
II	11.9	C	3.91	6.79	1.0
II	3.1	D	1.73	1.40	2.2
II	11.9	E	4.32	1.10	7.1
II	3.1	F	1.51	0.42	6.5

[a] See text.

in the experiments described herein was a GE Blacklite equipped with two 15-W bulbs with principal output at 355 nm; the intensity at the sample (stirred for polymer systems) was approximately 10^{-6} ein/min measured by ferrioxalate actinometry (7). The samples themselves were solutions or suspensions of the catalyst precursor in neat 1-pentene (Chemical Samples Co., > 99.9%) such that about $10^{-3}M$ Fe was present. Sample size was typically 1.0 mL; the aliquots were placed in 13 × 100-mm ampules, freeze-pump-thaw degassed, and hermetically sealed prior to illumination at 298 K. Analyses were all by gas chromatography using equipment described previously (7).

Polymer System II has not been previously described and its synthesis is described below. Tetrahydrofuran (THF) was distilled from potassium/benzophenone under nitrogen. All other solvents and chemicals were used as received, except for diphenylvinylphosphine, which was prepared by the reaction of chlorodiphenylphosphine with vinyl magnesium bromide according to the literature method (23). Styrene–1%-divinylbenzene beads were purchased from Bio-Rad Laboratories (SX-1, 200–400 mesh) and were brominated according to literature methods (24). Microanalytical analyses were performed by Schwarzkopf Microanalytical Laboratories, Woodside, New York. All reactions and manipulations involving phosphines were carried out under nitrogen.

Preparation of $C_6H_5P(H)CH_2CH_2P(C_6H_5)_2$. A neat solution of phenylphosphine (36.0 g, 0.33 mol) and diphenylvinylphosphine (36.8 g, 0.17 mol) containing AIBN (0.30 g) was irradiated under nitrogen for 2 hr with a 100-W Hanovia UV lamp. The excess diphenylphosphine was distilled from the resulting solution at atmospheric pressure and the 1-phenylphosphino-2-diphenylphosphinoethane was vacuum distilled at 190–195°C/0.05 torr, yield: 35.0 g (63%). The only other major byproduct was bis(2-diphenylphosphinoethyl)phenylphosphine produced by the reaction of two moles of diphenylvinylphosphine per mole of phenylphosphine. The ratio of the two products formed by this reaction is very sensitive to the ratio of the two reactants, with an excess of phenylphosphine favoring production of bis chelate phosphine.

Preparation of $P-P(Ph)CH_2CH_2PPh_2$. Lithium metal (5.0 g, 0.72 mol) was suspended in THF (25 mL) and a solution of 1-phenylphosphino-2-diphenylphosphinoethane (22.0 g, 68.3 mmol) dissolved in THF (50 mL) was added dropwise. A yellow solution initially formed and slowly became dark red. The reactants were stirred together for two days at room temperature and then refluxed for 1 day. The excess lithium metal was removed, and the solution was added dropwise to a brominated styrene–1%-divinylbenzene (11.9 g ~ 30% of the polystyryl rings brominated). The mixture was stirred for 2 days at room temperature and then refluxed for 1 day. The mixture was cooled and hydrolyzed slowly with 100 mL of deoxygenated water, filtered, and washed with 500 mL portions of the following deoxygenated solvents: water, acetone-water (1:1), acetone, benzene, and methanol. The phosphinated resin beads were dried in vacuo at room temperature for 24 hr. Analysis found 0.37% Br and 5.16% P, which corresponds to 11.9% of the polystyryl rings containing a bound $-P(Ph)CH_2CH_2PPh_2$ group.

A similar procedure was followed to prepare a phosphinated resin containing bound $-P(Ph)CH_2CH_2PPh_2$ groups with 3.1% of the polystyryl rings substituted.

Preparation of $P-[P(Ph)CH_2CH_2PPh_2]_x[Fe(CO)_n]_y$. $Fe_3(CO)_{12}$ (0.55 g, 3.28 mmol Fe) was dissolved in 25 mL of THF and reacted with 1.0 g of phosphinated resin containing bound $-P(Ph)CH_2CH_2PPh_2$ groups (11.9% of the rings substituted, 1.67 mmol P). The mixture was refluxed for 1 hr, cooled, and filtered. The resin beads then were washed repeatedly with alternate portions of dry, deoxygenated THF and methanol, and then dried in vacuo at 80°C for 24 hr. Analysis found: 3.91% P and 6.79% Fe corresponding to a P:Fe ratio of 1.0.

Three other polymer-attached catalyst resins were prepared similarly using different $Fe_3(CO)_{12}$:phosphinated resin ratios. Catalyst resin for Catalyst System D (see Table I) was prepared by the reaction of $Fe_3(CO)_{12}$ (0.10 g, 5.0 mmol Fe) with phosphinated resin (2.0 g, 3.1% rings substituted, 1.08 mmol P). Catalyst resin for Catalyst System E (see Table I) was prepared by the reaction of $Fe_3(CO)_{12}$ (0.032 g, 0.19 mmol Fe) with phosphinated resin (1.0 g, 11.9% rings substituted, 1.67 mmol P). Catalyst resin for Catalyst System F (see Table I) was prepared by the reaction of $Fe_3(CO)_{12}$ (0.023 g, 0.13 mmol Fe) with phosphinated resin (2.0 g, 3.1% rings substituted, 1.08 mmol P).

Results and Discussion

Polymer Systems Studied. The polymer used in this work is a commercially available (Bio-Rad Laboratories SX-1, 200–400 mesh) styrene–1% divinylbenzene, microporous resin. It was brominated and functionalized according to the procedures represented in Equations 1–3. The

$$P-\phi \xrightarrow[CCl_4]{Br_2, Fe} P-\phi-Br \quad (1)$$

$$P-\phi-Br \xrightarrow[THF]{LiPPh_2} P-\phi-PPh_2 \quad (2)$$
$$\text{I}$$

$$P-\phi-Br \xrightarrow[THF]{(Li)PhPCH_2CH_2PPh_2} P-\phi-P(Ph)CH_2CH_2PPh_2 \quad (3)$$
$$\text{II}$$

phosphinated material **I** and **II** were separately reacted thermally with $Fe_3(CO)_{12}$ in THF to incorporate $Fe(CO)_n$ ($n = 3, 4$) by attachment to the phosphine ligands (see Equations 4 and 5). Thermal reaction of $Fe_3(CO)_{12}$ with phosphine ligands is known to produce mononuclear iron carbonyl species (20). Elemental analyses (Schwarzkopf Microanalytical Laboratories) and IR spectra in the carbonyl region have been used to characterize these systems.

I $\xrightarrow[\text{THF}]{\text{Fe}_3(\text{CO})_{12}}$ (P)–PPh$_2$)$_{5-n}$Fe(CO)$_n$ (4)

$$n = 3, 4$$

II $\xrightarrow[\text{THF}]{\text{Fe}_3(\text{CO})_{12}}$ (P)–P(Ph)CH$_2$CH$_2$PPh$_2$)$_x$[Fe(CO)$_n$]$_y$ (5)

$$x = 1, 2;\ n = 3, 4;\ y = 1, 2$$

Table I summarizes the essential analytical results for the several polymer catalyst systems described herein, and Table II gives IR spectral data for polymer systems and other relevant complexes. The IR data (band positions and relative intensities) for anchoring system I reveal that there is a ratio of (P)–PPh$_2$Fe(CO)$_4$ to trans-(P)–PPh$_2$)$_2$Fe(CO)$_3$ between about 2–6 where the Fe(CO)$_3$ unit can serve as a cross-linking

Table II. IR Band Maxima in CO Stretching Region for Fe Carbonyl/Phosphine Complexes[a]

Complex	Band Maxima, cm^{-1} (ϵ)
Fe(CO)$_5$	2025(5470); 2000(11,830)
Fe(CO)$_4$PPh$_3$	2054(3500); 1978(2310); 1942(4870)
trans-Fe(CO)$_3$(PPh$_3$)$_2$	1893(5140)
(P)–PPh$_2$)$_{5-n}$Fe(CO)$_n$ ($n = 3, 4$) (A and B)	2045(—); 1968(—); 1932(—); 1876(—)
Fe(CO)$_4$P(Ph)$_2$CH$_2$CH$_2$PPh$_2$Fe(CO)$_4$[b]	2049(—); 1976(—); 1934(—)
(Ph$_2$PCH$_2$CH$_2$PPh$_2$)Fe(CO)$_3$[b]	1992(—); 1923(—); 1901(—)
(P)–P(Ph)CH$_2$CH$_2$PPh$_2$)$_x$[Fe(CO)$_n$]$_y$ ($x = 1, 2;\ n = 3, 4;\ y = 1, 2$)	
(C)	2055(—); 1982(—); 1934(—)
(D)	2050(—); 1975(—); 1935(—); 1880(—)
(E)	2050(—); 1980(—); 1935(—)
(F)	2040(—); 1975(—); 1935(—); 1885(—)

[a] Polymer systems measured as Nujol mulls; other complexes measured in hydrocarbon solution.
[b] Data from Refs. 25 and 26.

agent. The distribution of $Fe(CO)_n$ species for anchoring system **II** is more complex, owing to the possibility that the diphosphine anchor may or may not be a chelating reagent. Comparison of the IR spectra for catalyst systems IIC-IIF suggest that there are attached $Fe(CO)_4$ units. The band at 1880 or 1885 cm^{-1} in IIB and IIF suggests that $Fe(CO)_3$ serves as a cross-linking reagent as in system **I**, mimicking the model *trans*-$Fe(CO)_3(PPh_3)_2$ species. The data seem to rule out an important contribution from anchoring system **II** as a chelate to bind $Fe(CO)_3$ in the systems studied. Deliberate variation in the distribution of the coordination sphere of the attached iron in **I** and **II** is clearly achievable by careful control at the functionalization stages, but the work to date has focused mainly on first establishing some of the quantitative photocatalytic behavior in these systems.

Primary Photoprocesses in Iron Carbonyl Complexes. Equation 6 represents the primary chemical result of optically exciting $Fe(CO)_5$

$$Fe(CO)_5 \xrightarrow{h\nu} Fe(CO)_4 + CO \qquad (6)$$

(2, 11, 12). The reaction occurs efficiently and is believed to be the primary step in the $Fe(CO)_5$ photocatalyzed isomerization (21), hydrogenation (21), and hydrosilation (6) of alkenes. But in these cases it has been postulated that absorption of an additional photon by some $Fe(CO)_4(alkene)$ or $Fe(CO)_4(H)(R)$ (R = H, Si(alkyl)$_3$) is required to effect the alkene chemistry (6, 21).

The crucial question here with respect to polymer-anchored $Fe(CO)_n$ species is whether the Fe–P bond is photoinert. If the Fe–P is cleaved photochemically, then the anchored species can be released to the bulk solution and simply effect the same chemistry as that found beginning with $Fe(CO)_5$ in solution. Three lines of evidence support the conclusion that the Fe–P bond is relatively photoinert for the catalyst systems studied here. First, Equations 7 and 8 represent the chemistry occurring upon

$$Fe(CO)_4PPh_3 \xrightarrow[\substack{\Phi_{355\,nm} = 0.4 \pm 0.04 \\ \text{(benzene)}}]{h\nu,\ P(OMe)_3} Fe(CO)_3(PPh_3)(P(OMe)_3) + CO \qquad (7)$$

$$Fe(CO)_4PPh_3 \xrightarrow{h\nu,\ HSiEt_3} Fe(CO)_3(PPh_3)(H)(SiEt_3) + CO \qquad (8)$$

near-UV photoexcitation of $Fe(CO)_4PPh_3$ in the presence of $P(OMe)_3$ and $HSiEt_3$, respectively; in the former no $Fe(CO)_4(P(OMe)_3)$ was detected and in the latter no $Fe(CO)_4(H)(SiEt_3)$ was found (7). Thus, for $Fe(CO)_4(phosphine)$ the extrusion of CO is the principal result of optical excitation. Similar results were obtained for $Fe(CO)_3(PPh_3)_2$, but loss of PPh$_3$ here has an efficiency of about one-tenth that of CO loss. Second, irradiation of the polymer-anchored Fe(CO)$_n$ species does not yield detectable amounts of Fe(CO)$_n$ complexes in solution, even when the solution contains PPh$_3$ as a potential sequestering reagent. Finally, the catalytic chemistry for the polymer-anchored systems is inconsistent with that obtained from Fe(CO)$_5$; the data suggest the retention of at least one phosphine in the coordination sphere during actual catalytic reactions.

Qualitative Photocatalytic Behavior. Polymer-anchored Fe(CO)$_n$ species have been shown to serve as photochemical sources of catalytically active species that are capable of effecting 1-pentene isomerization, Equation 9, and 1-pentene reaction with HSiEt$_3$, Equation 10. The

$$\text{1-pentene} \xrightarrow[\text{[Catalyst Precursor]}]{h\nu} \text{cis-2-pentene} + \text{trans-2-pentene} \quad (9)$$

$$\text{1-pentene} + \text{HSiEt}_3 \xrightarrow[\text{[Catalyst Precursor]}]{h\nu} \text{products} +$$

$$\text{n-C}_5\text{H}_{11}\text{SiEt}_3 + \underset{\textbf{III}}{\text{CH}_3\text{CH=CHCH}_2\text{CH}_2\text{SiEt}_3} + \underset{\textbf{IV}}{\text{CH}_3\text{CH}_2\text{CH=CHCH}_2\text{SiEt}_3}$$

$$\underset{\textbf{V}}{\text{CH}_3\text{CH}_2\text{CH}_2\text{CH=CHSiEt}_3} \quad (10)$$

catalytic chemistry can be induced with near-UV (355 nm) irradiation at 25°C where there is no detectable dark reaction; no photoinduced reaction of the 1-pentene is found by irradiating suspensions of the anchoring material prior to attachment of the Fe(CO)$_n$ units. Solvent interactions are substantial but have not yet been fully evaluated. The importance of polymer solvation is reflected in the fact that 0.1M 1-pentene in an isooctane suspension of Catalyst System A (*see* Table I) undergoes much less than 1% isomerization after 12 hr of irradiation, whereas more than 25% isomerization results under the same conditions with benzene as the solvent. Benzene is an excellent swelling solvent. Therefore, reactions in benzene are suitable for comparison with homogeneous cases. For the work described here the solutions are generally neat 1-pentene or a neat 1:1 mol ratio of 1-pentene:HSiEt$_3$.

Catalytic action depends on the rigorous exclusion of O$_2$; reactions were run in hermetically sealed, freeze-pump-thaw-degassed ampules. Sustained catalysis requires continuous irradiation; i.e., when the light is turned off the reaction appears to stop, but it can be reinitiated by light.

any isomerization does take place on Catalyst Systems D, E, and F we see a trans:cis product ratio in accord with the evidence of more than one phosphine group in the Fe coordination sphere.

Table V shows the variation in isomerization quantum yields as the 1-pentene concentration is varied for Catalyst System B using benzene as cosolvent. There is more than a ten-fold change in the initial quantum yields with variation of 1-pentene concentration. Such an effect has been reported previously using $Fe(CO)_5$ in homogeneous solution (21). There is a modest change in the initial trans:cis ratio with variation of 1-pentene concentration. This effect may result from differences in steric effects owing to change in the degree of polymer swelling by benzene and 1-pentene. Alternatively, the variation in trans:cis ratio may indicate some effect from pentene not coordinated to iron, perhaps associatively directing decay of the proposed (21) π-allyl hydride intermediate for isomerization.

Table V. Dependence of Quantum Yield on 1-Pentene Concentration[a]

1-Pentene Concentration (M)	% Conversion	Φ[b]	(trans:cis)[c]
0.3	9.0	0.3	1.6
0.3	20.7	0.4	1.6
1.0	14.7	0.9	1.8
1.0	19.8	0.8	1.8
2.5	8.4	1.2	1.7
2.5	10.3	1.0	1.8
5.0	8.5	1.7	1.5
9.1 [neat]	6.4	4.8	0.8
9.1 [neat]	10.6	4.0	1.0

[a] Irradiation of 5 mg of Catalyst System B suspended in 1.0 mL of degassed benzene/1-pentene solution. Solutions are stirred throughout the irradiation period. Reaction is carried out at 25°C using a GE Blacklite (355 nm) irradiation source.
[b] Observed quantum yield for 1-pentene disappearance; number of 1-pentene molecules consumed per photon incident on the sample.
[c] Ratio of trans- to cis-2-pentene products.

Photocatalyzed Reaction of 1-Pentene with $HSiEt_3$. Table VI shows the distribution of Si-containing products resulting from the various photocatalysts. Except for some minor variations, the distribution of products seems to be essentially independent of the catalyst precursor or the extent of conversion. The amount of n-pentane in each case is initially equal to the total (pentenyl)$SiEt_3$ yield. These data testify to the ability to effect synthetic catalysis of some consequence using the polymer-anchored catalyst precursors. The interesting products here are the (pentenyl)$SiEt_3$ species, which are not generally formed in catalyzed

Table VI. Product Distribution in

Catalyst Precursor	% Conversion
$Fe(CO)_5$	2
	80
$Fe(CO)_4PPh_3$	8
	30
$Fe(CO)_3(PPh_3)_2$	6
	40
(P)–$PPh_2)_{5-n}Fe(CO)_n$ [b] ($n = 3, 4$) (A)	20
	50
(P)–$P(Ph)CH_2CH_2PPh_2)_x[Fe(CO)_n]_y$ (C) [c] ($x = 1, 2; n = 3, 4; y = 1, 2$)	27

[a] Numbers given are the percent of all Si-containing products. Products generated from irradiation of neat 1:1 alkene:silane solutions containing $10^{-3}M$ catalyst precursor. Cf. text, Equation 10, for structure of (pentenyl)SiEt$_3$ isomers **III**, **IV**, and **V**.

reactions of alkenes with silicon hydrides (22). Since the product distribution does not vary greatly with any of the precursors, we cannot unequivocally conclude that the phosphines are retained in the Fe coordination sphere, but there is not any evidence that the phosphine is lost. The catalytic reaction simply may not be too sensitive to catalyst structure; the Si-containing products are likely generated by competitive β- and reductive-elimination processes (4, 6), which may not be as sensitive to the steric environment as is the decay of the π-allyl hydride intermediate (21) for isomerization.

Summary

Polymer-anchored $Fe(CO)_n$ ($n = 3, 4$) units are effective photocatalysts for alkene isomerization and reaction with a silicon hydride. The catalytic chemistry parallels that found for appropriate homogeneous model complexes. Primary photoprocesses for the models and the distribution of photocatalytic products reveal that the anchoring bond attaching the catalyst to the polymer is relatively photoinert. While quantum yields for catalytic reactions substantially exceed unity, long-lived catalytic activity in the dark does not occur, even in polymer-anchored systems. The high turnover numbers ($> 2 \times 10^4$) indicate good catalyst lifetime for synthetic applications.

Photocatalyzed Reaction of 1-Pentene and HSiEt$_3$[a]

(n-Pentyl)SiEt$_3$[a]	(Pentyl)SiEt$_3$[a]		
	III	IV	V
16.5	21.3	52.4	9.8
17.5	16.1	51.2	15.2
8.1	16.2	62.0	13.6
10.5	16.0	58.1	15.4
19.8	14.8	50.5	14.8
11.1	17.3	57.9	13.6
8.4	21.1	58.7	11.8
14.7	15.5	55.4	14.4
11.3	16.4	55.0	17.3

[b] Similar product distributions were found for catalysis using Catalyst System B.
[c] Experiments with C were conducted in m-xylene solutions of 1.1M 1-pentene and 1.1M HSiEt$_3$.

Acknowledgments

We thank the Office of Naval Research for support of this research at the University of Alabama (ONR Inorganic Polymers Program) and the Massachusetts Institute of Technology. MSW acknowledges support as a Dreyfus Teacher–Scholar grant recipient, 1975-1980.

Literature Cited

1. Cotton, F. A.; Wilkinson, G.; "Advanced Inorganic Chemistry," 3rd ed.; Interscience: New York, 1972; pp. 770–800.
2. Wrighton, M. *Chem. Rev.* **1974**, *74*, 401.
3. Wrighton, M. S. *Top. Curr. Chem.* **1976**, *65*, 37.
4. Austin, R. G.; Paonessa, R. S.; Giordano, P. J.; Wrighton, M. S. *Adv. Chem. Ser.* **1978**, *168*, 189.
5. Wrighton, M.; Ginley, D. S.; Schroeder, M. A.; Morse, D. L. *Pure Appl. Chem.* **1975**, *41*, 671.
6. Schroeder, M. A.; Wrighton, M. S. *J. Organomet. Chem.* **1977**, *128*, 345.
7. Sanner, R. D.; Austin, R. G.; Wrighton, M. S.; Honnick, W. D.; Pittman, C. U., Jr. *Inorg. Chem.* **1979**, *18*, 928.
8. Pittman, C. U., Jr.; Hanes, R. M. *Ann. N.Y. Acad. Sci.* **1974**, *239*, 76.
9. Pittman, C. U., Jr.; Smith, L. R. In "Organotransition-Metal Chemistry"; Ishii, Y., Tsutsui, M., Eds.; Plenum: New York, 1975; pp. 143–156.
10. Grubbs, R. H.; Kroll, L. C. *J. Am. Chem. Soc.* **1971**, *93*, 3062.
11. Poliakiff, M. *J. Chem. Soc. Dalton Trans.* **1974**, 210.
12. Poliakoff, M.; Turner, J. J. *J. Chem. Soc. Dalton Trans.* **1974**, 2276.
13. Perutz, R. N.; Turner, J. J. *J. Am. Chem. Soc.* **1975**, *97*, 4791.

14. Gubitosa, G.; Boldt, M.; Brintzinger, H. H. *J. Am. Chem. Soc.* **1977**, *99*, 5174.
15. Grubbs, R.; Lau, C. P.; Cukier, R.; Brubaker, C., Jr. *J. Am. Chem. Soc.* **1977**, *99*, 4517.
16. Jarrell, M. S.; Gates, B. C.; Nicholson, E. D. *J. Am. Chem. Soc.* **1978**, *100*, 5727.
17. Pittman, C. U., Jr.; Ng, Q. *J. Organomet. Chem.* **1978**, *153*, 85.
18. Jayalekshmy, P.; Mazur, S. *J. Am. Chem. Soc.* **1976**, *98*, 6710.
19. Scott, L. T.; Rebek, J.; Ovsyanki, L.; Sims, C. L. *J. Am. Chem. Soc.* **1977**, *99*, 625.
20. Angelici, R. J.; Siefert, E. E. *Inorg. Chem.* **1966**, *5*, 1457.
21. Wrighton, M. S.; Schroeder, M. A., *J. Am. Chem. Soc.* **1976**, *98*, 551.
22. Harrod, J. F.; Chalk, A. J. "Organic Synthesis via Metal Carbonyls"; Wender, I., Pino, P., Eds.; Wiley: New York, 1977; Vol. 2, pp. 673–704.
23. Berlin, K. D.; Butler, G. G. *J. Org. Chem.* **1961**, *26*, 2537.
24. Pittman, C. U., Jr.; Smith, C. R.; Hanes, R. M. *J. Am. Chem. Soc.* **1975**, *97*, 1742.
25. Reckziegel, A.; Bigorgne, M. *J. Organomet. Chem.* **1965**, *3*, 341.
26. Manuel, T. A. *Inorg. Chem.* **1963**, *2*, 854.

RECEIVED October 2, 1978.

Photoeffects on Reactions over Transition Metals

B. E. KOEL[1] and J. M. WHITE

Department of Chemistry, University of Texas, Austin, TX 78712

J. L. ERSKINE and P. R. ANTONIEWICZ

Department of Physics, University of Texas, Austin, TX 78712

> *This chapter reviews experiments and theory related to photodesorption and photoreactions over transition metals. At wavelengths above 300 nm, photoinduced desorption of CO from Ni appears to be thermally induced. Around 250 nm there appears to be a true quantum effect in the photodesorption of CO from W and perhaps from Ni, although the latter experiments may be dominated by background effects. The theory that can be applied to the photodesorption problem relies heavily on analogies between photodesorption and electron stimulated desorption. Existing models invoke Auger transitions and electron tunneling as the key steps. These models are in the very early stages of development. For reactions, the only well established case of photoenhancement involving small molecules is the CO oxidation over Pd at pressures of around 1 torr. At pressures of around 10^{-6} torr the photoeffect is no longer measurable.*

Photochemistry is an old subject and yet it remains very interesting. The use of light to promote chemical reactions and to probe the character of chemical species is among the most powerful and incisive tools available to the scientist. As the title of this volume suggests, there is now a high level of interest in using light to control the rate and

[1] NSF Trainee.

specificity of interfacial processes. Nowhere in industry do interfacial processes play a larger role than in heterogeneous catalysis. Because of the tremendous economic consequences, there is an ongoing effort to improve catalytic activity, efficiency, and specificity. One potential route for improvement would use light instead of heat to promote specific reactions. On a laboratory scale the use of light to promote reactions on transition metals has proved difficult and frustrating, as will be noted in the text of this chapter.

The purpose of this chapter is to review some experimental photodesorption work done at the University of Texas and compare it with other experimental work, to describe some of the mechanisms that may be used in understanding and describing photodesorption, and finally, to suggest some directions for future research.

It is probably fair to say that the detailed requirements for observing photoeffects as the result of direct atomic level excitations are not well understood either experimentally or theoretically. However, the analogy with electron stimulated processes can serve as a good beginning point. Thus, the latter part of this paper relies heavily on electron stimulated desorption.

Desorption of CO from Ni

Since surface reaction rates involve the complex interplay of adsorption, migration, bond cleavage, bond formation, and desorption, some simplification has often been sought and systems have been chosen in which desorption is central. For example, the effect of light on the photodesorption of CO from Ni has been studied frequently, beginning with Lange and Riemersma (1). Their data (Figure 1) shows that the yield of CO molecules per photon incident on Ni foil is 10^{-8}–10^{-9} and varies with wavelength by a factor of twenty in passing from 460 to 330 nm. Subsequently, Adams and Donaldson (2) reported similar results. Both groups interpret their results in terms of direct photoinduced bond cleavage, i.e., a quantum effect.

A moments consideration reveals the difficulty of establishing on metals that an observed desorption arises from direct photon-absorption-induced-chemisorption bond cleavage rather than local or bulk lattice heating. Mechanistically, photodesorption via a quantum effect may be defined as the absorption of a photon by the electron density at the adsorbate–substrate interface followed by adsorbate desorption without equilibration of the energy in the surrounding metal lattice. At the other extreme, lattice heating, the photon energy is absorbed by the metal and heats the lattice locally and, by conduction, throughout. Because of their large UV–visible extinction coefficients, metals absorb light very near the

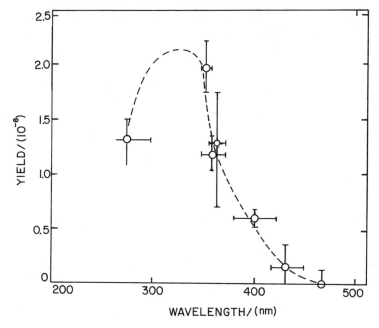

Figure 1. *Wavelength dependence of photodesorption of CO from Ni. The results are plotted with probable errors (1).*

surface. If the photon energy absorbed by the metal is not rapidly distributed throughout the lattice, a significant temperature gradient can exist in the surface region and photodesorption may occur via a local heating effect.

Because a significant amount of weakly bound CO is present in the Ni–CO chemisorption system, the separation of thermal and quantum effects is particularly troublesome. In the preliminary work mentioned above (1, 2), adequate account was not taken either of the power absorbed or of the rate at which the target was heated. Since the arc lamps used are rich in UV photons (300–400 nm) and this is just the region where the yield tends to maximize, the observed effect may more nearly reflect the energy distribution of the light source than the quantum yield distribution. McAllister and White (3) have shown that this is indeed the case over the range 320–600 nm. Using a filtered high pressure mercury arc and accounting for wavelength variations in the reflection coefficient of Ni, the photodesorption of CO from a polycrystalline Ni foil was studied at constant absorbed power from various wavelengths (3). The foil was prepared either by heating in vacuum to 975 K or by Ar$^+$ bombardment and annealing. No Auger electron spectroscopy was done, but there is evidence that heating in vacuum to 975 K will give a carbon- or sulfur-contaminated surface while Ar^+ bombardment will give a clean surface (4, 5). Figure 2 shows four separate determinations (at four wavelengths) of the energy efficiency at constant absorbed power (19.2 mW cm^{-2}) for a foil prepared by heating in vacuum. Within experimental error ($\pm$ 10%) there is no variation of energy efficiency with wavelength and although the quantum yield does increase with decreasing wavelength as shown in the lower part of Figure 2, this variation is not, by itself, evidence for a quantum effect. Further evidence that quantum photodesorption plays a minimal role in these carbon-covered Ni foils saturated with about 1.5×10^{12} CO molecules cm^{-2} is shown in Figure 3. Here heating rates, as measured by an iron–constantan thermocouple spot-welded to the dark side of the substrate, were systematically varied either with neutral density filters in the light beam or by varying the resistive heating current through the substrate. Over a factor of six, there is no discernible distinction in the initial desorption rate (determined from the mass spectrometrically determined CO pressure vs. time curves) produced by photons (365 or 680 nm) and resistive heating. Thus, on contaminated Ni the major pathway for CO desorption is thermal and, since the bulk temperature is measured in the above experiments, local heating does not predominate.

As shown in Figures 4 and 5 the same situation holds for Ar^+-bombarded substrates where the amount of adsorbed CO is about 3×10^{14} molecules cm^{-2} and is tightly held (25 kcal mol^{-1}). Further experiments on the Ar^+-bombarded substrates showed that over the ambient pressure range (10^{-8}–10^{-6} torr) of CO, 320-nm light and resistive heating gave the same desorption rates. As the ambient pressure increased, the concentration of weakly held CO increased and the initial desorption rates increased. It was also shown over this pressure range that the total number of CO molecules desorbed in transients induced by 320-nm light and by resistive heating was identical.

On the basis of this data we estimate that any photodesorption quantum effects in the CO–Ni system over the range 300–600 nm have cross sections of less than 10^{-22}–10^{-23} cm^2, if the effect exists at all. This conclusion is in good agreement with that of Brainard (6) who investigated the photodesorption of CO from polycrystalline Ni surfaces prepared by Ar^+ bombardment and interpreted the results in terms of bulk heating with no significant quantum desorption. Moreover, we conclude that the work mentioned above and the extensions thereof to other systems like Fe (2), Zr (2), and W (7, 8, 9) are probably strongly controlled by thermal effects for any wavelength above 300 nm.

This conclusion is also in good agreement with that of Genequand (10) and Paigne (11), who both found extremely low yields of CO above 300 nm using pulsed light sources to distinguish thermal and quantum

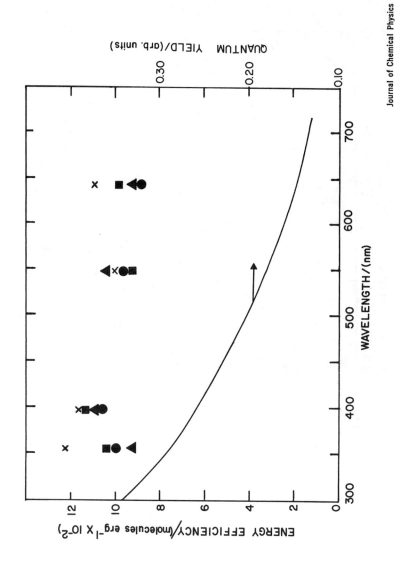

Figure 2. The variation of the energy efficiency with wavelength at a fixed absorbed radiant power of 19.2 mW cm^{-2}. The target was prepared by high-temperature outgassing (3).

Journal of Chemical Physics

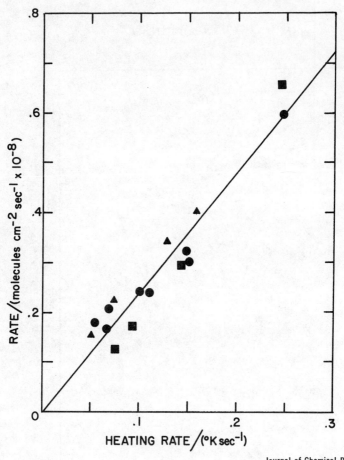

Figure 3. The variation of the initial rate of desorption with the bulk Ni heating rate: (▲) 3650 Å light; (■) 6800 Å light; (●) electrical current. The target was prepared by high temperature outgassing (3).

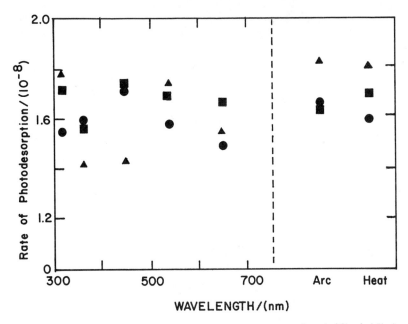

Figure 4. The variation of the initial rate of photodesorption with wavelength at a fixed absorbed power of 20.6 mW cm^{-2}. The target was prepared by Ar$^+$ bombardment. Right side of figure gives the rates of desorption produced by equivalent bulk heating with the attenuated full arc (arc) and with electrical current (heat) (3).

effects. They estimated maximum photodesorption cross sections of 10^{-23}–10^{-24} cm^2 above 300 nm. Below 300 nm, Genequand found a sharply rising yield of CO, which agrees very nicely with the earlier work of Adams and Donaldson (2). For comparison, we note that Kronauer and Menzel (7) found a sharply rising photodesorption probability for CO off polycrystalline W as the wavelength dropped below 360 nm. They reported a yield of 4×10^{-7} at 250 nm that is attributed to a quantum effect on the basis of carefully compensated photon-induced temperature increases. On the basis of this data it appears (but see below) that a small quantum photodesorption effect exists below 300 nm in both the Ni–CO and W–CO systems.

Keeping in mind the very small yield (10^{-9}–10^{-7}), it is pertinent to consider possible spurious sources of desorption in both these systems. This is particularly relevant in light of the significant work of Lichtman and co-workers (12) showing that around 250 nm the quantum photodesorption of CO from stainless steel becomes quite strong. As shown in Figure 6 the yield is about 5×10^{-3} at 185 nm and drops by at least 2.5 orders of magnitude as the wavelength increases to 300 nm. This

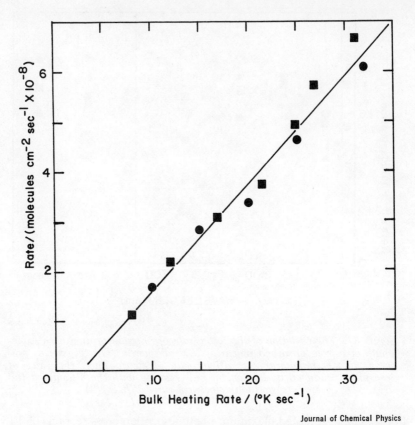

Journal of Chemical Physics

Figure 5. Variation of the initial rate of desorption of CO with bulk Ni heating rate: (●) 320-nm light; (■) electrical current. The substrate was prepared by Ar^+ bombardment (3).

remarkable quantum photodesorption property of stainless steel is attributed to the presence of a layer of chromium oxide at the surface. Thus, the surface is more like a semiconductor than a metal, and semiconductors have well known photoproperties related to the band gap and band bending at the surface (13–17). The relatively high yield of CO from stainless steel raises the possibility that much of the disagreement about photodesorption in CO–metal systems may arise from background effects due to oxide surfaces, particularly when the metal is stainless steel. This would not seem to influence the results of Kronauer and Menzel (7), who used a glass system and tested rather thoroughly for background effects. Neither would it influence the results of McAllister and White (3) who did no experiments below 300 nm, where stainless steel background effects become important.

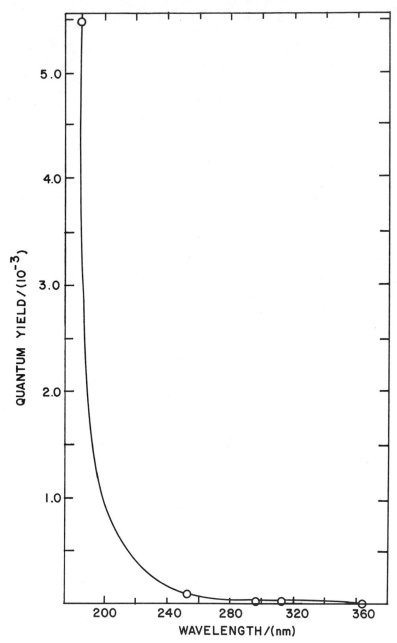

Figure 6. Variation of total yield with wavelength for photodesorption from 304 stainless steel. Photon flux density less than 290 mW cm^{-2} at all wavelengths (12).

On the basis of this work several conclusions may be drawn for the CO–Ni and CO–W systems: (1) The quantum photodesorption yield is negligibly small ($\lesssim 10^{-8}$) above 300 nm; (2) Around 250 nm there appears to be a threshold for quantum photodesorption in both systems, which suggests that future experiments should be done at wavelengths into the vacuum UV in an effort to isolate structure in the yield function. Such structure would be useful in guiding the development of models for the photodesorption process; (3) The work of Lichtman (12) is very interesting and suggests that enhanced quantum photodesorption might be observed on metal surfaces that are partially oxidized or otherwise modified to mimic the properties of semiconductor or insulator surfaces. This work also makes obvious the absolute necessity of accounting for desorption from sources other than the substrate.

Pd-Catalyzed Oxidation of CO

The number of examples of photoinduced chemical reactions on metal surfaces is extremely small and mechanisms are not understood. While there are a few isolated reports involving benzaldehyde on Cu (18), acetone on Ni (19) and ethylene on Ni and Pt (20), perhaps the best data is available on the oxidation of CO over Pd at pressures in the torr regime. Baddour and Modell (21) reported that the rate of CO oxidation over polycrystalline Pd wire could be photoenhanced tenfold at 420 K using 6.1 torr CO, 14.7 torr O_2, and 740 torr He in a recirculating loop reactor. This effect was wavelength dependent with a threshold at about 320 nm, which is reminiscent of the photodesorption of CO from Ni, W, and stainless steel. Subsequently, Chen, Close, and White (22) undertook a study of this process at low (10^{-5} torr) and high pressures with particular emphasis on separating the thermal and photoinduced (254 nm) rates. At low pressures no photoenhancement was detectable for a wide variety of experimental conditions of temperature, partial pressure, and pressure ratios. These conditions ranged from situations of high and low surface coverage, rate control by O_2 adsorption, and rate control by CO collision frequency (23, 24). Radiation from 185–600 nm was used.

At high pressures, however, the results of Chen, Close, and White (22) accord with and extend slightly the work of Baddour and Modell (21). Figure 7 shows the temperature dependence of the thermal and photoinduced rates and indicates a sharp change in the temperature dependence of both rates between 410 and 450 K. Between 333 and 413 K the photorate has an apparent activation energy of 45 kJ mol^{-1} (10.8 kcal mol^{-1}) but above 413 K the rate deteriorates, suggesting that the steady-state population of some species intimately involved in the photo-reaction is declining. The data of Figure 7 involved fixed pressures of

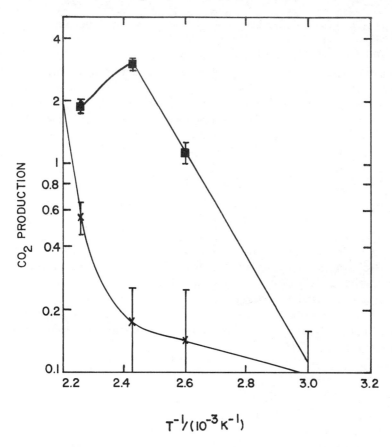

Figure 7. CO production (arbitrary units) during a 12-hr experiment vs. the inverse of the temperature (22). (■) Photorate, (×) thermal rate.

CO and O_2, 5 and 15 torr, respectively. Under these conditions the thermal rate is inhibited strongly by CO (25) and the surface involves significant amounts of rather weakly held CO that can be removed by heating or by lowering the CO partial pressure. IR spectroscopic measurements suggest two types of adsorbed CO (26, 27, 28); the more weakly held of these species is removed by heating into the temperature range 413–493 K. These IR data correlate the temperature dependence of the photorate with the loss of weakly held CO. The origin of the photoeffect itself is then related to some other step in the oxidation reaction. The most likely candidates appear to be the photoenhancement of the rate of O_2 dissociation on sites that normally do not dissociate O_2 or a photocatalyzed reaction between weakly held O_2 and adsorbed CO. At present there is no basis for distinguishing these two paths.

Mechanistic Considerations

Photon and Electron Stimulated Processes. There are several mechanisms that can lead to photon induced desorption. These mechanisms may be separated into the two general classes already indicated: one in which the desorption is thermally activated, and the other in which atomic-level electronic excitations produce desorption. Mechanisms falling into the electronic-excitation category are intrinsically more interesting because they are more likely to lead to useful spectroscopic probes of adsorbates. To date, photodesorption experiments have used, for the most part, photon energies of the same order of magnitude as the binding energies of gases on metal surfaces (0.5–6 eV). These photon energies give results that would be expected from heating of the substrate, and no relationship between the heat of adsorption and the photon energy for desorption is found (3). For energies up to 5–6 eV neutral molecule desorption appears to dominate. However, at electron energies in the neighborhood of 20 eV, positive ions are observed to desorb (29, 30). The electron stimulated desorption of ions is well known (29–32), whereas photon stimulated desorption of ions is not. However, the production of a particular desorption product should not depend on the excitation mode (33).

Even though ion desorption represents a very small fraction of the desorbed species in a photon or electron desorption experiment, measurements of neutral species have not been pursued extensively because of the relative difficulty of detection. Using electrons, excited neutral CO and O have been detected by time-of-flight methods by Redhead (34) and Newsham and Sandstrom (31). Desorbed ion energy distributions (30) and angular distributions (32, 35, 36) have been observed using electron stimulated desorption, and this technique shows promise of yielding direct information of local adsorbate geometry.

As noted before, it would be helpful to study photodesorption for energies greater than 6 eV. Peavey and Lichtman (8) have carried out some photodesorption experiments using synchrotron radiation, which extend the wavelength range into the vacuum ultraviolet region, and these experiments indicate the existence of an intrinsic quantum desorption mechanism. Desorption of CO_2, CO, and H_2 was observed from a W sample when exposed to the integrated light flux (unmonochromatized) from the Stoughton, Wisconsin storage ring. The energy distribution of the synchroton radiation (240-MeV electron beam energy) extended from approximately 20 nm into the IR with a maximum at approximately 50 nm ($\sim$ 25 eV). A rough estimate of the total beam power at the sample was 5 μW cm^{-2}, which was judged low enough to rule out thermally activated desorption processes. No ion desorption was observed; it is likely that stray radiation interacting with stainless steel gave rise to a significant fraction of the observed neutrals.

As noted above, the phenomena of photodesorption and electron stimulated desorption are closely related. Clearly, the excitation mechanisms are different for photons and electrons, but the dynamics that lead to desorption are very similar for the two cases. Both processes excite the adsorbed molecule–substrate complex to an excited (electronic) state, which may then lead to a final-state molecule or ion desorbed from the surface. The details of the excitation process and the states initially populated may differ, but the subsequent desorption process in each case is identical. This is because the motion of the excited molecule is much slower than the electronic relaxation to those excited states that lead to desorption.

One of the ways the two excitation methods differ is that a photon must give up all of its energy in an interaction process whereas an electron can suffer several inelastic collisions before or after it excites the adsorbate. There is also a distinction between electron excitation of a metal and an insulator. In a metal, the electron mobility is sufficiently large that the electron that stimulates desorption does not remain in the vicinity of the excited molecule whereas in an insulator there may be charging effects. Momentum transfer to an adsorbate by a low energy electron is negligible because of the large mass ratio. Therefore, atomic level desorption by either photons or electrons involves energy transfer to the electronic structure of the adsorbate. There are several different channels that the excited molecule may take. It can desorb from the metal in its excited state and radiatively decay or it can relax in the vicinity of the surface and desorb as a ground-state molecule. If the excited state is ionic, it can desorb either as an ion or a neutral (after being neutralized by an electron tunneling from the substrate) (33). For high enough excitation energies, secondary electrons may also contribute to the excitation of the adsorbates. Finally, there exists the possibility that the excitation energy is degraded to heat the surface (3).

Photon–Surface Interaction. In both thermal and quantum photodesorption processes, the first step involves the interaction of the incident beam with the surface. Monolayer coverages of adsorbed atoms represent a very small perturbation on the reflection and transmission processes. These perturbations can be analyzed by comparison with a clean surface. The incident and reflected fields combine to form a standing wave above the surface and it is this combined field that acts on an adsorbate atom. It is important to use this field if one is interested in comparing the desorption efficiencies. Depending on the reflection coefficient, which is determined by the wavelength, angle of incidence, polarization and optical constants of the material, the total electric field near the surface can be reduced considerably from the incident field value.

The transmitted component of electric field decays exponentially with distance into the bulk; the loss of intensity being accompanied by heat production in the material. The power absorbed by a material is given by

$$P(\hbar\omega) = \frac{1}{2} \int \text{Re} \{J \cdot E dV\} = \frac{1}{2} \sigma^{(1)}(\hbar\omega) \int E^2 dV$$

where E is the electric field vector in the material, J is the current density, and $\sigma^{(1)}(\hbar\omega)$ is the real part of the optical conductivity. The associated increase in temperature of the sample can lead to thermal desorption. This mechanism must produce a nonlinear effective desorption cross section as a function of incident beam intensity (for a fixed wavelength). The dominant nonlinear dependence is the exponential temperature dependence associated with thermal activation processes.

Two limiting cases can be analyzed, one where the incident intensity is constant and a steady state condition is established and a second where a short light pulse is used. Fabel, Cox, and Lichtman (12) have analyzed the temperature rise of a thermally isolated sample under steady state conditions. In this case, heat input from the light beam is equal to heat radiated by black-body radiation; to maintain temperature changes of less than a few degrees above ambient temperature (assumed to be 300 K), the total input power must be kept below 1 mW cm^{-2}.

The analysis for a pulsed light source is more difficult and must take into account the heat capacity of the material and thermal conduction from the optical penetration layer into the bulk material. The estimate of Fabel, Cox, and Lichtman (12) does not account for the possibility of a higher temperature at the surface because of the small penetration depth of light into the surface; therefore, the analysis only applies rigorously to relatively thin, thermally isolated films. To our knowledge, no comprehensive analytical treatment of photon-induced thermal desorption has been carried out and compared with experimental data.

Microscopic Desorption Models. The basic microscopic model used to discuss electron and photodesorption phenomena is accredited to Menzel, Gomer and Redhead (MGR) (37, 38, 39). In the MGR model an excitation causes a Frank–Condon transition from the ground state of the adsorbate–surface system to a neutral or ionic antibonding state of the complex (adsorbed species plus the substrate). As a consequence of this excitation the neutral or ionic particle starts to move away from the surface. If de-excitation does not occur, either ions or neutrals can be desorbed, but often the excited state is recaptured by electron tunneling to fill the orbital emptied by photon absorption. The antibonding states from which desorption occurs must be quite high in energy at the position of the adsorbed molecule since ions are observed to desorb with kinetic energies of up to 10 eV. The antibonding states were postulated and subsequent results and elaborations start with this premise. To our knowledge there have been no theoretical calculations of the antibonding states.

In semiconductors, the model proposed by Baidyaroy, Bottoms, and Mark (*40, 41*) and by Shapira, Cox, and Lichtman (*42*) accounts for most of the observed features. The primary property of semiconductors is a photodesorption threshold at the band gap energy. When an acceptor-like adsorbate is on the surface of an n-type semiconductor, charge transfer between the surface and adsorbate sets up a potential that helps bond the adsorbate to the surface. When photons having energy above the band gap interact with the surface, electron–hole pairs are created near the surface. In the presence of the surface potential, holes migrate to the surface where they can neutralize an adsorbate; this leads to desorption of a neutral atom. This model accounts for the primary feature (band gap radiation threshold) but additional work is needed to investigate the dynamics leading to desorption after neutralization.

A fundamentally new mechanism of electron stimulated desorption has been proposed recently by Knotek and Feibelman (*43*). They show that Auger decay of a core hole explains both the desorption thresholds and the large charge transfer involved in the electron stimulated desorption of positive ions from maximal-valence transition-metal oxides. The MGR model does not readily account for these features. The proposed Auger process will proceed in the same way upon photon creation of a core hole, but this has not been verified experimentally.

To understand this process, consider the specific case of TiO_2. Since the Ti ion is in a maximal-valence state, there are no valence electrons available for intraatomic Auger transitions (normally the dominant relaxation channel). When a $3p$ electron is ejected from the core, interatomic Auger transitions can take place from the valence levels of an O ion. In this process the charge on the O atom changes by two or three electrons, leaving a neutral or positively charged atom. If this O atom or positive ion is at the surface, it finds itself in a Madelung potential for a negative ion (and possibly in an antibonding state) and is therefore repelled from the surface.

This mechanism leads to positive ion desorption (possibly neutral desorption also, but this apparently has not yet been observed experimentally) and more importantly to a specific threshold for the process to occur. The desorption threshold is associated with the energy necessary to create a core hole in the maximal-valence positive ion. In the case of TiO_2, the $3p$ shell of the Ti atom occurs at 34 eV. This is the energy range where synchrotron radiation provides the most reasonable approach for experimental studies. The mechanism could lead to an atom-specific, valence-sensitive probe of surfaces.

Another new desorption mechanism has been suggested recently (44). In the proposed model a neutral atom interacts with the substrate via a potential energy curve that has a steeply repulsive part as the distance between the atom and the substrate becomes small and an attractive part that decreases as z^{-3}, where z is the distance from the substrate. The position of the minimum in the potential curve is approximately given by the sum of the atomic radii of the adsorbate and substrate atoms.

In the desorption sequence an adsorbed atom or molecule is ionized by an electron or photon. The minimum energy for this step is the energy required to transfer an electron from the adsorbate to the Fermi level of the substrate. Due to the fact that the ion is much smaller than the adsorbate atom, the ion–substrate potential energy curve has its minimum nearer the substrate. The force on the ion is toward the substrate, imparting kinetic energy to the ion. As the ion comes nearer the surface the probability for an electron to tunnel from the substrate to neutralize the ion increases and the ion finally is neutralized. In the meantime, the atom has acquired kinetic energy and is nearer the substrate than the initial equilibrium position and therefore higher up on the repulsive part of the potential energy curve. If the sum of the new potential energy plus the acquired kinetic energy is more than the desorption energy, the atom will desorb. This sequence can be seen in Figure 8.

This model works equally well for explaining the production of positive ions. In this case one needs three potential energy curves: one for the neutral atom, one for the singly ionized ion, and one for the doubly ionized ion. As in the previous case an ion is created (now with two electrons missing) and is attracted toward the substrate. This ion can pick up one or two electrons. If it picks up one electron then a singly ionized ion comes off the surface. If it picks up two electrons then a neutral atom is desorbed.

This model predicts that the threshold for producing positive ions from a surface is equal to the energy necessary to produce a doubly ionized ion on the surface. This means that one needs sufficient energy to excite an Auger process that will leave an ion with two valence electrons missing. Note that formation of a singly charged ion leads to neutral desorption while a doubly charged ion leads to both positive ion and neutral desorption.

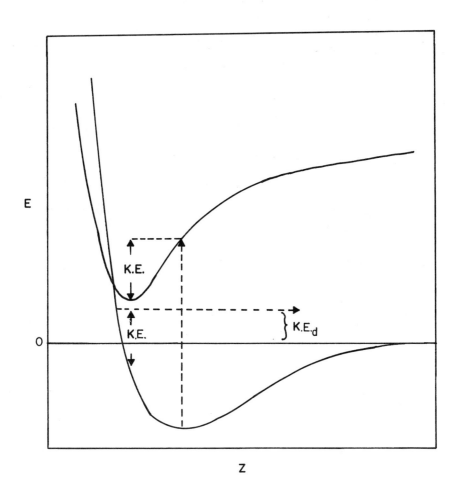

Figure 8. Depiction of an atomic desorption event. The potential energy curves represent the total energy of the system consisting of the substrate in the ground state and a stationary adsorbate. After excitation to the upper state, the adsorbate is accelerated toward the substrate gaining kinetic energy; the adsorbate is de-excited to the lower curve by interaction with the substrate. It still has the kinetic energy gained in the above and now its total energy is greater than the binding energy, consequently leading to desorption with kinetic energy KE_d.

Summary

Using wavelengths above 300 nm, photodesorption of CO from Ni and W is dominated by thermal desorption. If a quantum effect exists it must have a cross section below 10^{-23} cm^2. Beginning around 250 nm and extending to shorter wavelengths, there appears to be some quantum effect. The results suggest that wavelengths into the vacuum UV could be used profitably in future studies designed to search for quantum photodesorption. Spurious background effects due to scattered light must be carefully avoided as clearly indicated by the excellent work of Lichtman (8). This work also indicates that oxidized surfaces may have much stronger photodesorption properties than the metals themselves.

Electron stimulated desorption of ions and neutrals is well known and there are analogies that can be drawn between photoinduced and electron stimulated desorption, particularly at energies above 10 eV. Recently, fairly detailed theoretical models have been proposed for certain types of electron induced desorption. These involve Auger transitions and tunneling.

While considerable electron stimulated desorption data exists, related photon experiments and theoretical models are in a very rudimentary stage of development.

Very few photoenhanced reactions occurring over transition metals are known and none are well understood. There is evidence from two laboratories that the CO oxidation reaction over Pd can be photoenhanced when the reactant pressures are of the order of a few torr. The mechanism of this photoenhancement is not understood.

Acknowledgment

We gratefully acknowledge the Office of Naval Research for their support in this research.

Literature Cited

1. Lange, W. J.; Riemersma, H. *Trans. Natl. Vac. Symp.* **1961**, *8*, 167.
2. Adams, R. O.; Donaldson, E. E. *J. Chem. Phys.* **1965**, *42*, 770.
3. McAllister, J. W.; White, J. M. *J. Chem. Phys.* **1973**, *58*, 1496.
4. Holloway, P. H.; Hudson, J. B. *Surf. Sci.* **1972**, *33*, 56.
5. Madden, H. H.; Küppers, J.; Ertl, G. *J. Chem. Phys.* **1973**, *55*, 3401.
6. Brainard, J. "Desorption by Photons; Carbon Monoxide Adsorbed on Nickel," Ph.D. Thesis, Washington State University, Pullman, Washington, 1968.
7. Kronauer, P.; Menzel, D. In "Adsorption–Desorption Phenomena"; Ricca, F., Ed.; Academic: New York, 1972.
8. Peavey, J.; Lichtman, D. *Surf. Sci.* **1971**, *27*, 649.
9. Lange, W. J. *J. Vac. Sci. Technol.* **1965**, *2*, 74.
10. Genequand, P. *Surf. Sci.* **1971**, *25*, 643.

11. Paigne, J. *J. Chim. Phys. Phys.-Chim. Biol.* **1972**, *69*, 1.
12. Fabel, G. W.; Cox, S. M.; Lichtman, D. *Surf. Sci.* **1973**, *40*, 571.
13. Hauffe, K. *Rev. Pure Appl. Chem.* **1968**, *18*, 79.
14. Wolkenstein, T. *Adv. Catal.* **1973**, *23*, 157.
15. Murphy, W. R.; Veerkamp, T. F.; Leland, T. W. *J. Catal.* **1976**, *43*, 304.
16. Sancier, K. M.; Morrison, S. R. *Surf. Sci.* **1973**, *36*, 622.
17. Shapira, Y.; McQuistan, R. B.; Lichtman, D. *Phys. Rev. B* **1977**, *15*, 2163.
18. Terenin, A. N. *Probl. Kinet. Katal.* **1955**, *8*, 17.
19. Terenin, A. N. *J. Chim. Phys. Phys.-Chim. Biol.* **1957**, *54*, 114.
20. Aleksandrowicz, J., Ph.D. Thesis, Massachusetts Institute of Technology, Cambridge, MA, 1970.
21. Baddour, R. F.; Modell, J. *J. Phys. Chem.* **1970**, *74*, 1392.
22. Chen, B.-H.; Close, J. S.; White, J. M. *J. Catal.* **1977**, *46*, 253.
23. Golchet, A.; White, J. M. *J. Catal.* **1978**, *53*, 266.
24. Bonzel, H. P.; Ku, R. *Surf. Sci.* **1972**, *33*, 91.
25. Cochran, H. D.; Donnely, R. G.; Modell, M.; Baddour, R. F. In "Colloid and Interface Science"; Kerker, M., Ed.; Academic: New York, 1976; Vol. III.
26. Eischens, R. P.; Francis, S. A.; Pliskin, W. A. *J. Phys. Chem.* **1956**, *60*, 194.
27. Baddour, R. F.; Modell, M.; Huesser, U. K. *J. Phys. Chem.* **1968**, *72*, 3621.
28. Palazov, A.; Change, C. C.; Kokes, R. J. *J. Catal.* **1975**, *36*, 338.
29. Madey, T. E.; Yates, J. T., Jr. *J. Vac. Sci. Technol.* **1971**, *8*, 525.
30. Nishijima, M.; Propst, F. M. *Phys. Rev. B* **1975**, *2*, 2368.
31. Newsham, I. G.; Sandstrom, D. R. *J. Vac. Sci. Technol.* **1973**, *10*, 39.
32. Czyzewski, J. J.; Madey, T. E.; Yates, J. T., Jr. *Phys. Rev. Lett.* **1974**, *32*, 777.
33. Gersten, J. I; Janow, R.; Tzoar, N. *Phys. Rev. B* **1975**, *11*, 1267.
34. Redhead, P. A. *Nuovo Cimento* **1967**, *Suppl. 5*, 586.
35. Madey, T. E.; Czyzewski, J. J.; Yates, J. T., Jr. *Surf. Sci.* **1975**, *49*, 465.
36. Ibid. **1976**, *57*, 580.
37. Menzel, D.; Gomer, R. *J. Chem. Phys.* **1964**, *41*, 3311.
38. Redhead, P. A. *Can. J. Phys.* **1964**, *42*, 886.
39. Menzel, D.; Gomer, R. *J. Chem. Phys.* **1964**, *41*, 3311.
40. Baidyaroy, S.; Bottoms, W. R.; Mark, P. *Surf. Sci.* **1971**, *28*, 571.
41. Ibid. **1972**, *29*, 165.
42. Shapira, Y.; Cox, S. M.; Lichtman, D. *Surf. Sci.* **1975**, *50*, 503.
43. Knotek, M. C.; Feibelman, P. J. *Phys. Rev. Lett.* **1978**, *40*, 964.
44. Antoniewicz, P. R., unpublished.

RECEIVED October 2, 1978.

4

Photochemistry of Surfactant and Hydrophobic Compounds in Organizates and at Interfaces

Effects of Environment and Solute Penetration on Excited State Quenching and Reactivity[1]

DAVID G. WHITTEN, JANET A. MERCER-SMITH, RUSSELL H. SCHMEHL, and PAUL R. WORSHAM

Department of Chemistry, University of North Carolina, Chapel Hill, NC 27514

> *This chapter reviews several photoreactions occurring at interfaces and in "organizates" such as monolayer films, micelles, and multilayer assemblies. The first section reports a study of some functionalized surfactants whose photoreactions have been carried out in films at an air–water interface, micelles, and assemblies. The reactions observed are environment-sensitive and suggest that the molecules studied, surfactant ketone and styrene derivatives, could be useful environmental probes in other organizates. A second section deals with penetration of multilayer assemblies by various solutes at an assembly–solution interface, while the third section deals with interfacial photoredox reactions that occur as a consequence of such penetration phenomena.*

This chapter deals predominantly with photoreactions occurring at various types of interfaces. In particular there is a focus on phenomena involving functionalized surfactant molecules incorporated into molecular "organizates." The types of organizates formed from these molecules include monolayer films at an air–water interface, micelles, liposomes or vesicles, lipid bilayers, and multilayer assemblies; the latter are usually

[1] Photochemical Reactivity in Organized Assemblies, Paper XIII. Paper XII, D. G. Whitten, *Augewandte Chemie*, 1979, 18, 440.

constructed by consecutive deposition of several monolayers onto a rigid support. The chapter is divided into three parts: the first section compares photoreactivity observed for some functionalized surfactants anchored in monolayer films, miceles, and multilayer assemblies. This section emphasizes the utility of these molecules and their reactivity as environmental probes and also points out some possible synthetic applications. The second and third sections deal with interfacial phenomena involving reactions between a surfactant molecule anchored in an assembly and a solute penetrating, contacting, or otherwise interacting from an adjacent liquid or vapor phase.

There have been numerous reports of reactions proceeding at rates different from those observed in homogeneous solution when one or more of the reagents is incorporated into an organizate such as a micelle (*1–4*) or monolayer film at an air–water interface (*5–9*). In the case of micelles this has triggered much interest in their use as models for enzyme systems (*11–16*). More recently there has been considerable interest in photophysical and photochemical phenomena occurring in these phases. Investigations of photochemical reactions occurring in micelles have suggested important roles for surface or host charge interactions, selective solubilization, local concentrations, and viscosity–orientation effects (*17–20*). While some of these factors undoubtedly play major roles in the more highly condensed monolayer films and multilayer assemblies, investigations of photochemical and photophysical phenomena in these phases have indicated that the tight packing into highly organized structures in which there is little mobility is often of critical importance (*22–26*).

Properties of Organizates as Probed by the Photoreactions of Functionalized Surfactant Molecules

Numerous studies of fluorescence behavior in organizates have provided information about the location and mobility of "probe" molecules in various organizates (*21, 27, 28*). In a number of elegant investigations Kuhn and co-workers have shown that the regular spacing in a known and controlled molecular architecture (i.e., multilayer assemblies formed by transferral of monolayer films) is well suited for studies of distance dependence of energy transfer as well as a number of other photophysical processes (*22, 23*). Investigations from our own laboratories have demonstrated that the microenvironment obtained by monolayer incorporation can produce striking effects on photoreactivity with polar or hydrophilic chromophores (*24–30*). As pointed out above, these effects can be attributed to high effective concentrations of reactants, restricted diffusion, and to tight packing into sites in which molecular motion is somewhat restricted. In recent investigations we have synthesized a series of molecules such as 1 and 2 in which the reactive unit is in the hydrophobic portion of a surfactant molecule, such that the anchoring carboxyl groups

$$R-(CH_2)_{13}COOH$$

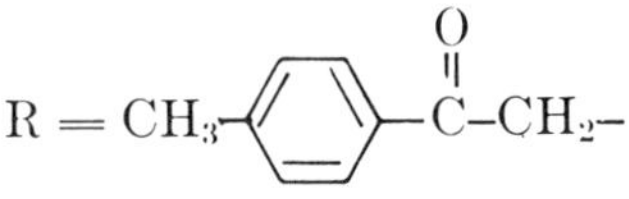

$$R = CH_3-\text{C}_6H_4-\overset{O}{\underset{\|}{C}}-CH_2- \quad \quad R = CH_3-\text{C}_6H_4-\overset{H}{\underset{H}{\overset{|}{C}}}=C\diagdown$$

1 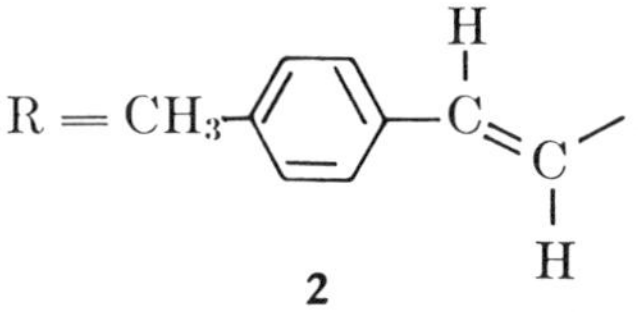   2

and reactive site are well separated. A frequent and often legitimate criticism of the use of structural probes in organizates is that the probes themselves either suffiicently disrupt the system, leading to spurious results, or reside in special sites so only a particular area is sampled (31). In 1 and 2 as well as in other systems presently under investgiation we have attempted to design probe molecules compatible with simple surfactants such as saturated fatty acids. By using relatively small reactive groups with molecular dimensions and polarity similar to a hydrocarbon chain, it was anticipated that minimal disruption of the system would be produced by the probe. As will be pointed out, our results suggest that both 1 and 2 are fairly good probes for organizates in this respect.

For 1 the chief intramolecular photoreactions, anticipated on the basis of solution studies of structurally related ketones, would be the Type I and II photoelimination processes (Equations 1 and 2, respectively). Intermolecular processes could include numerous reactions originating via

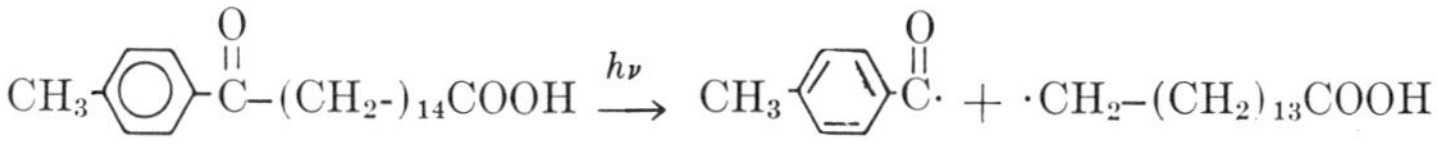

$$CH_3\text{-}C_6H_4\text{-}\overset{O}{\underset{\|}{C}}\text{-}(CH_2\text{-})_{14}COOH \xrightarrow{h\nu} CH_3\text{-}C_6H_4\text{-}\overset{O}{\underset{\|}{C}}\cdot + \cdot CH_2\text{-}(CH_2)_{13}COOH \quad (1)$$

1

$$CH_3\text{-}C_6H_4\text{-}\overset{O}{\underset{\|}{C}}\text{-}(CH_2)_{14}COOH \xrightarrow{h\nu} CH_3\text{-}C_6H_4\text{-}\overset{O}{\underset{\|}{C}}\text{-}CH_3 + CH_2\text{=}CH\text{-}(CH_2)_{11}COOH \quad (2)$$

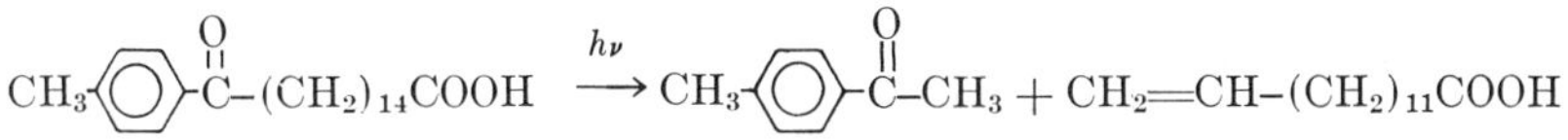

hydrogen atom abstraction; in general, however, for ketones possessing an abstractable γ hydrogen the dominant process is the Type II elimination, which would proceed via formation of diradical 3 in the case of 1. The Type II process has been observed for different ketones in a number of diverse media including polymers, solids, melts, and micelles (32, 33). Reduction in mobility in polymers, for example below the glass transition

CH₃—⟨phenyl⟩—C(·O-H)(CH₂)(CH₂)···CH–(CH₂)₁₁COOH

3

temperature or in rigid crystals, drastically reduces or eliminates this process (34, 35, 36). These features suggested that incorporation of 1 into various organizates and study of its photoreactivity therein could furnish useful structural information. This was further suggested by the observation that the fate of the diradicals formed by γ-hydrogen abstraction is extremely solvent-environment sensitive (33, 37), and thus with 1 might give an indication of the polarity of the ketone microenvironment. Therefore, we have studied photoreactions of 1 in solution, in deposited multi-layer assemblies, and in anionic micelles (38).

Irradiation of 1 in degassed benzene solution leads to the production of p-methylacetophenone, the anticipated Type II product, as the only volatile product with a reasonable quantum efficiency ($\phi_{II} = 0.2$). Although 1 is water insoluble, it can be incorporated into aqueous sodium dodecyl sulfate (SDS) solutions above the critical micelle concentration. Irradiation of 1 in SDS micelles also leads to the Type II elimination product, p-methylacetophenone, as the only detectable volatile product. However, in the SDS micelles the efficiency for the process is enhanced over that obtained in benzene ($\phi_{II} = 0.8$). Ketone 1 does not form good monolayer films by itself; films formed by the ketone show increases in pressure at relatively large areas. As these films are further compressed a gradual rise in the pressure–area curve typical of expanded or liquid monolayers is observed (39). This type of behavior is probably due to the polarity of the carbonyl groups that are attracted to the water interface in these films of pure ketone since we have observed this for other carbonyl compounds. When 1 is mixed with arachidic acid (n-$C_{20}H_{40}O_2$) in ratios of 1:5 or greater, the corresponding films exhibit typical condensed monolayer behavior with very steep pressure–area isotherms. In these films the measured area per molecule of 1 is found to be 20 Å², which is nearly the same as that for arachidic acid molecules in pure condensed films (39).

Irradiation of multilayer assemblies formed from 1–arachidic acid mixtures leads to relatively slow loss of the carbonyl group ($\phi_{-1} = 0.06$) as indicated by monitoring the carbonyl transitions in the IR (6360 nm) or long wavelength UV (290 nm). In contrast to the behavior of 1 in solution or micelles, irradiation in the multilayer assemblies yielded very small amounts of volatile products. Only a trace of p-methylacetophenone and no Type I products (toluene or p-methylbenzaldehyde) were detected. The quantum yield for the Type II process in multilayers can be calculated as 0.03–0.0001 and thus the bulk of the photoreaction must involve paths other than type I or Type II cleavage. Some nonvolatile surfactant products could be extracted from the assemblies; preliminary analyses suggest that these are alcohols and olefins resulting from radical coupling and disproportionation following an intermolecular hydrogen abstraction.

Comparison of the photobehavior of 1 in the different media indicates that there must be striking differences in the various microenvironments. The sharp reduction in the quantum efficiency of the Type II fragmentation upon going from solution to the multilayer assemblies suggests that the tightly packed, condensed molecular environment provided by the latter does not permit the excited ketone to assume the cyclic six-membered transition state needed for formation of biradical 3. The area per molecule measured for 1 is about the same as that obtained for a fatty acid in monolayers and very close to the intermolecular spacing measured for crystalline linear paraffins (40). In the latter the chains are packed in a regular "zigzag" arrangement and it has been established that such an arrangement persists in monolayer films and assemblies constructed from straight-chain saturated fatty acids, esters, and alcohols. Both the measured area per molecule and photobehavior observed for 1 in the multilayers are consistent with an arrangement wherein 1 is also extended in a highly regular structure. The lack of much Type II reaction further indicates that little, if any, rearrangement can occur during the excited state lifetime. Based on these results the most consistent picture of the hydrophobic region of the assemblies is a region in which relatively little motion can occur. These results suggest that the degree of order and/or restriction of motion in the multilayers investigated thus far is greater than that in most polymers. Also, Guillet and co-workers have found that several carbonyl-containing polymers give the Type II reaction both above and below the glass transition temperature, although with reduced efficiency in the latter case (34, 35, 36). Since different surfactants and mixtures thereof can form a variety of monolayer films, which differ greatly in compressibility and presumably in crystallinity of the hydrophobic region (39), it might be expected that corresponding changes in the photoreactivity of 1 in these environments should be observed. We are currently examining the photoreactivity of 1 in assemblies formed using different hosts, pure and in mixtures, as a probe of these environments. We are also examining the behavior of 1 in assemblies contacted with various aqueous and nonaqueous solutions. As will be pointed out, several investigations have established that solute and solvent molecules can penetrate assemblies from solution contacting the assembly without dissolving or producing major apparent disruptions of the multilayer structure (22, 23). In preliminary studies, we have found that irradiation of assemblies containing 1 and arachidic acid in contact with organic solvents such as methylcyclohexane leads to pronounced enhancement of Type II reactivity (41). Although quantum yields have not yet been measured for these systems, it seems clear that the observed enhancement indicates a change from solid-like to more fluid properties for the hydrophobic region of the assemblies. This occurs with the penetration of the organic solvent.

In contrast to the results obtained for 1 in assemblies, the high quantum yields observed in SDS micelles indicate a highly mobile, liquid-like environment in the latter phase. This is not at all unreasonable in view of other investigations, which also suggest a relatively low viscosity for the interior of micelles (16, 20, 21). What is perhaps surprising is that the quantum yield for the SDS micelles is enhanced by a factor of four over the value measured for benzene solutions. Studies of solvent effects on the Type II reaction of other aryl ketones have indicated that quantum efficiencies are generally enhanced in polar, hydrogen-bonding solvents (32–37); this has been attributed to a hydrogen-bonding stabilization of the biradical analogous to 3. This stabilization evidently favors product formation vs. back reaction and leads to enhanced efficiencies, approaching unity in alcohols or water–alcohol mixtures. In previous studies of the nonsurfactant ketones octanophenone and valerophenone solubilized in cationic micelles, relatively high quantum efficiencies for the Type II process were obtained (37). It was concluded from these results that the carbonyl group was not exclusively in a hydrocarbon region of the micelle, but rather in either the Stern layer or near the micellar surface (37). The efficiencies measured for 1 are close to those obtained for the nonsurfactant ketones and thus suggest that 1 must be in a similar, relatively polar environment in the SDS micelles. While the results could be interpreted as indicating that considerable penetration of water into the interior micelle occurs, a second interpretation might be that the polymethylene chain in 1 is mostly bent back so that both ends lie near the surface. Such folding has been suggested to occur in other cases (42) and would probably be reasonable in view of the high surface-to-volume ratio in micelles, which strongly favors surface sites for even very weakly surface-active molecules (*43*). Such a picture of a largely folded configuration of **1** in micelles would also be consistent with the observed behavior of **1** when spread in pure films at a water–air interface.

The photochemical behavior of **2** also has been investigated in a variety of media including solutions, monolayer films at the air–water interface, anionic micelles, and multilayer assemblies. The pictures of the microenvironment provided by the photoreactivity of **2** generally are congruent with those obtained by the study of ketone **1**. The chief reaction obtained for **2** upon irradiation in solution, the solid state, and all of the organizates in which it has been studied thus far is dimerization. In all cases the dimers appear to be cyclobutane structures formed by cycloaddition across the ethylene units (*44*). The reaction has been studied most extensively with **2** in monolayer films at an air–water interface and in multilayer assemblies. Unlike ketone **1**, the pure hydrocarbon **2** readily forms monolayer films at an air–water interface, exhibiting the same condensed behavior usually found with linear saturated fatty acids such as arachidate (*45*). The measured area per molecule in pure and in mixed films is ca. 18–19 $Å^2$ in the compressed (20 dyne/cm) films. This area is close to that measured for **1** in mixed films and the same as observed for most fatty acids. This suggests that the polymethylene chains of **2** are packed in a regular arrangement and that the chromophore packs without causing any distortion. Irradiation at 254 nm of compressed films of pure **2** at the air–water interface leads to rapid formation of a single (>95%) dimeric product. When the irradiation is carried out at constant area, there is no discernable change in surface pressure; the converse is true when the irradiation is carried out in the constant pressure mode. The lack of pressure or area change with photodimerization suggests that little rearrangement can occur during the process and that the packing of monomers in the monolayer film controls the stereochemistry of the product formed.

Multilayer assemblies can be formed readily by transferral of the compressed films of **2** to quartz slides. Examination of these assemblies with a polarizing microscope indicates that the hydrophobic surface is highly regular. Irradiation of these multilayers leads to formation of the same product obtained by irradiation of the films; microscopic examination of multilayer assemblies following irradiation indicates no detectable disruption of the hydrophobic surface. The same product is formed in multilayer assemblies where there is no hydrophobic contact between layers of **2** and where there are adjacent layers of **2**. In both cases the reaction may be followed spectroscopically and the same isosbestic points are observed during the conversion (*45*). The NMR spectrum is most consistent with **4** as the structure of the predominant photodimers formed in the films and assemblies (*45*). The observation of a weak fluorescence from the photoproduct similar to that of the intramolecular benzene excimer formed in other fused systems (*46*) is also consistent with the assignment of a trans–syn–trans structure (**4**).

$$\underset{\underset{4}{\text{COOH} \quad \text{COOH}}}{\underset{(CH_2)_{13} \quad (CH_2)_{13}}{\text{[cyclobutane-linked bis(4-methylphenyl) structure]}}}$$

Irradiation of **2** in solution leads to the formation of several dimeric products that can be partially separated by high pressure liquid chromatography (HPLC). The presence of more than the four dimers resulting from simple combination of two *trans*-**2** units suggests that irradiation in acetonitrile solution yields some trans–cis isomerization in addition to dimerization. Product **4** is among the dimers produced in solution but it does not appear to be the predominant product. Irradiation of solutions containing **2** incorporated in SDS micelles also leads to dimer formation as the chief photoreaction; here, too, there are several dimeric products formed but it appears that, in contrast to the solution, **4** is the major product (*44*).

The photobehavior of **2** in the monolayer films and assemblies indicates a regular solid-like structure in the hydrophobic regions that once again exerts a remarkable control over reactivity. In this case the effect takes the form of a topological control (For a discussion of topological control in solid state photochemistry, *see* Ref. *47*); thus while the same type of reactivity (dimer formation) as in solution is observed, the reaction in the films and assemblies is stereoselectively controlled. This is similar to some crystal photoreactions where specific monomer packing in the crystal leads to selective product formation (For a discussion of topological control in solid state photochemistry, *see* Ref. *47*). Reaction of **2** in SDS micelles leads to several products, which indicates that there is much less order and/or restriction of mobility in the hydrophobic interior of the micelles. However, the indication that there is some preference for forming product 4 in the micelles compared with in solution suggests that there may be either some preferential association of the styrene chromophore in 2 prior to excitation or that the chains may have some tendency to extend so that the chromophores of neighboring 2 molecules align. If the latter is true, this would indicate rather different sites for the reactive chromophores in 1 and 2 in the SDS micelles. While this may be reasonable in view of the rather different polarities of the styrene and aryl ketone chromophores, such a result underlines the importance of interpreting with care experiments using reactive molecules as probes of the microenvironment in loosely organized units such as micelles.

Reactions in Multilayer Assemblies: Penetration of the Assembly by Various Solutes at an Assembly–Solution Interface

The ability to incorporate reactive molecules in an assembly having a regular, specified, and controllable structure suggests numerous possibilities for carrying out selective reactions, particularly interfacial processes in which there is some control of the reactive solute's access to the anchored reagent. Therefore, it is of particular importance to determine what sort of penetration can occur with different types of solutes and what effects penetration of solute or solvent molecules produces on the assemblies. One of the earliest studies of phenomena of this sort was by Kuhn and co-workers (*22, 23, 48*) using an azo dye coupling reaction as a probe of the ability of organic solutes to penetrate and react in an assembly constructed from multilayers of fatty acid. In this study a surfactant diazonium ion was placed in assemblies and covered by varying numbers of layers of cadmium arachidate. The layers were immersed in an aqueous solution containing 8-hydroxyquinoline, which forms an azo dye by coupling with the diazonium salt in solution; conversion of the assembly-bound diazonium salt to azo dye proceeded at the same rate for the various parts of the sample, indicating rapid penetration of the assembly by the organic solute. Little or no alteration of the assembly was produced by the penetration or reaction; this was indicated by the subsequent deposition of a fluorescent dye on the assembly and a study of fluorescence quenching by the newly formed azo dye (*22, 23*). The extent of fluorescence quenching observed showed the expected dependence on the distance between the donor and acceptor, which is determined by the number and length of intervening fatty acid units.

In contrast to their findings with neutral organic solutes, Kuhn and co-workers (*22, 23*) found that larger, charged molecules such as the dye 5 would not readily penetrate multilayer assemblies. In the case of 5, exposure of an aqueous solution of the dye to a multilayer assembly with a hydrophobic outer face led to adsorption of 5 onto the surface of the assembly. This was easily checked by fluorescence quenching experiments that demonstrated a fairly uniform coverage of the surface but no penetration in an assembly constructed from cadmium arachidate. In additional studies with 5 it was found that some penetration could occur when different fatty acids or counter ions were used (23). Extended studies with charged surfactants have demonstrated that for these compounds exposure of multilayer systems can lead to the formation of new monolayers on the surface. This occurs when the surfactant has an organization at the molecular level similar, but not identical, to layers transferred by the Langmiur–Blodgett method (49–53).

5

Our studies of penetration of multilayer assemblies have involved predominantly the anchoring of porphyrins and their metal complexes in the assemblies. We have monitored reactions of the anchored porphyrins with reagents entering the assembly from a solution–assembly or gas–assembly interface. Several studies have indicated that gaseous reagents such as CO, N_2, O_2, and NO can readily penetrate even multiple layers of fatty acid such as cadmium arachidate and react with porphyrins or metalloporphyrins incorporated in the assemblies (26, 45, 54, 55). Porphyrins located at either hydrophilic or hydrophobic sites can readily react with these neutral reagents. However, in photooxidation reactions of free base protoporphyrin IX derivatives, we have observed the formation of rather different product distributions in films, micelles, and assemblies compared with those formed in fluid solution (56). For example, the diester 6 gives predominantly the hydroxyaldehyde 7, with only a trace of diformyl deuteroporphyrin 8, upon irradiation in methylene chloride or chloroform. In contrast, irradiation in the organizates yields 8 as the predominant product to the extent of as much as 93% in dry multilayers (56). The source of the difference in product distribution is presently under investigation; several factors may be important here, including the polarity and degree of congestion in the microenvironment. Studies of ligand exchange with anchored metal porphyrins indicate that small, neutral organic ligands such as pyridine and piperidine can readily penetrate the assemblies from an aqueous solution. This agrees with the findings of Kuhn and co-workers (22, 23).

We have recently examined penetration of multilayer assemblies by metal ions by investigating formation of metalloporphyrins from anchored free base porphyrins in assemblies (57). In this study we have used several different porphyrins whose substitution patterns control the type of site occupied by the chromophore in films and multilayer assemblies; thus in some cases the porphyrin ring is folded into a relatively hydrophobic region of the assemblies while in others the porphyrin ring lies at the hydrophilic interface. Although in several cases the rates of incorporation of different metal ions differ in films and assemblies from those observed in solution, one of the more striking results of this study was the finding that only porphyrins located in hydrophilic sites would incorporate metal ions (57). This suggests that the metal ions, present in the form of their aquo complexes, either do not penetrate hydrophobic regions or are not present in appreciable concentrations. A study of the assembly structure rate dependence of Cu^{2+} incorporation into porphyrin **9**, which is indicated to lie at a hydrophilic interface, has furnished additional information concerning the mechanism and extent of metal ion penetration into these assemblies (45, 57). For porphyrin **9**, the $\alpha,\alpha,\alpha,\alpha$ isomer which has all of the hydrophobic groups on one side of the porphyrin ring, we find Cu^{2+} readily incorporated when the porphyrin is contained in films at an air–water interface or in assemblies with the porphyrin at an outer

9

hydrophilic surface. The porphyrin also incorporates Cu^{2+} when multilayer assemblies are immersed in aqueous Cu^{2+} solutions. In these assemblies the metal ions must enter the asemblies: (A) via direct diffusion through the layers by dissolving alternately in hydrophilic and hydrophobic portions, (B) by passing through defect channels ("Swiss cheese effect") or (C) by entering the assemblies at an edge and penetrating along the hydrophilic–hydrophilic interface. A study of assemblies containing **9** covered with varying numbers of layers of cadmium arachidate suggests that both Paths B and C operate for the Cu^{2+} ions but that C is the major path for assemblies where the metal ions are separated from the interfacial solution by several monolayers. It was found (*see* Table I) that with fewer than three covering layers of fatty acid the rate of metallation was dependent upon the number of covering layers. However, with greater numbers of layers the rate was constant, indicating that metal

Table I. **Pseudo First-Order Rate Constants for Metallation of Porphyrin 9 in Multilayer Assemblies of Different Architecture Contacted with Aqueous Cu^{2+}**

Experiment	Assembly Configuration[a]		$k_{meas.}$[b] ($\times\ 10^2\ sec^{-1}$)	k_{rel}
A	G · 5Cd–Ar · 10	9–Cd–Ar	5.6	4.0
A	G · 5Cd–Ar · 6	9–Cd–Ar · 4Cd–Ar	1.4	1.0
A	G · 5Cd–Ar · 6	9–Cd–Ar · 6Cd–Ar	1.4	1.0
B	G · 5Cd–Ar · 9	9–Cd–Ar · 1Cd–Ar	7.9	3.0
B	G · 5Cd–Ar · 9	9–Cd–Ar · 3Cd–Ar	2.5	0.9
B	G · 5Cd–Ar · 9	9–Cd–Ar · 5Cd–Ar	2.65	1.0

[a] G = glass, layers subsequently listed in sequence deposited, thus the first entry is glass coated with 5 layers of cadmium arachidate followed by 10 layers of 9–arachidate in a 1:5 mixture.
[b] Temperature 50°C; $[Cu^{2+}] = 0.001M$.

incorporation proceeded as outlined in Path B. A study of absorption and fluorescence changes in assemblies deposited on glass slides also indicated that in assemblies with six covering layers of cadmium arachidate spectral, changes consistent with metal incorporation occurred first at the edges of the slide and only subsequently near the center.

To summarize the results of several investigations, it is clear that small, neutral molecules (O_2, NO, CO N_2) can diffuse rapidly through both hydrophobic and hydrophilic regions of multilayer assemblies without causing any noticeable changes in the assembly. Moderately large, neutral organic molecules can penetrate both regions of assemblies, but diffusion is relatively slow and probably involves phase changes or other transient disruptions of the assembly without promoting gross rearrangement. Charged molecules exhibit different behavior depending upon size and perhaps several other factors; positively charged metal ions can enter assemblies through edges or defect channels and then move rapidly, in the case of anionic hosts, in the plane formed by a hydrophilic–hydrophilic interface. In contrast larger positive ions may either penetrate to some extent or simply adsorb on the surface, depending upon the system.

Interfacial Photoinduced Redox Phenomena

Light-induced electron transfer processes in solution have been the subject of intensive investigation (58–84). For a number of substrates, including dyes, aromatic hydrocarbons, and transition metal complexes, quenching of excited states concurrent with one electron transfer can be a prominent, rapid, and efficient process with a number of potential donors and acceptors. The general pattern of quenching and back electron transfer process is summarized in Equations 3–6 where the quenching processes (Equations 3 and 5) can involve considerable conversion of excitation energy into high energy products. Although these products might be

$$S^{n*} + A^{ox} \rightarrow S^{n+1} + A^{red(\bar{\cdot})} \qquad (3)$$

$$S^{n+1} + A^{red(\bar{\cdot})} \rightarrow S^n + A^{ox} \qquad (4)$$

$$S^{n*} + D^{red} \rightarrow S^{n-1} + D^{ox(\dot{+})} \qquad (5)$$

$$S^{n-1} + D^{ox(\dot{+})} \rightarrow S^n + D^{red} \qquad (6)$$

stable upon separation, in solution the normal fate of the product pair is energy-wasting back electron transfer (Equations 4 and 6). There has been much recent interest in developing ways to avert the back electron transfer processes in both solution and other media so that one or both of the high energy products can be diverted for useful chemistry or be stored as a fuel (84–88). Among the methods employed have been the use of organizates such as micelles where the shielding of one of the reagents in a charged micellar environment can result in significant alteration in rates of both quenching and back reactions (18, 19, 85, 86). Given the capabilities of a relatively loose organizate such as a micelle, it might be expected that there would be a number of possible ways to design organized multilayer assemblies or solution–multilayer systems in which appropriate combinations of hosts and donor–acceptor pairs might promote efficient permanent redox chemistry.

We have previously investigated light-induced electron transfer reactions in multilayer assemblies using surfactant analogs of tris(2,2′-bipyridine)ruthenium(II)$^{2+}$ and 1,1′-dialkyl-4,4′bipyridinium^{2+} ("Paraquat," PQ^{2+}) (10a) as excited donor and electron acceptor oxidant, respectively (89). In our studies we found that efficient quenching of the ruthenium complex luminescence occurs when the complex and acceptor are arranged in assemblies such that there is near-molecular contact. Subsequent experiments by Seefeld, Möbius, and Kuhn (90) using the same compounds and also a variety of surfactant cyanine dyes such as 11 as excited substrates have indicated quenching of excited states by 10a can occur over distances of 10–75 Å. The distance for half quenching increases monotonically with the estimated exothermicity of the electron transfer process. These results are most consistently explained by a mechanism involving electron tunneling from the excited substrate to 10a (90, 91).

Essentially all of the luminescence of 11 can be quenched by 10a in monolayer assemblies, which emphasizes the utility of the assemblies for observing and/or obtaining electron transfer processes not readily occurring in solution. In solution, quenching of the short-lived fluorescence

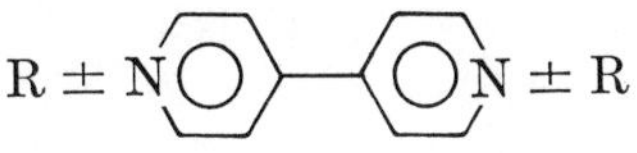

R ± N⟨○⟩—⟨○⟩N ± R

10 a R = $C_{18}H_{37}$
b R = CH_3

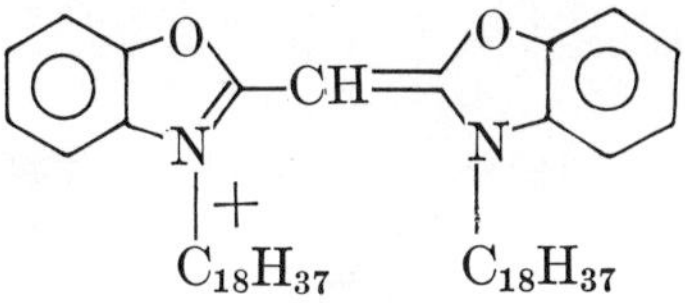

[benzoxazole structure] —CH= [benzoxazole structure]
                    |+
               $C_{18}H_{37}$ $C_{18}H_{37}$

11

(92) of **11** is barely observable with the highest obtainable concentrations of **10b** (the Stern–Volmer constant, $K = 3.2M^{-1}$ corresponding to about 6% quenching with 0.02M **10b**). The high degree of quenching observed with normal mixing ratios is probably because of the effective high quencher concentration in the assemblies, which is far in excess of those obtainable in solution; this suggests that even rather slow processes or reactions involving short-lived nonemitting excited states as substrates (93) can be very important in these systems.

Recently we have studied electron transfer processes in multilayers and at multilayer–solution interfaces using fluorescent dyes such as **11** and palladium porphyrins as substrates. For the former, we have focused primarily on interfacial quenching phenomena, while for the latter we have examined permanent redox chemistry occurring upon photolysis under a variety of different conditions in assemblies.

Permanent photoreduction reactions have been observed previously for a number of metalloporphyrins in solution (94–99). The general course of reaction involves consecutive reduction of two double bonds of the porphyrin to yield first a chlorin and then a dihydrochlorin or isobacteriochlorin derivative (Equation 7). In studies with several palladium(II) and platinum(II) complexes in solution, we have found reductive addition occurs with a variety of amines including triethylamine and N,N-dimethylaniline (DMA) (100). The reaction proceeds from the metalloporphyrin triplet. The indicated mechanism is outlined in Equations 8–10 for palladium(II) porphyrins and DMA. We have studied the

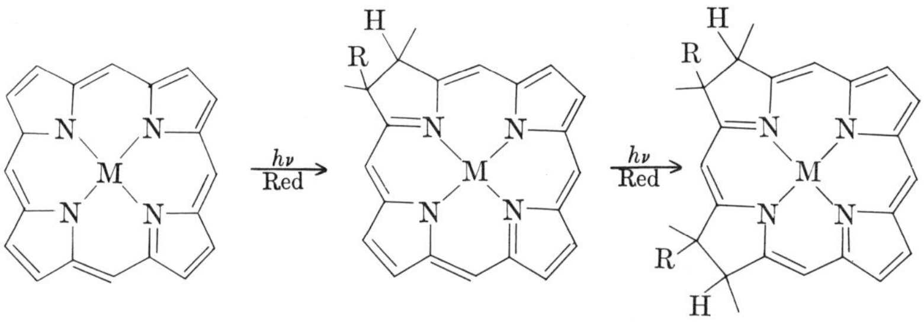

$$\text{porphyrin M} \xrightarrow{h\nu / \text{Red}} \text{chlorin} \xrightarrow{h\nu / \text{Red}} \text{isobacteriochlorin} \quad (7)$$

12   **13**

$$PdP^{3*} + (CH_3)_2N{-}\langle\underline{}\rangle \longrightarrow PdTPP^{\dot{-}} {-}{-}{-} \overset{CH_3}{\underset{CH_3}{\overset{\backslash+}{\cdot N}}}{-}\langle\underline{}\rangle \quad (8)$$

$$PdP^{\dot{-}} {-}{-}{-} \overset{CH_3}{\underset{CH_3}{\overset{\backslash+}{\cdot N}}}{-}\langle\underline{}\rangle \longrightarrow PdPH\cdot {-}{-}{-} \overset{CH_3}{\underset{\cdot CH_2}{\overset{\backslash}{N}}}{-}\langle\underline{}\rangle \quad (9)$$

$$PdPH\cdot {-}{-}{-} \overset{CH_3}{\underset{\cdot CH_2}{\overset{\backslash}{N}}}{-}\langle\underline{}\rangle \longrightarrow 12, \ R = {-}CH_2{-}\overset{CH_3}{\underset{}{N}}{-}\langle\underline{}\rangle \quad (10)$$

reaction using two surfactant porphyrins, the palladium(II) complex of 9 (14) and the palladium(II) complex of the tetra(octadecyl)ester of α,β,γ,δ-tetra(p-carboxyphenyl) porphine (15). Both of these porphyrins can be incorporated readily into films and assemblies; 15 is indicated to reside in a hydrophobic region in assemblies while, as noted earlier, 14 lies at the hydrophilic interface.

Irradiation of both 14 and 15 in monolayer assemblies contacted with an aqueous solution of DMA under argon leads to spectroscopic changes consistent with the formation of reductive adducts (*see* Equation 7). No reduction is observed when the assemblies are contacted with water alone or irradiated dry. For 15 a weak phosphorescence is observed in monolayer assemblies; this emission is strongly quenched by admission of oxygen and a similar quenching is observed upon exposure of the assembly to an aqueous solution of DMA. We found that 80% of the emission is quenched by immersion of a slide containing 26 layers of 15 in aqueous DMA; thus the quenching is not confined to the outer layers. However, the reaction proceeds more rapidly (ca. 2×) in assemblies in which the porphyrin is present only in an outer layer having hydrophilic contact with the aqueous DMA. In the case of 15 generally only the first step of the reduction proceeds; slow decomposition of the material becomes a problem before isobacteriochlorin formation occurs to an appreciable extent. Irradiation of 14 leads to reduction; in this case irradiation of the assemblies at an aqueous DMA–assembly interface leads to formation of both reductive addition products (*see* Equation 7). Chromatographic analysis indicates that the products formed are the same products formed upon irradiation of 14 in organic solvents containing DMA. The photoreduction proceeds slowly and some decomposition is observed during the irradiation; however, assemblies where the porphyrin layers are covered with 0, 2, 4, 6, and 8 outer layers of cadmium arachidate reduce at essentially the same rate yielding identical product distributions. The results obtained with **14** and **15** and DMA agree with the expectation that relatively small, neutral organic molecules should easily penetrate monolayer assemblies in concentrations high enough to permit photoreaction. Although we have not yet measured precise quantum efficiencies, they are indicated to be relatively low both in solution and in the assemblies. It is not possible to determine whether reaction is actually facilitated in the assemblies.

The reductive photoaddition of **14** using the surfactant analog (**16**) of DMA in films and assemblies also has been investigated. Upon irradiation of films formed from cadmium arachidate, **14**, and **16** in a

$$\text{C}_6\text{H}_5-\text{N}\begin{smallmatrix}\diagup \text{C}_{18}\text{H}_{37}\\ \diagdown \text{C}_{18}\text{H}_{37}\end{smallmatrix}$$

16

2.5:1:1 mixture at an air–water interface, chlorin-type products formed in very low yield. Irradiation of multilayer assemblies formed by transfer of these films, either dry (under argon) or in argon-saturated water, results in a limited amount of reduction to form first chlorin and then isobacteriochlorin. Reduction is by no means complete before a general decomposition sets in; the major product formed is the doubly-reduced isobacteriochlorin. Assemblies constructed with adjacent layers containing **14** and **16** in hydrophilic contact give similar results; considerable reduction occurs and the predominant product is the doubly-reduced isobacteriochlorin. In both cases reaction with **16** as the reductant is substantially slower than with aqueous DMA. A reasonable inference from the studies with **16** is that the assemblies contain the porphyrin in a number of different sites, some containing one or more molecules of **16** close enough to react and some where the porphyrin is more isolated. Since **14** phosphoresces only weakly in solution and not at all in assemblies, it has not been possible to correlate quenching with reaction in this case. We are extending our studies of interfacial photoredox reactions of the porphyrins with other diffusable solute quenchers. Reasonably clean, although as yet inefficient, interfacial redox processes occur with relatively hydrophobic systems. This suggests the possibility of carrying out conversions in which the porphyrin (or other reagent) can function as a catalyst, mediating net redox processes between solutes in the contacting solution (95, 101).

As pointed out above, quenching of the fluorescence of 11 by the potential surfactant electron acceptor 10a has been studied in assemblies of varying molecular architecture (90). We have examined quenching of the fluorescence of 11 by the nonsurfactant acceptor 10b and by Cu^{2+} ions that are introduced by contacting the assemblies with aqueous solutions. Table II indicates the different configurations used in this study and the quenching observed. There was no appreciable quenching of the fluorescence of 11 immediately upon immersion of the assemblies into aqueous solutions of 10b; this is in agreement with the expectation that the lifetime of 11 is too short to permit appreciable dynamic quenching at the concentrations used. However, there is a slow quenching whose rate and extent depends on the assembly configuration; which reaches as high as 40%. This quenching can be reversed by replacing the aqueous solution of 10b with a solution of Cd^{2+} ($0.025M$); in several cases nearly complete restoration of the original fluorescence can be obtained. Similar quenching processes can be obtained using Cu^{2+} ions as the interfacially introduced quencher.

These results suggest that most of the interfacial electron transfer quenching of 11 occurs by prior association of the quencher solute with the assembly. The slow quenching process is significant even for assemblies where the fluorescing molecule is buried under several cadmium arachidate layers. It almost certainly occurs through penetration of the charged quencher into the assemblies via the hydrophilic interface similar to that observed in the studies of Cu^{2+} incorporation into porphyrins (57). That the penetration can occur and be reversed in essentially an ion-exchange process suggests that schemes in which penetration, subsequent reaction with an anchored catalyst, and exit occur sequentially might be usefully carried out in these assemblies. In these cases significant quenching occurs with solute concentrations much too low to give appreciable

Table II. Fluorescence Quenching of Cyanine Dye 11 in Multilayer Assemblies with 5mM Aqueous Paraquat Dichloride

Assembly Arrangement	I/I_o		
	(immed)	(10 min)	(20 min)
CdA⊐ (1:5, 11:CdA)	1.00	0.63 ± 0.01	0.59 ± 0.01
CdA⊐	0.89	0.60	0.52
CdA⊐ •–	0.98	0.69 ± 0.01	0.59 ± 0.02
CdA⊐ •– –• •– –• •–	0.98	0.87 ± 0.05	0.84 ± 0.05

dynamic quenching in homogeneous solutions; systems could be designed to observe and perhaps exploit a number of processes not observable in solution.

In both cases described above we have seen no evidence that anchoring of one or both of the reagents used in a light-induced electron transfer process can greatly retard the back electron transfer responsible for energy wasting in these processes. It is reasonable that in at least some cases the retarded diffusion of solutes penetrating the layers facilitates back reactions. However, it would be reasonable to expect that suitably designed systems might offer advantages over homogeneous solution, especially for excited substrates contained in an outer layer at or near a hydrophobic surface. For such a substrate dynamic quenching by electron donation to e.g., a dicationic acceptor should lead to ionic products that would have little probability of re-encounter, especially if reagents are present in solution and assembly to oxidize or otherwise further react the acceptor and reduce the donor (*101*). Experiments designed to accomplish this process are currently in progress.

Acknowledgment

We are grateful to the National Science Foundation (Grant No. CHE76-01074), Army Research Office (Grant No. DAAG29-77-G-0063), and National Institutes of Health (Grant No. GM15238-11) for support of this work. We thank H. Kuhn and D. Möbius for helpful discussion and a sample of surfactant dye 11. We are especially grateful to the W. R. Grace Company for an award that assisted investigations by R. H. Schmehl.

Literature Cited

1. Fendler, J. H.; Fendler, E. J. "Catalysis in Micellar and Macromolecular Systems"; Academic: New York, 1975; Chap. 4, p. 86 and references therein.
2. Hartley, G. S. *Trans. Faraday Soc.* **1934**, *30*, 444.
3. Bunton, C. A.; Robinson, L.; Schaak, J.; Stam, M. F. *J. Org. Chem.* **1971**, *36*, 2346.
4. Dunlap, R. B.; Cordes, E. H. *J. Phys. Chem.* **1969**, *73*, 361.
5. Kögl, F.; Havinga, E. *Rec. Trav. Chem.* **1940**, *59*, 600.
6. Fosbinder, R. J.; Rideal, E. K. *Proc. R. Soc. London, Ser. A* **1933**, *143*, 61.
7. Gee, G.; Rideal, E. K. *Proc. R. Soc. London, Ser. A* **1935**, *153*, 116.
8. Gaines, G. L., Jr. "Insoluble Monolayers at Liquid–Gas Interfaces"; Wiley–Interscience: New York, 1966; p. 308.
9. Hopf, F. R.; Möbius, D.; Whitten, D. G. *J. Am. Chem. Soc.* **1976**, *96*, 1584.
10. Cordes, E. H.; Dunlap, R. B. *Acc. Chem. Res.* **1969**, *2*, 329.
11. Fendler, E. J.; Fendler, J. H. *Adv. Phys. Org. Chem.* **1970**, *8*, 271.

12. Piszkiewicz, D. *J. Am. Chem. Soc.* **1976,** *98,* 3053.
13. Ibid., p. 7695.
14. Ibid., p. 1550.
15. Ibid., p. 7695.
16. Tanford, C. "The Hydrophobic Effect"; Wiley–Interscience: New York, 1973.
17. Hautala, R. R.; Letsinger, R. L. *J. Org. Chem.* **1971,** *36,* 3762.
18. Grätzel, M.; Thomas, J. K. *J. Am. Chem. Soc.* **1973,** *95,* 6885.
19. Scheever, R.; Grätzel, M. *J. Am. Chem. Soc.* **1977,** *99,* 865.
20. Kalyanasundarum, K.; Thomas, J. K. *J. Am. Chem. Soc.* **1977,** *99,* 2039.
21. Hautala, R. R.; Schore, N. E.; Turro, N. J. *J. Am. Chem. Soc.* **1973,** *95,* 5508.
22. Kuhn, H.; Möbius, D. *Angew Chem., Intl. Ed. Engl.* **1971,** *10,* 620.
23. Kuhn, H.; Möbius, D.; Bucher, H. In "Physical Methods of Chemistry"; Weissburger, A., Rossier, B., Ed.; Wiley: New York, 1972; Vol. I, Part 3b, p. 577.
24. Whitten, D. G. *J. Am. Chem. Soc.* **1974,** *94,* 594.
25. Quina, F. H.; Whitten, D. G. *J. Am. Chem. Soc.* **1977,** *99,* 877.
26. Whitten, D. G.; Hopf, F. R.; Quina, F. H.; Sprintschnik, G.; Sprintschnik, H. W. *Pure Appl. Chem.* **1977,** *49,* 379.
27. Geiger, M. W.; Turro, N. J. *Photochem. Photobiol.* **1976,** *26,* 221.
28. Galla, H.; Sackman, E. *Biochem. Biophys. Acta.* **1974,** *339,* 103.
29. Quina, F. H.; Whitten, D. G. *J. Am. Chem. Soc.* **1975,** *97,* 1602.
30. Hopf, F. R.; Möbius, D.; Whitten, D. G. *J. Am. Chem. Soc.* **1976,** *98,* 1584.
31. Menger, F. M.; Jerkunica, J. M.; Johnston, J. C. *J. Am. Chem. Soc.* **1978,** *100,* 4676.
32. Wagner, P. J. *Acc. Chem. Res.* **1971,** *4,* 168, and references therein.
33. Wagner, P. J. *Pure Appl. Chem.* **1977,** *49,* 259.
34. Gaillet, J. E. *Pure Appl. Chem.* **1977,** *49,* 249.
35. Hartley, G. H.; Guillet, J. E. *Macromolecules* **1968,** *1,* 165.
36. Slivinskas, J. A.; Guillet, J. E. *J. Polym. Sci., Polym. Chem. Ed.* **1973,** *11,* 3043.
37. Turro, N. J.; Liu, K.-C.; Chow, M.-F. *Photochem. Photobiol.* **1977,** *26,* 413.
38. Worsham, P. R.; Eaker, D. W.; Whitten, D. G. *J. Am. Chem. Soc.* **1978,** *100,* 7091.
39. Gaines, G. L., Jr. "Insoluble Monolayers at Liquid–Gas Interfaces;" Wiley–Interscience: New York, 1966; p. 156.
40. Ibid., p. 162.
41. Worsham, P. R., unpublished data.
42. Muller, N.; Birkhahn, R. H. *J. Phys. Chem.* **1967,** *71,* 957.
43. Fendler, J. H.; Fendler, E. J.; Infante, G. A.; Shih, P.-S.; Patterson, L. K. *J. Am. Chem. Soc.* **1975,** *97,* 89.
44. Worsham, P. R.; Russell, J.; Whitten, D. G., unpublished data.
45. Whitten, D. G.; Eaker, D. W.; Horsey, B. E.; Schmehl, R. H.; Worsham, P. R. *Ber. Bunsenges. Phys. Chem.* **1978,** *82,* 858.
46. Birks, J. B. "Photophysics of Aromatic Molecules"; Wiley-Interscience: London, 170; p 324.
47. Schmidt, G. M. J. "Solid State Photochemistry"; Verlag Chemie: Wienheim-New York, 1976.
48. Bruning, R., Ph.D. Dissertation, University of Marburg, 1969.
49. Polymeropoulos, E. E.; Sagiv, J. *J. Chem. Phys.* **1978,** *69,* 1836.
50. Bigelow, W. C.; Pickett, D. L.; Zisman, W. A. *J. Colloid Sci.* **1946,** *1,* 513.
51. Brockway, L. O.; Karle, J. *J. Colloid Sci.* **1947,** *2,* 277.
52. Bigelow, W. C.; Brockway, L. O. *J. Colloid Sci.* **1956,** *11,* 60.
53. Bartell, L. S.; Ruch, R. J. *J. Phys. Chem.* **1959,** *63,* 1045.

54. Hopf, F. R.; Whitten, D. G. *J. Am. Chem. Soc.* **1976**, *98*, 7422.
55. Horsey, B. E.; Hopf, F. R.; Schmehl, R. H.; Whitten, D. G. "Porphyrin Chemistry Advances," Longe, F. R., Ed.; Ann Arbor: Ann Arbor, 1979; p. 17.
56. Horsey, B. E.; Whitten, D. G. *J. Am. Chem. Soc.* **1978**, *100*, 1293.
57. Schmehl, R. H.; Shaw, G. L.; Whitten, D. G. *Chem. Phys. Lett.* **1978**, *58*, 579.
58. Leonhardt, H.; Weller, A. *Ber. Bunsenges. Phys. Chem.* **1963**, *67*, 791.
59. Knibbe, H.; Rehm, D.; Weller, A. *Ber. Bunsenges. Phys. Chem.* **1968**, *72*, 257.
60. Rehm, D.; Weller, A. *Ber. Bunsenges. Phys. Chem.* **1969**, *73*, 834.
61. Rehm, D.; Weller, A. *Isr. J. Chem.* **1970**, *8*, 259.
62. Grellmann, K. H.; Watkins, A. R.; Weller, A. *J. Phys. Chem.* **1972**, *76*, 469.
63. Ibid., p. 3132.
64. Kawai, K.; Yamamoto, N.; Tsubomura, T. *Bull. Chem. Soc. Jpn.* **1969**, *42*, 369.
65. Yamashita, H.; Kokubun, H.; Kuizumi, M. *Bull. Chem. Soc. Jpn.* **1968**, *41*, 2312.
66. Bonneau, R.; Fornier-de-Violet, P.; Joussot-Dubien, J. *Photochem. Photobiol.* **1974**, *19*, 129.
67. Vogelmann, E.; Kramer, H. E. A. *Photochem. Photobiol.* **1976**, *23*, 383.
68. Gafney, H. D.; Adamson, A. W. *J. Am. Chem. Soc.* **1972**, *94*, 8238.
69. Bock, C. R.; Meyer, T. J.; Whitten, D. G. *J. Am. Chem. Soc.* **1974**, *96*, 4710.
70. Bock, C. R., Meyer, T. J.; Whitten, D. G. *J. Am. Chem. Soc.* **1975**, *97*, 2909.
71. Navon, G.; Sutin, N. *Inorg. Chem.* **1974**, *13*, 2159.
72. Lawrence, G. S.; Balzani, V. *Inorg. Chem.* **1974**, *13*, 2976.
73. Harbour, J. R.; Tollin, G. *Photochem. Photobiol.* **1974**, *19*, 147.
74. Young, R. C.; Meyer, T. J.; Whitten, D. G. *J. Am. Chem. Soc.* **1975**, *97*, 4781.
75. Ibid., **1976**, *98*, 6536.
76. Creutz, C.; Sutin, N. *Inorg. Chem.* **1976**, *15*, 496.
77. Lin, C. T.; Böttcher, W.; Chou, M.; Creutz, C.; Sutin, N. *J. Am. Chem. Soc.* **1976**, *98*, 6536.
78. Lin, C. T.; Sutin, N. *J. Am. Chem. Soc.* **1975**, *97*, 3543.
79. Creutz, C.; Sutin, N. *J. Am. Chem. Soc.* **1977**, *99*, 241.
80. Toma, H. E.; Creutz, C. *Inorg. Chem.* **1977**, *16*, 545.
81. Balzani, V.; Moggi, L.; Manfrin, M. F.; Bolletta, F.; Lawrence, G. S. *Coord. Chem. Rev.* **1975**, *15*, 321.
82. Juris, A.; Gandolfi, M. T.; Manfrin, M. F.; Balzani, V. *J. Am. Chem. Soc.* **1976**, *98*, 1047.
83. Sutin, N.; Creutz, C. "Coordination Chemistry, Volume 1," *Adv. Chem. Ser.* **1978**, *168*, 1.
84. DeLaive, P. J.; Giannotti, C.; Whitten, D. G. "Coordination Chemistry, Volume 1," *Adv. Chem. Ser.* **1978**, *168*, 28.
85. Maestri, M.; Grätzel, M. *Ber. Bunsenges. Phys. Chem.* **1977**, *81*, 504.
86. Meisel, D.; Matheson, M. S.; Rabani, J. *J. Am. Chem. Soc.* **1978**, *100*, 117.
87. DeLaive, P. J.; Lee, J. T.; Abruña, H.; Sprintschnik, H. W.; Meyer, T. J.; Whitten, D. G. "Coordination Chemistry, Volume 1," *Adv. Chem. Ser.* **1978**, *168*, 28.
88. Gray, H. B.; Mann, K. R.; Lewis, N. S.; Thich, J. A.; Richman, R. M. "Coordination Chemistry, Volume 1," *Adv. Chem. Ser.* **1978**, *168*, 44.
89. Sprintschnik, G.; Sprintschnik, H. W.; Kirsch, P. P.; Whitten, D. G. *J. Am. Chem. Soc.* **1977**, *99*, 4947.

90. Seefeld, K.-P.; Möbius, D.; Kuhn, H. *Helv. Chim. Acta.* **1977**, *60*, 2608.
91. Kuhn, H. *Pure Appl. Chem.* **1979**, *51*, 341.
92. Schafer, F. P.; Röllig, K. *Z. Phys. Chem., n.F.* **1964**, *40*, 198.
93. Young, R. C.; Nagle, J. K.; Meyer, T. J.; Whitten, D. G. *J. Am. Chem. Soc.* **1978**, *100*, 4773.
94. Fuhrhop, J.-H.; Lumbantobing, T. *Tetrahedron Lett.* **1970**, 2815.
95. Harel, Y.; Manassen, J. *J. Am. Chem. Soc.* **1977**, *99*, 5817.
96. Whitten, D. G.; Yau, J. C.; Carroll, F. A. *J. Am. Chem. Soc.* **1971**, *93*, 2291.
97. Seely, G. R.; Calvin, M. *J. Chem. Phys.* **1955**, *23*, 1068.
98. Suboch, W. P.; Losev, A. P.; Gurinovitch, G. P. *Photochem. Photobiol.* **1974**, *20*, 183.
99. Seely, G. R.; Talmadge, K. *Photochem. Photobiol.* **1964**, *3*, 195.
100. Mercer-Smith, J.; Sutcliffe, C. R.; Schmehl, R. H.; Whitten, D. G. *J. Am. Chem. Soc.* **1979**, *101*, 3997.
101. DeLaive, P. J.; Gianotti, C.; Whitten, D. G. *J. Am. Chem. Soc.* **1978**, *100*, 7413.

RECEIVED October 2, 1978.

Characterization and Chemical Reactions of Surfactant Monolayer Films

STEVEN J. VALENTY

General Electric Company Corporate Research and Development, Schenectady, NY 12301

The study of chemical reactions of suitably functionalized surfactant monolayer films is suggested as a corollary approach toward understanding the chemistry of bonding modifiers to electrode and catalytic surfaces. Liquid- and vapor-phase chromatography have been used in conjunction with absorption, fluorescence, and IR spectrometry to detect and quantitate the chemistry occurring in such films containing aldehyde or ester molecules. Carboxylic acid surfactant derivatives of ruthenium(II)tris(2,2'-bipyridine) have been used to assess solid surface acidity. The observation of methylene blue adsorption to photogalvanic electrodes is related to the orientation and absorption spectra of its surfactant analogs in monolayer films. The effect of counterion on luminescence properties of derivatives of ruthenium(II)tris(2,2'-bipyridine) are unique to the positively charged interface.

The study of interfacial photoprocesses emphasizing energy conversion and chemical synthesis is predicated upon the expectation that constraining specially tailored reaction centers to a surface will produce reaction products or kinetics different from what is observed in either of the adjoining homogeneous phases alone. In some experimental approaches, a surface is modified by covalently attaching specialized molecules whose varied functions include selective electron transfer with a specific redox couple in an adjoining electrolyte solution, prevention of electrode corrosion, and preferential binding of reactant molecules. It is clear that the chemistry involved in coupling the modifying molecule to the surface and the nature of the covalent bond so formed are important to both the mechanism of the modifier's action and its operational lifetime. A variety of techniques, including electron spectroscopy for chemical analysis (ESCA), atomic emission spectroscopy (AES), IR, optical spectroscopy, and electroanalytical methods, have been used to characterize a surface before and after chemical modification.

Where applicable the study of chemical reactions of suitably functionalized surfactant monolayer films at the gas–water interface offers another approach towards understanding this chemistry. Operating at the air–water interface and, hence, within the constraint of aqueous-based chemistry, this technique allows the preparation of a variety of functionalized surfaces by forming a monolayer from selected synthetic surfactant monomers. Further, the use of this particular methodology allows experimental control, to varying degrees, of the resulting two-dimensional films' composition, orientation, and electrical charge. Because a monomolecularly thin layer is being dealt with here, a chemically modified film can be recovered from the interface and separated into its molecular constituents, each of which can then be quantitated and characterized. Such is not the case where attachment is made to a bulk material.

Since the monolayer film can be transferred from the gas–water interface onto a variety of solid supports, the well-known surface analysis techniques mentioned earlier can be used for characterizing these films as well. In some instances, such a procedure would allow quantitative calibration of the surface analysis because the exact amount of material transferred to the solid surface is known. The efficiency of subsequent procedures to physically adsorb or covalently attach similar but nonsurfactant modifiers to this solid surface by other means could be determined directly by reference to the monolayer experiment.

Further, the response of a specially tailored surfactant, say to pH as observed by optical spectroscopy, in a monolayer at the well-characterized gas–water interface could be used to probe the same surface property of less known materials (in this case, necessarily transparent to the optical radiation used) upon transferring the monolayer to that surface.

In support of the preceeding statements, I should like to describe work done in characterizing monolayer films and chemical reactions occurring in them at the gas–water interface and on solid surfaces by classical techniques as well as by applying newer analytical methods to detect and quantitate the small amounts of material contained therein. While no new photochemistry mediated by monolayers will be discussed, a striking example of how constraining a molecule to an interface alters its photophysical behavior will be presented.

Experimental

The synthesis and characterization of Compounds I–X have been described (1, 2, 3). Stearaldehyde (StAld) (Supelco, 99+%), stearyl alcohol (StOH) (Sigma, 98%), poly-l-lysine (PL) (as HBr salt, Sigma), 3-methyl-2-benzothiazolinone hydrazone hydrochloride (reagent, Aldrich), sodium borohydride (Aldrich), inorganic reagents (analytical grade), and organic solvents (Burdick and Jackson) were used without further purification. Triply distilled water from a quartz still was used in the preparation of the monolayer subphase solutions.

Monolayers of StAld and methylene blue/ruthenium bipyridyl surfactant complexes were spread from dilute n-hexane and chloroform solutions, respectively.

Surface pressure–area isotherms (1), surface viscosity (4), optical spectrometry of glass-supported monolayers (1), emission spectrometry of monolayers at the air–water interface (1), high-performance liquid chromatographic (HPLC) analyses (5), vapor-phase chromatography (VPC) analyses for H_2 dissolved in water (6) and for StAld/StOH (4) were done using apparatus and procedures previously described. The Fourier transform IR (FTIR) spectra of monolayers of I on hydrophilic germanium attenuated total reflectance (ATR) plates (Herrick Scientific, $50 \times 20 \times 1$ mm, $\Theta = 45°$, single-pass parallel-piped, 50 reflections) were recorded with a Nicolet 7199. The ATR–IR spectra of the StAld–PL system were obtained as described earlier (4). The field desorption mass spectra (FDMS) were recorded on a Varian-Mat 731 mass spectrometer. The construction details and operation of the multicompartment trough have been discussed elsewhere (7).

Results and Discussion

Chemical Reactions and the Multicompartment Monolayer Trough. The mechanics of performing a chemical reaction on a monolayer film require several operations: (a) the film must be spread at the gas–water interface and compressed to the desired surface pressure; (b) the reaction must occur under conditions where the surface pressure can be monitored and held constant if desired; (c) the reaction must be quenched to obtain a chemically stable system such that later analysis will reflect the composition of the film while it was still at the gas–water interface; and (d) the film must be recovered as quantitatively as possible from the interface. In order to satisfy these requirements, a multicompartment trough has been used to allow a series of chemical reactions to be performed upon a well-defined monolayer.

The multicompartment trough used here is shown in Figure 1. The trough subphase is divided into a number of individual compartments by submerged hydrophilic glass barriers that extend the width of the trough and are covered by ca. 0.5 cm water. The several compartments can be filled and emptied independently. Surfactant monolayers are constrained between the hydrophobic sides of the trough and two motor-

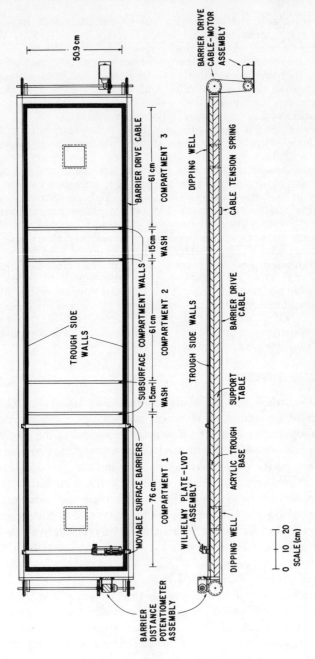

Figure 1. Multicompartment monolayer trough

ized hydrophobic surface barriers which can be driven independently to control the surface pressure (as measured by a Wilhelmy balance) or simultaneously to laterally transport the film at constant area from the surface of one compartment to another over the submerged barrier.

Borohydride Reduction of StAld. The reduction of StAld to StOH by aqueous $NaBH_4$ illustrates the use of this trough. A monolayer of StAld (3.2 µmol, 0.86 mg) was formed on Compartment 1 containing $5.0 \times 10^{-3} M$ borate buffer (pH 9.0) and compressed to $\Pi = 10$ dyn/cm. The film was moved to Compartment 2 (transfer time, 1 min), which contained freshly prepared $1.0 \times 10^{-2} M$ $NaBH_4$ in aqueous borate buffer. After 30 min, the film was moved back to Compartment 1, compressed until it collapsed into threads, scraped off the surface and dissolved in chloroform. VPC analysis indicated StOH corresponding to 84 mol % of the initially spread StAld with 1 mol % unreduced aldehyde. Control experiments have shown 85 ± 2 mol % recovery of either the original StAld spread and manipulated in the absence of $NaBH_4$ or the StOH spread and manipulated in the presence of $NaBH_4$.

Sequential Chemical Reactions. A more complex series of reactions is shown by the covalent attachment of a dye to a preformed monolayer film. In the presence of an aldehyde and ferric chloride, 3-methyl-2-benzothiazolinone hydrazone (MBTH) forms a deep-blue-colored cation via oxidative coupling of the initially formed aldimine with a second mole of MBTH (8, 9).

A monolayer of StAld (3.0 μmol) was formed on $5 \times 10^{-3} M$ borate buffer (pH 9) in Compartment 1 and compressed ($\pi = 10$ dyn/cm). The film was shifted onto the surface of Compartment 2 containing MBTH ($1.1 \times 10^{-2} M$ in pH 9 borate buffer) and allowed to stand for 30 min. The film was returned to Compartment 1, compressed to collapse the film into white threads, and scraped off the surface to give a homogeneous solution in CCl_4. VPC analysis showed less than 0.5 mol % unreacted StAld and two major components eluting at much longer retention times. VPC–MS (mass spectrometry) showed that these two components have the same molecular ion ($m/e = 420$) and fragmentation patterns similar to the formation of isomeric imines. In a second reaction using identical conditions, the film is shifted from the MBTH containing subphase directly (no intervening wash compartment) onto a subphase freshly prepared with $FeCl_3$ dissolved in distilled water ($1.8 \times 10^{-3} M$). After standing 30 min, the film is collapsed (a blue coloration is apparent) and scraped off the surface to give a deep-blue homogeneous solution in $CHCl_3$. VPC analysis indicated 3 mol % StAld unreacted and the visible absorption spectrum showed a maximum at 673 nm with a shoulder at 630 nm (1.00:0.83 o.d. ratio) as reported for similar compounds (8, 9). Using $\epsilon^{673} = 5.2 \times 10^4$ from the literature (8), the yield of the reaction is 9 mol % based on StAld. The blue color can be bleached by dithionite and recovered by Fe^{3+} oxidation. Since the imine appears to be produced in high yield in the initial reaction and remains in the blue-colored film by VPC analysis, its conversion to the relatively larger structure of the dye cation is limiting overall reaction yield. Increasing the surface area available per molecule by using one-third the amount of StAld increased the yield of dye cation to ca. 15%. The dye-cation-containing monolayer film can be transferred with the two-dimensional structure intact to a glass slide and its absorption spectrum can be recorded. The visible spectrum of the dye cation in the monolayer is compared with that in homogeneous solution in Figure 2.

Polycondensation Chemistry. In an experimental approach seeking to synthesize planar, ultrathin polymer films possessing rubber elasticity and functional sites that might prove useful in modelling biological membrane structure and selective transport capability, polycondensation reactions in monolayer films were studied (4). The synthetic strategy to obtain such a film grafts surfactant monomers of StAld formed in a monolayer array onto linear, but randomly coiled, water-soluble PL chains. A sufficient number of unreacted lysyl amino groups are left to form the interchain crosslinkage upon further condensation with a dialdehyde such as glutaraldehyde. Thus, the two dimensional surface active network is achieved.

The measurement of intrinsic viscosity provides a convenient indication of bulk-phase polymerizations. Most of the literature claiming to

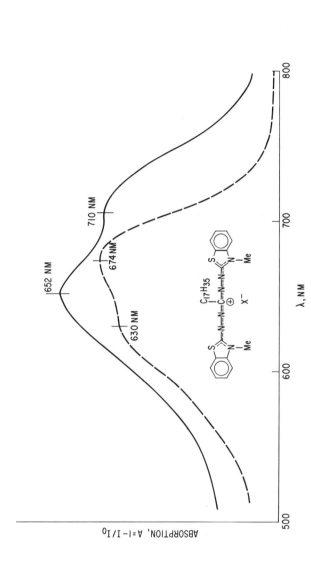

Figure 2. Absorption spectra of StAld–MBTH reaction product as a monolayer coating both sides of a hydrophilic glass slide (———, transferred from 1.8×10^{-3}M $FeCl_3$ at $\Pi = 30$ dyn/cm, $A^{652} = 5.6 \times 10^{-2}$, $A^{710} = 4.5 \times 10^{-2}$); and in $CHCl_3$ solution (– – –, $\varepsilon^{674} = 5.2 \times 10^4$, intensity arbitrarily scaled).

have prepared monomolecular polymer films has presented some surface viscosity measurement in support (4). When a StAld monolayer is formed and compressed to $\Pi = 10$ dyn/cm on a subphase containing 4.2 μg/mL PL at pH 7.0 (phosphate), η_s^{rel} increases linearly with time as illustrated in Figure 3. (Note that η_s^{rel} has a low constant value for StAld without PL in the subphase and that it is ca. zero for PL alone and it is not detectable for a monolayer of n-octadecanol with PL in the subphase.) The observation of a time-dependent relaxation in the surface pressure–area isotherm of the reacted film upon compression–expansion cycling where none was observed for a StAld film alone also indicates a change in film structure. The viscosity changes during film reaction have been shown to be dependent upon PL concentration, PL molecular weight, pH, temperature, and StAld surface pressure. The measurement of surface viscosity has

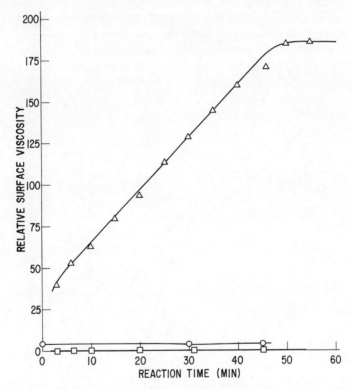

Figure 3. A comparison of relative surface viscosity vs. time for StAld alone, (○); PL (15K) alone, (□); and StAld–PL (15K), (△). All measurements were obtained on a subphase containing pH 7.0, 1.0×10^{-3}M phosphate buffer at 20°C. Where used, the PL (15K) concentration was 4.2 μg/mL. The measurements for StAld and StAld–PL (15K) were done at a constant $\Pi = 10$ dyn/cm.

been used as a rapid means of surveying reaction kinetics assuming that a rise in the observed viscosity is related in some manner to the degree of reaction.

The ordinate of Figure 3 would be more useful to the chemist if it were also labelled in percent reaction of starting material or percent formation of product. For this purpose, the multicompartment trough was used to study the grafting reaction between StAld and PL as shown in Figure 4. A monolayer of StAld (3.2 μmol, 0.86 mg) was formed and compressed to 10 dyn/cm on aqueous borate buffer (pH 9.0, $5.0 \times 10^{-3}M$) in Compartment 1. The compressed film is moved laterally over borate buffer (the same as in Compartment 1) in Compartment 2 onto aqueous PL (1.0 μg/mL, total ca. 2.8 mg, ~ 15,000 mol wt) in Compartment 3. After a period of time, the film was passed back over Subphase 2 to wash out any unattached PL. The reaction is quenched by reduction with aqueous $NaBH_4$ ($10^{-2}M$, pH 9.0, borate buffer, 30 min.) in Compartment 1 and transferred with washing across Compartment 2 onto the surface of $10^{-2}M$ NaOH in Compartment 3. At this stage, or any intermediate one, the film may be transferred to a solid support (glass slide, CaF_2 and Ge IR plates) via the Langmuir–Blodgett technique for spectral observation or collapsed, scraped from the surface, and dissolved in an appropriate solvent for chromatographic and spectral characterization.

Figure 5 shows the results of VPC analysis of reduced films scraped from the water surface and extracted with $CHCl_3$ to dissolve the StOH (reduction product of StAld noted previously). As the reaction time lengthens, more H_2O- and $CHCl_3$-insoluble white solid is obtained. The chloroform-soluble fraction shows only StOH and a small amount of unreduced StAld ($\leq 2\%$). The disappearance of StAld occurs in a nearly linear fashion (0.42%/min or 1.3×10^{-2} μmol/min) until 180 min (ca.

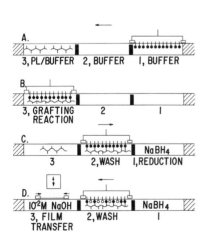

Figure 4. Schematic description of operations involved in grafting a monolayer of StAld onto PL and subsequent $NaBH_4$ reduction using the multicompartment trough. Operations: (A) form monolayer of StAld and compress to desired Π on surface of buffer in Compartment 1; (B) shift film at constant area across buffer in second compartment onto PL containing subphase to start reaction; (C) shift film at constant area across wash buffer in Compartment 2 onto alkaline $NaBH_4$ in Compartment 1 to reduce surface film; (D) transfer film across wash compartment onto 10^{-2}M NaOH in Compartment 1 for film removal from surface.

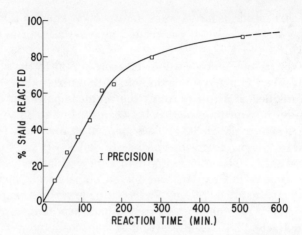

Figure 5. VPC analysis for unreacted StAld as StOH after $NaBH_4$ reduction as a function of StAld–PL reaction time. Conditions: 3.2 μmol StAld, 1.0 μg/mL PL (15K), pH 9.0, $5.0 \times 10^{-3}M$ borate buffer, $22 \pm 2°C$, 10 dyn/cm.

66% consumption) when the reaction slows markedly leaving some 10% unreacted StAld after 8.5 hr. The precision of the analysis is ±2% in the linear portion of the trace. If the grafted film is collapsed and scraped from the interface before $NaBH_4$ reduction, the precision of the VPC analysis for StAld drops to ±15–20% although the scatter of points very nearly reproduces the more precise data of Figure 5. A number of the films were extracted into CCl_4, the intensity of the carbonyl stretch frequency of StAld (1730 cm^{-1}) monitored in a 1.0-cm path length cell, and the StAld concentration obtained from an empirical calibration curve. The resultant data for disappearance of StAld follows the VPC-derived information. The greater precision for the VPC analysis after reduction results from "freezing" the composition of the condensation equilibria while the film is still at the air–water interface.

Grafted but unreduced films were also transferred ($\pi = 25$ dyn/cm, 10 monolayers per side) to a hydrophobic Ge ATR plate and their IR spectra were obtained. Of major importance was the disappearance of the StAld carbonyl stretch frequency (1708 cm^{-1}) with reaction time and the simultaneous appearance of a broader band at 1645 cm^{-1}, which can be assigned as the carbonyl amide I band of polypeptides.

The disappearance of reactants (PL assay described elsewhere (4)), the formation of a H_2O- and CH_3Cl-insoluble surface film, and the increase in surface viscosity with increasing contact time between an insoluble StAld monolayer and a solution of PL suggest covalent attachment of the octadecyl hydrocarbon chain to the PL backbone. Preliminary quantitative analyses indicate ca. two lysyl groups per attached octadecyl hydrocarbon chain or one free amino group for each lysyl amino–StAld linkage. Continued work will be required to directly assay the number of free lysyl amino groups present in the grafted film and to determine its overall structure and rheological properties. VPC analysis has proved to be a most useful analytical tool in monitoring surface film reactions and has shown clearly that condensation chemistry continues long after the relative surface viscosity has increased to a value exceeding the range of the constant angular velocity torsion pendulum device used here.

Hydrolysis of a Positively Charged Surfactant Ester. Chromatography also has been indispensable for the detection and quantitation of nonvolatile surfactant substances. HPLC is the method of choice when comparing the hydrolysis kinetics of the surfactant diester derivative of ruthenium(II) tris(2,2′-bypyridine), **I**, to that of its water soluble analog, **II** (7).

It had been observed previously that the area of a monolayer of **I** spread on alkaline subphase and held at constant surface pressure decreases with time whereas that of a monolayer spread on a subphase of

$$\left[(bpy)_2 Ru \begin{matrix} N\bigcirc\text{-COOR} \\ N\bigcirc\text{-COOR} \end{matrix} \right]^{2+} \xrightarrow[-ROH]{OH^-} \left[(bpy)_2 Ru \begin{matrix} N\bigcirc\text{-COOR} \\ N\bigcirc\text{-COO}^- \end{matrix} \right]^{1+} \xrightarrow[-ROH]{OH^-} \left[(bpy)_2 Ru \begin{matrix} N\bigcirc\text{-COO}^- \\ N\bigcirc\text{-COO}^- \end{matrix} \right]^{0}$$

I, R = R = $C_{18}H_{37}$ **IV**, R = $C_{18}H_{37}$
II, R = R = C_2H_5 **V**, R = C_2H_5 **III**

pH < 5 did not (1). By analogy to the chemistry observed for **I** in alkaline 50% aqueous tetrahydrofuran (THF) and **II** in alkaline aqueous solution (5), the diester surfactant was expected to hydrolyze to the water-soluble dicarboxylate, **III**, and *n*-octadecanol via **IV**.

Figure 6 shows a schematic of the trough operations involved in the hydrolysis of **I**. The initial step requires the formation of a well-defined monolayer film on a nonhydrolyzing subphase ($10^{-3} M$ NaCl). Any instability in the film that might occur if the surfactant were slightly water soluble or underwent counterion exchange with a bulk subphase can be detected and distinguished from any changes accompanying the hydrolysis reaction itself. Furthermore, the film can be compressed to the desired surface pressure prior to bringing it into contact with alkaline subphase. This eliminates the interval of time (ca. 5 min) on single-compartment troughs starting when the initial drop of spreading solution touches the interface and continuing through solvent evaporation until the desired surface pressure is reached. The film then is transported laterally across

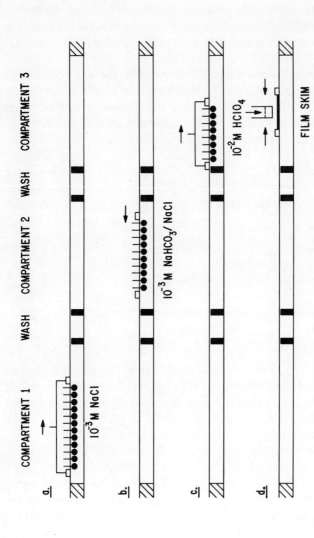

Figure 6. Schematic description of operations involved in the hydrolysis reaction of **I** in monolayer films at the air–water interface. Operations: (A) Apply CHCl$_3$ solution of **I** to surface of 1.0×10^{-3}M NaCl, allow CHCl$_3$ to evaporate and compress film to desired surface pressure; (B) shift film at constant area across 1.0×10^{-3}M NaCl in first wash compartment onto alkaline subphase and maintain constant area across 1.0×10^{-3}M NaHCO$_3$/NaCl in second wash compartment II by varying barrier separation; (C) shift film at constant area across 1.0×10^{-3}M NaCl in second wash compartment onto 1.0×10^{-2}M HClO$_4$ to stop hydrolysis; (D) collapse film while on perchloric acid, skim off resulting thick film and dissolve in CHCl$_3$.

the subsurface walls at constant area onto the alkaline subphase ($10^{-3}M$ NaHCO$_3$/NaCl, pH 8.2) by simultaneously moving the two surface barriers. By driving only one of the surface barriers while using the other as a fixed platform for the surface pressure detector, the film area can be adjusted to maintain a constant II. The hydrolysis reaction is quenched by moving the film across the wash compartment to minimize the volume of entrained alkali onto an acid-containing subphase ($10^{-2}M$ HClO$_4$). Previous studies had shown clearly that the hydrolysis reaction in bulk solution is stopped immediately by acid quenching (5). In addition to supplying protons, aqueous perchloric acid was chosen for the perchlorate anions' strong condensing effect on films of **I** and because **IV** is known to form stable films on perchlorate-containing subphases (1). Films of **I** on aqueous perchloric acid can be collapsed into visibly reddish threads while attempts to do so on aqueous hydrochloric acid lead to the subphase overflowing the side walls of the trough with resultant material loss. The collapsed, thicker film of **I** is easily skimmed off using a spatula sandwiched between two surface barriers and dissolved in an appropriate solvent for subsequent HPLC analysis. Control experiments show the absolute recovery for StOH, **I**, and **IV** to be 90, 78 ± 5, and 90 mol %, respectively.

Rate constants derived for the sequential hydrolyses, II → V → III, in homogeneous solution are in quantitative agreement with simple consecutive pseudo-first order reaction kinetics ($k_{-II} = 5.0 \times 10^{-3}$ min^{-1}, $k_{-V} = 7.0 \times 10^{-3}$ min^{-1} in $10^{-3}M$ NaHCO$_3$/NaCl, pH 8.2°). The rate laws governing the monolayer hydrolyses, **I** → **IV** → **III**)see Figure 7), are not as simple. A plot of log [**I**] vs. time (II = 10 dyn/cm, $1.0 \times 10^{-3}M$ NaHCO$_3$/NaCl) shows the disappearance of **I** as quantitatively described by first-order rate constants, $k_{-I} = 5.3 \times 10^{-2}$ min^{-1} (early) and 1.1×10^{-2} min^{-1} (late), whose straight lines intersect at $t \approx 45$ min. The disappearance of **I** according to two rate constants may reflect the higher surface charge density found in the early portion of the sequential hydrolysis reactions ($t < 20$ min), where the doubly charged **I** predominates, over that found at longer reaction times ($t > 40$ min), where **I** is only a minor component in the mainly singly charged film. As the positive surface charge increases, the concentration of hydroxide anions in the layer of subphase adjacent to the ester monolayer also increases.

The appearance or disappearance of **IV** in this same reaction cannot be fitted to a simple pseudo first-order consecutive reaction rate expression over any interval of time. However, the disappearance of **IV** ($t > 40$ min) can be fitted (only fair match, $r^2 = 0.76$) to first-order kinetics, $k_{-IV}^{obs} = 9.0 \times 10^{-3}$ min^{-1}. A separate series of experiments starting with **IV** gives an almost identical first-order rate constant, $k_{-IV}^{obs} = 1.1 \times 10^{-2}$ min^{-1}. These observed rate constants must be corrected for the disso-

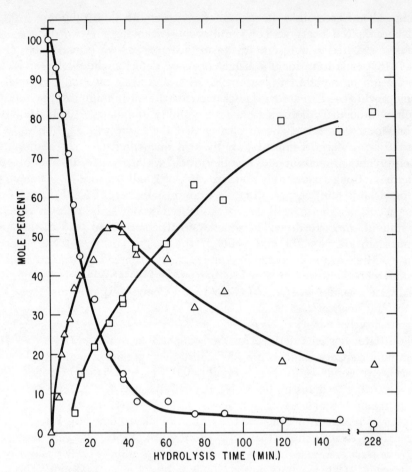

*Figure 7. Relative concentrations of reactant (**I**, ○) and products (**IV**, △; **III**, □) during hydrolysis on $1.0 \times 10^{-3}M$ $NaHCO_3/NaCl$ (10.0 ± 0.5 dyn/cm, 22–23°C, initial amount of **I** is 3.0×10^{-7} mol at a surface concentration of 6.7×10^{13} molecules/cm^2) as determined by HPLC. The amounts of **III** were calculated from $100 - (I + IV)$.*

lution of **IV** into the subphase during the hydrolysis as estimated by the first-order film shrinkage (k_{-IV}^{sol}) observed on $1.0 \times 10^{-3}M$ NaCl ($\pi = 10$ dyn/cm): $k_{-IV} = k_{-IV}^{obs} - k_{-IV}^{sol} = (1.1 - 0.4) \times 10^{-2}$ min^{-1} $= 7.0 \times 10^{-3}$ min^{-1}. HPLC analysis of films recovered from various subphases clearly confirms dissolution of **IV** from the monolayer in the case of $1.0 \times 10^{-3}M$ NaCl.

A comparison based only on film shrinkage would suggest a faster reaction proceeded at the higher surface pressure whereas quantitative analysis of the recovered films show little, if any, difference. HPLC analysis indicates that **IV** has a very high rate of dissolution at 33.3 dyn/cm and a constant, equilibrium concentration of **IV** in the surface film is attained very rapidly in the hydrolysis reaction starting with **I**.

It is clear that the rate of hydrolysis of **I** constrained to the interface proceeds ten times faster than that of its counterpart, **II**, in homogeneous solution at the same bulk pH. Alexander and Rideal noticed no difference when the hydrolysis rates of neutral esters were compared in both environments (*10*). However, the rate enhancement obtained by having a positively charged surface film rather than a neutral film is striking. To achieve experimentally reasonable rates for the hydrolysis of ethyl octadecanoate, Alexander and Rideal were required to employ $1N$ NaOH (pOH $= 0!$) in the subphase. The hydrolysis rate constant observed, $k = 2.09 \times 10^{-2}$ min^{-1} (21.2°C, 3 dyn/cm) lies between $k_{-\mathrm{I}}$ and $k_{-\mathrm{IV}}$ obtained in this study on a subphase having a much lower hydroxyl ion concentration, pOH 5.8.

Monolayer Probe for Surface Acidity. The knowledge that monolayers of **I** hydrolyze readily on the surface of alkaline subphases can be used to develop a technique for assessing the pH of unknown hydrophilic solid surfaces to which the surfactant can be adsorbed. For instance, monolayers of **I** can be transferred from $10^{-3}M$ NaCl onto soda–lime glass slides which have been pretreated by sonication in alkaline Alconox or strongly acidic aqueous solution. Soaking the slide in acetonitrile containing MeSO$_3$H ($0.015M$) and HOAc (0.5% V/V) readily desorbs the surfactant and allows quantitation by HPLC of **I** and **IV** in the resulting solution (98 ± 2% mol recovery, pmol sensitivity). When the slide was dried in a dessicator immediately after transferring the monolayer, 0.5 mol % **IV** was found on the alkaline-treated glass and 2–3 mol % **IV** on the acid-washed glass. However, if the monolayer coated slides are immersed in triply distilled water, then steadily increasing amounts of **IV** (and **III**) are seen with time; at 4-hr immersion 6 mol % **IV** is desorbed from alkaline cleaned glass and 14–17 mol % from acid-cleaned glass. Correlations between the hydrolysis rates of **I** on aqueous subphases of known bulk pH and on solid surfaces would allow estimates of the acidity of that surface. The resulting accuracy is limited only by our present ability to calculate surface pH from bulk solution pH values.

If the solid is visually transparent and a surfactant employed whose absorption spectrum is pH dependent, then measurements of the optical spectrum of the solid adsorbed monolayer will likewise provide a surface acidity probe. For instance, the absorption and emission spectra of **IV** are pH dependent. Figure 8 shows a comparison between the absorption spectra of **IV** and its ethyl ester analog in 20% aqueous THF and on acid or alkaline pretreated slides. Other surfactant pH indicators are known and could be employed in this way (*11, 12*).

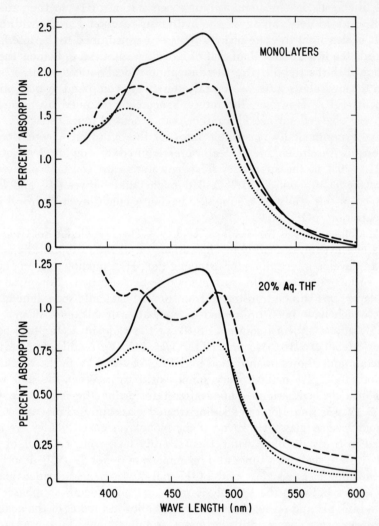

Figure 8. Upper traces: Typical absorption spectra of **IV** and Ru-$(COOC_{18})(COOEt)$ as single monolayers on both sides of a hydrophilic glass slide transferred from 10^{-3}M $NaClO_4$: (· · ·) $Ru(COOC_{18})(COOEt)$ on acid cleaned glass; (———) **IV** on acid cleaned glass, 99 mol % pure; and (– – –) **IV** on alconox cleaned glass, also contains 5 mol % **I** and 10 mol % n-octadecanol. Lower traces: Absorption spectra of **IV** and Ru-$(COOC_{18})(COOEt)$ as ca. equivalent number of molecules in 20% aqueous THF: (· · ·) $Ru(COOC_{18})(COOEt)$, 2.9×10^{-7}M, pH ~ 6; (———), **IV**, 5.5×10^{-7}M, pH ~ 6; and (– – –) **IV**, 5.5×10^{-7}M, pH ~ 2.

Orientation of Methylene Blue Adsorption. The adsorption of thionine and methylene blue to electrode surfaces and the resulting electrochemical selectivity may be critical when these dyes are used to absorb and convert light to electrical energy in photogalvanic devices (*13, 14*). Little is known about the orientation of this adsorption. One of the advantages of the monolayer model in studying surface phenomena is the experimental control that can be exercised over orientation by synthesis

$$CH_3-N(R_1)-\text{[phenothiazine]}^+-N(R_2)-CH_3 \quad X^-$$

	R_1	R_2
VI	$-CH_3$	$-CH_3$
VII	$-C_{18}H_{37}$	$-CH_3$
VIII	$-C_{18}H_{37}$	$-C_{18}H_{37}$

of the appropriate surfactant. In this case, monolayer films showing different molecular orientations were obtained from mono-octadecyl, **VII**, and di-octadecyl, **VIII**, derivatives of methylene blue, **VI**.

Surface pressure–area curves of **VII** and **VIII** on aqueous $10^{-3}M$ $HClO_4$ are shown in Figure 9. From Corey-Pauling-Koltun (CPK)

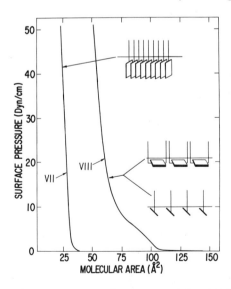

*Figure 9. Π–A curves for **VII** and **VIII** on $10^{-3}M$ $HClO_4$ recorded at 23°C and with a compression rate of 6.6 $A^2/molecule/min$*

molecular models, the dimensions of the rectangular methylene blue ring system (including the exocyclic N,N-dimethyl amino substituents) are approximately 14.5 Å (long) × 7.6 Å (wide) × 5.5 Å (thick). The three basic orthogonal cross-sectional areas can be obtained from these values: ca. 110 Å^2, in the plane of the ring system; ca. 80 Å^2, along the long axis; and ca. 42 Å^2, along the short axis.

A comparison of the experimentally observed areas at which Π starts to rise with the model derived cross-sectional areas suggests that the chromophore of **VIII** (108 Å^2) lies nearly coplanar with the air–water interface while that of **VII** (30 Å^2) has its long axis nearly normal to this same interface. It appears that the molecular areas (26–30 Å^2) observed for films formed from **VII** over the entire range of Π approximates that area (24 Å^2) obtained with the molecular models by interleaving the exocyclic N,N-dimethyl amino substituents. The limiting area measured for **VII** at high Π suggests an orientation in which both the plane and short axis of the chromophore are more nearly normal to the interface (area determined from close packed models is 60–65 Å^2).

The absorption spectra of glass-supported monolayer arrays of **VII** and **VIII** show differences based on the number of hydrocarbon chains attached to the chromophore, the glass pretreatment, subphase counterion, and surface pressure at which the film was transferred off the subphase onto the glass surface. For instance, Figure 10 compares the ab-

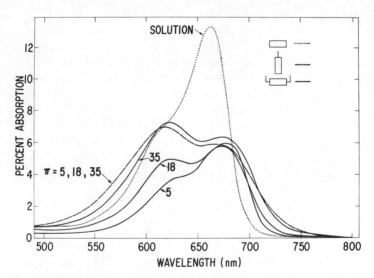

Figure 10. Absorption spectra of (– – –) **VII**, and (———) **VIII**, transferred from 10^{-3}M HCl as a single monolayer on both sides of an acid cleaned hydrophilic glass slide; Π as labelled in dyn/cm. Absorption spectrum of (· · ·) **VI**, in 10^{-3}M HCl, 5 × 10^{-7}M, 1.0 cm^2 cuvette.

sorption spectra of monolayers of **VII** and **VIII** transferred from $10^{-3}M$ HCl onto acid cleaned glass to ca. equivalent number of molecules of **VI** in homogeneous $10^{-3}M$ HCl solution (there is no anion effect on this set of spectra). It is obvious that the absorption spectrum of **VII** is insensitive to increasing π while that of **VIII** changes substantially over the same range. The absorption spectrum of **I** in aqueous solution exhibits a concentration dependence that has been interpreted as resulting from dimerization and further aggregation of the chromophores (15). The long wavelength band is assigned to the monomer, the short to the dimer, and further broadening to higher aggregation.

It has been proposed that the dimerization of thiazine dyes involves the face-to-face apposition of the two chromophores lying in adjacent parallel planes in which van der Waals–London interactions provide the attractive force (e.g., see Rabinowitch (15), Sheppard (16), and Tomita (17)). The data reported here can be interpreted using this dimer geometry. At low π, films containing **VII** already show strong spectral interaction between the chromophores (long axis normal to glass surface) while films of **VIII** reflect chromophores lying coplanar with the glass surface that must be compressed to high π before the chromophores are induced to tilt and take part in the face-to-face interaction.

While spectroscopic studies using polarized light to precisely define chromophore orientation on the glass surface have not been done, the present information should be useful in the study of photogalvanic devices by: (1) providing optical densities of monolayer dye films which will allow quantitation of the amount of dye adsorbed from homogeneous solution; (2) suggesting possible orientations for the dye adsorption; and (3) supplying monolayer absorption spectra obtained under a variety of conditions for correlation with photoelectrochemical action spectra.

Counterion Effects. π–A Isotherms. To the surface chemist, monolayers of **I** are novel because they contain the first known example of a doubly charged cationic surfactant. The hydrolysis behavior of **I** at the air–water interface demonstrated one aspect of this unique characteristic. Largely to avoid the complications of hydrolysis while still retaining the doubly charged nature of the surfactant molecule, a dinonadecyl analog of **I** was synthesized in which the alkyl side chains were connected to the bipyridyl ligand by carbon–carbon single bonds (3).

When monolayers of **I** and **IX** are formed on aqueous subphases (pH < 5 for **I**), the films exhibit stable and generally reversible π–A curves that are markedly dependent on the counterions present in the subphase solution (see Figure 11) (1,3). On subphases giving the more expanded films (e.g. Cl⁻, AcO⁻), the π–A curves are exactly reversible and do not vary with compression rate or with the time the film is held at high surface pressure (at least up to 30 dyn/cm). With the more

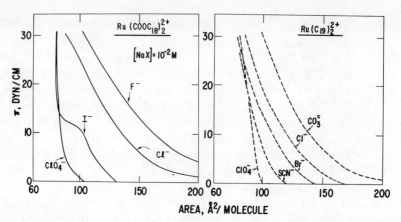

Figure 11. Π–A curves for **I** and **IX** on aqueous subphases 0.01M in NaX, 22–24°C

strongly condensing counterions (e.g. ClO_4^-), some surface pressure decay is observed after rapid compression to high Π and there is a small amount of compression–expansion hysteresis in cycling. This hysteresis is largely reversible, since a second compression curve close approximates the initial compression. For the ions exhibiting intermediate behavior (SCN^-, I^-), there is also some hysteresis, but the shape of the compression curve persists with expansion and recompression cycles. Although their molecular areas are approximately twice as large, the ruthenium surfactants have Π–A curves similar in shape to and showing the same counterion specificity as those of such singly charged cationic surfactants as docosyl trimethylammonium bromide (*18*).

When films of **I** are spread on mixed chloride–perchlorate solutions, a high selectivity is observed. Thus, the curve on $10^{-2}M$ $Cl^- + 10^{-3}M$ ClO_4^- is identical to that on ClO_4^- alone; for $10^{-2}M$ $Cl^- + 10^{-4}M$ ClO_4^- a small expansion at Π < 5 dyn/cm occurs, but at higher Π the curve again

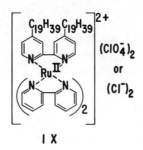

I X

matches that on ClO_4^-; on $10^{-2}M$ $Cl^- + 10^{-5}M$ ClO_4^-, the compression curve agrees with that obtained on Cl^- alone, but a hysteresis loop is observed on expansion in the range 30 > Π > 17 dyn/cm.

The specificity of these doubly charged complexes for ClO_4^- over Cl^- can also be demonstrated by FTIR analysis of monolayer films of **I** transferred off mixed anion subphases onto hydrophilic germanium ATR plates. As shown in Figure 12, even at $ClO_4^-:Cl^- = 1:1000$, the absorption at 1100 cm^{-1} attributable to ClO_4^- is observed. Little difference in relative intensities of the ClO_4^- is noted for $ClO_4^-:Cl^- > 1:100$.

The more tightly bound nature of ClO_4^- relative to Cl^- is emphasized in the field desorption mass spectra (FDMS) of solid **IX** as both the perchlorate and chloride salt (*see* Figure 13). The molecular ion of the chloride salt shows the loss of both anions; the molecular ion of the perchlorate compound shows only a loss of one anion.

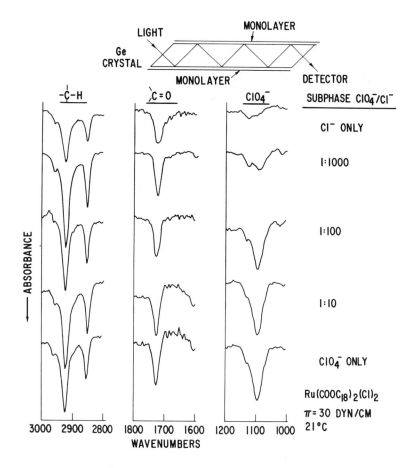

*Figure 12. Partial FTIR of **I** transferred from $10^{-3}M$ NaCl, $10^{-3}M$ NaCl/$10^{-6}M$ NaClO$_4$, $10^{-3}M$ NaCl/$10^{-5}M$ NaClO$_4$, $10^{-3}M$ NaCl/$10^{-4}M$ NaClO$_4$ and $10^{-3}M$ NaClO$_4$ aqueous subphases at $\Pi = 30$ dyn/cm, 21°C as a single monolayer on both sides of a hydrophilic Ge ATR plate.*

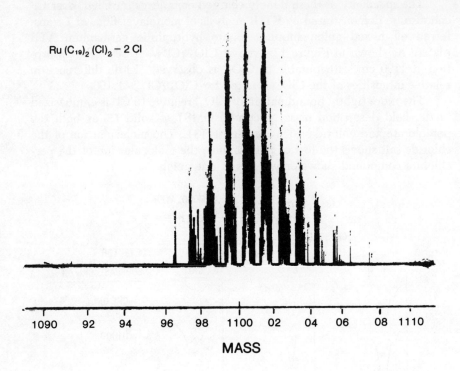
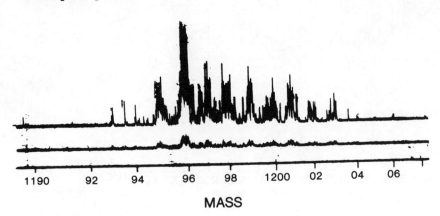

Figure 13. FDMS of the perchlorate and chloride salts of solid **IX**

LUMINESCENCE. The most striking effect of counterions is their major influence on monolayer luminescence at the air–water interface (*see* Figure 14). Such differences are unique to the interface and are not found either in homogeneous aqueous organic solutions of **I** (or **IX**) with added salts or in wholly aqueous solutions of **II** and **III** at levels of 4M KCl and NaClO$_4$. Because of the inability to estimate the quantum yield of luminescence in the monolayer fluorimeter, it is not possible to decide whether the difference in luminescence intensity results from a reduction in yield by ions such as Cl$^-$, or an enhancement by ClO$_4^-$, as compared with yields in other environments. The large solvent effects on luminescence yield of Ru(bpy)$_3^{2+}$ derivatives (*19, 20, 21*) suggest that the environment change induced by interaction with the strongly condensing ClO$_4^-$ ions might enhance luminescence. Another case of anion alteration (increased lifetime, reduced nonradiative decay rate) of luminescence properties for a transition metal bipyridyl complex has been reported recently (*22*), although since in that situation a different type of excited state is affected and the effect is observed in aqueous solution, the observations may be unrelated.

Förster theory calculations based on the small overlap between absorption and emission spectra of **I** in CHCl$_3$ solution (using a luminescence quantum yield, $\Phi_L = 0.18$) leads to an energy transfer distance, $R_0 = 15$ Å (If this value is corrected for the monolayer environment by introducing $\Phi_L = 0.045$, as measured for aqueous solutions of the dimethyl ester (*21*), $R_0 = 12$ Å). In a close packed array occupying 85 Å^2/mole-

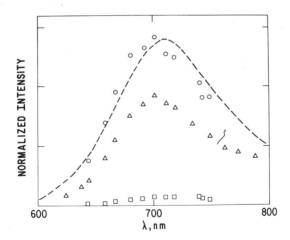

*Figure 14. Spectral distribution of luminescence from **I** in monolayers on aqueous (○) 0.01M NaClO$_4$; (△) 0.01M NaBF$_4$; and (□) 0.01M NaCl. All at Π = 4–6 dyn/cm. Luminescence spectrum of monolayer transferred to hydrophilic glass slide from (– – –) 10^{-3}M NaClO$_4$, arbitrary intensity scale.*

cule, the center-to-center distance is 10.1 Å. Therefore it might be expected that a substantial amount of energy migration could occur in these films. Experimental results give no evidence of concentration quenching as the chromophores are brought closer together in monolayers on chloride subphases.

Photochemical reactions involving the excited state of **I** or **IX** with counterion would also lead to a reduction in luminescence yield. The oxidation of chloride anion to atomic chlorine is one such possibility that would appear to be eliminated by observing exactly the same luminescence intensity when chloride is replaced by the more easily oxidized bromide. While the presence of iodide anion completely quenches the luminescence of **I**, this observation is clouded by the possibility of an alternative reaction with molecular iodine, which might have been formed in the freshly prepared but open aqueous subphase.

PHOTOLYSES. Single monolayers of **I** on hydrophilic glass have been irradiated with white (300–900 nm) light in contact with a variety of aqueous solutions. The choice of Cl^- or ClO_4^- as counterion in the photolyte had no effect. The average disappearance quantum yield for **I** in these experiments is $5 \pm 2 \times 10^{-6}$. Other than small amounts (1–3 mol %) of **IV**, which are found in both the photolysis and dark, aqueous control, chromatographic analysis under currently employed conditions shows no component(s) that could account for the amount of **I** consumed in the photolysis. Absorption spectra of the dry monolayers taken before and after photolysis show the development of a broad band extending throughout the visible spectrum to ca. 700 nm with the original maxima still at 415 nm and 490 nm but superimposed on the new band and of diminished absorption. Neither HPLC analysis nor absorption spectroscopy of the aqueous photolysis solution show any trace (< 2 mol %) of **III** or any other water soluble $Ru(II)(bpy)_3^{2+}$ derivative. In no case was molecular hydrogen observed by VPC in these photolyses (if produced, $\Phi_{H_2} \lesssim 1 \times 10^{-4}$).

NEUTRAL MONOLAYER. To obtain a neutral surfactant molecule with structural and photophysical characteristics similar to **I** and **IX**, one of the bipyridine ligands in **IX** was replaced by two cyano ligands to

$$\left[\begin{array}{c} C_{19}H_{39}\ C_{19}H_{39} \\ \text{CN--Ru--CN} \end{array} \right]^0$$

X

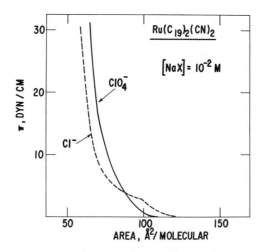

*Figure 15. Π–A curves for **X** on aqueous subphases 10^{-3}M in NaCl and $NaClO_4$, 23°C*

give **X** (3). This compound is expected to have properties similar to the water soluble compound, Ru(II)(bpy)$_2$(CN)$_2$, studied by Demas et al. (23). **X** is a chloroform-soluble material that gives a strong isotope pattern at mass 998 (^{102}Ru) in FDMS for its molecular ion. Stable monolayer films of **X** can be formed on an aqueous subphase and show little selectivity between ClO_4^- and Cl^- in Π–A isotherms (see Figure 15). Sigle monolayer films can be transferred to hydrophilic quartz slides. The absorption and uncorrected emission spectra of **IX** and **X** are shown in Figure 16. The emission intensity of quartz-supported monolayers of **X** does not depend on either the counterion (ClO_4^-, Cl^-) in the subphase

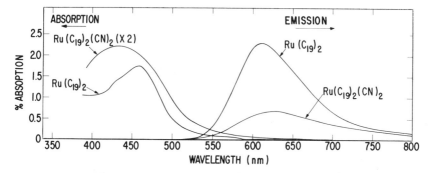

*Figure 16. Absorption spectra and uncorrected emission spectra of **IX** and **X** transferred from 10^{-3}M $NaClO_4$ as a single monolayer on both sides of an alkaline sonicated hydrophilic quartz slide at $\Pi = 30$ dyn/cm, 22°C. Excitation at 436 nm and emission band pass is 4.2 nm.*

from which the monolayer was transferred or the counterion in the aqueous solution in contact with the monolayer during the measurement. Under similar conditions, both I and IX would show a two- to threefold intensity increase when ClO_4^- replaced Cl^-. Hence, whatever the exact nature underlying the effect of differing counterions upon the photophysics of ruthenium(II)tris(2,2'-bpy) complexes, it is unique to an interfacial environment carrying a positive charge.

Conclusion

Some aspects of the directed synthesis of catalytically active or selective surfaces by chemically attaching modifying functionalities can be modelled by the study of chemical reactions in monolayer films. Applications of commercially available instrumentation have made it possible to obtain as complete an experimental description of the chemistry occurring in films of monomolecular thickness as for reactions occurring in homogeneous solution.

Acknowledgments

The author wishes to thank G. L. Gaines, Jr. for his general encouragement, valuable discussions, and experimental collaboration in the work on the counterion effects. He acknowledges with pleasure the synthetic contributions made by P. E. Behnken in the preparation of the surfactant ruthenium tris(bipyridyl) complexes and his many colleagues at the Corporate Research and Development Center who provided helpful discussion and experimental assistance. This work was partially supported by the Division of Basic Energy Sciences, Department of Energy, and RANN, the National Science Foundation.

Literature Cited

1. Gaines, G. L., Jr.; Behnken, P. E.; Valenty, S. J. *J. Am. Chem. Soc.* **1978**, *100*, 6549.
2. Valenty, S. J. *J. Colloid Interface Sci.* **1979**, *68*, 486.
3. Valenty, S. J.; Behnken, P. E.; Gaines, G. L., Jr. *Inorg. Chem.* **1979**, *18*, 2160.
4. Valenty, S. J. *Macromolecules* **1978**, *11*, 1221.
5. Valenty, S. J.; Behnken, P. E. *Anal. Chem.* **1978**, *50*, 834.
6. Valenty, S. J. *Anal. Chem.* **1978**, *50*, 669.
7. Valenty, S. J. *J. Am. Chem. Soc.* **1979**, *101*, 1.
8. Hünig, S.; Fritsch, K. H. *Justus Liebigs Ann. Chem.* **1957**, *609*, 172.
9. Sawicki, E.; Hauser, T. R.; Stanley, T. W.; Elbert, W. *Anal. Chem.* **1961**, *33*, 93.
10. Alexander, A. E.; Rideal, E. K. *Proc. R. Soc. London, Ser. A* **1937**, *163*, 70.
11. Gaines, G. L., Jr. *Anal. Chem.* **1976**, *48*, 450.
12. Fernández, M. S.; Fromherz, P. *J. Phys. Chem.* **1977**, *81*, 1755.

13. Albery, W. J.; Archer, M. D. *Electrochim. Acta* **1976**, *21*, 1155.
14. Albery, W. J.; Bowen, W. R.; Fisher, F. S.; Foulds, A. W.; Hall, K. J.; Hillman, A. R.; Orchard, A. F.; Egdell, R. G. "Extended Abstracts," 2nd International Conference Photochemical Conversion and Storage of Solar Energy, University of Cambridge, Cambridge, United Kingdom, August 1978.
15. Rabinowitch, E.; Epstein, L. F. *J. Am. Chem. Soc.* **1941**, *63*, 69.
16. Sheppard, S. E.; Geddes, A. L. *J. Am. Chem. Soc.* **1944**, *66*, 2003.
17. Tomita, G. *Z. Naturforsch.* **1968**, *23b*, 922.
18. Goddard, E. D.; Kao, O.; Kung, H. C. *J. Colloid Interface Sci.* **1968**, *27*, 616.
19. Van Houten, J.; Watts, R. J. *J. Am. Chem. Soc.* **1976**, *98*, 4853.
20. Seefeld, K. P.; Möbius, D.; Kuhn, H. *Helv. Chim. Acta* **1977**, *60*, 2608.
21. Harriman, A. *J. Am. Chem. Soc., Chem. Commun.* **1977**, 777.
22. Henry, M. S. *J. Am. Chem. Soc.* **1977**, *99*, 6138.
23. Demas, J. N.; Addington, J. W.; Peterson, S. H.; Harris, E. W. *J. Phys. Chem.* **1977**, *81*, 1039.

RECEIVED October 2, 1978.

6

Effect of the Condensed Phase, Homogeneous Solution, and Micellar Systems on Photoionization

J. K. THOMAS and P. PICIULO

Chemistry Department and Radiation Laboratory, University of Notre Dame, Notre Dame, IN 46556

> *Several physical techniques are used to investigate the effect of condensed-phase alkanes, alcohols, and micellar systems on the photoionization of arenes and aromatic amines. In alkanes the ionization threshold is reduced by approximately 2 eV compared with that in the gas phase; the decrease can be as large as 4.0 eV in micellar systems. The negative surface of anionic micelles helps to separate efficiently the ion pair formed in the initial photoionization stage. Other micellar systems where photoassisted electrons transfer to an acceptor other than water (e.g., quinone) are discussed. In many ways these systems offer advantages over direct photoionization for storage of energy.*

Increasing concern with alternate sources of energy over the past few years has focused attention on the possible utility of several photophysical and photochemical systems for energy storage (1). This chapter presents various aspects of photoionization of molecules as this process could, under suitable conditions, lead to the conversion of light into electrical energy. Considerable data is already available for photoionization of molecules in the gas phase. Unfortunately, the energies required to eject electrons from molecules in the gas phase are greater than 6 eV and the corresponding wavelength of light lies in the vacuum UV part of the spectrum. However, there is evidence (2) that the ionization potential, or onset of ionization (I_p), of a molecule is reduced in condensed phases compared with that in the gas phase. A 2-eV drop in I_p is experienced in solutes dissolved in alkane liquids compared with the gas phase (3, 4).

Indeed, in anionic micellar systems the onset of photo-ionization (I_p) is dropped even further, (5, 6, 7) and the micelle also helps to efficiently separate the ions produced. This latter process occurs as the electron is ejected into the water phase while the cation is stabilized by the anionic charge of the micelle, which also repels the electron. This latter event is observed most clearly in the effect of cationic micelles on exciplex systems such as pyrene–dimethylaniline (8, 9). The exciplex dissociates into pyrene anions and dimethylaniline cations at the micelle surface. The cation is repelled from the cationic micelle surface while the anion is retained. In this way the micelle promotes efficient separation of ions and leads to greatly enhanced ion lifetimes compared with homogeneous solution.

The condensed phase and, in particular, micelles enhance the photo-production of ions and also lead to stabilization of the ionic products. The object of this chapter is to reflect on experiments that measure the factors influencing the decrease in I_p of a molecule in condensed phases compared with the gas phase (the ultimate object being the design of systems that might be useful in the storage of solar energy). The format of the chapter is to proceed to investigate I_p's of molecules in alkanes, alcohols, and aqueous micellar solutions and to relate the observed data to a coherent picture of the photoejection process. Finally, the added utility of promoting electron transfer to a second solute rather than to the solvent is discussed.

Experimental

Materials. Aromatic compounds such as pyrene and anthracene were recrystallized from alcohol; solvents such as alkanes were Phillips 66 research grade and purified further by passage down a 3-ft column of activated alumina. Alcohols were distilled from the sulfate salt of dinitrophenyl hydrazine; water was simply distilled from an alkaline potassium permanganate solution, followed by distillation from acidic potassium dichromate solution. Surfactants such as sodium lauryl sulfate (NaLS), and cetyltrimethylammonium bromide (CTAB), were purchased from BDH as pure grade, and also were recrystallized from alcohol.

Techniques. The I_p's in insulators such as alkanes were measured by the photoproduced current generated in these systems (3, 4). Solutions were placed in 1-cm³ quartz cells containing parallel electrodes of platinum separated by 5 mm. A high voltage was applied across the electrodes and the resultant carrier was measured by a Keithley picoammeter. Light from a 1000-W xenon lamp was projected between the electrodes after passage through a Bausch and Lomb f/3.5 monochromator and suitable filters to cut out stray light. The I_p was detected by an abrupt increase in photocurrent upon varying the excitation wavelength.

This technique does not work for conducting fluids such as alcohols. Here a different technique developed by Lipsky was attempted (10).

The solution was excited with light of various wavelengths in an Aminco spectrofluorimeter and the yield fluorescence at a selected emission wavelength was noted. This produces a fluorescence activation spectrum of the molecule. The experiment was repeated in the presence of varying concentrations of an electron scavenger such as $CHCl_3$. In hydrocarbons the fluorescence yield decreased below the wavelength associated with I_p. This is explained in the following fashion: excitation of the molecule A leads to the excited state which is observed via its fluorescence.

$$A \xrightarrow{h\nu} A^* \rightarrow \text{fluorescence} \qquad (1)$$

The molecule $CHCl_3$ has little effect on this process. At short wavelengths photoionization occurs.

$$A \xrightarrow{h\nu} A^+ + e^- \qquad (2)$$

This is followed by recombination to give the excited state of A.

$$A^+ + e^- \rightarrow A^* \qquad (3)$$

However, in the presence of $CHCl_3$ the electron (e^-) reacts giving a product that does not lead to excited A.

$$e^- + CHCl_3 \rightarrow CHCl_2 + Cl^- \qquad (4)$$

These processes have been identified in photolysis of solutions of N,N'-tetramethylparaphenylenediamine (TMPD) (10) and in laser photolysis of pyrene solutions (11). The $CHCl_3$ only reduces the fluorescence yield if photoionization occurs. Thus the effect of $CHCl_3$ on the activation spectrum can be used to measure I_p of the molecule. This technique was utilized in alcoholic solutions.

Aqueous micellar solutions present special problems regarding measurement of I_p's. The technique used consists of photolyzing the molecule in micellar solution in the presence of SF_6. If hydrated electrons are produced in the photoionization event they react with SF_6 to give six fluoride ions (12). The yield of fluoride ion, measured by a fluoride-sensitive electrode, is also a measure of photoionization. The excited singlet or triplet states of some of these molecules may react with SF_6, presumably via electron transfer, to give F^-. However, the rate constants for such reactions are slow ($k < 10^7$ mol^{-1} sec^{-1}). Also such reactions are eliminated by introducing 10^{-4} mol O_2 into the system at [SF_6] of $10^{-4}M$. Under these conditions O_2 efficiently removes the triplet state ($k = 2.4 \times 10^{10}$ mol^{-1} sec^{-1}) while 50% of the hydrated electrons react with O_2 and 50% with SF_6 ($k_{e^-+O_2} = 1.8 \times 10^{10}$ mol^{-1} sec^{-1}, $k_{e^-+SF_6} = 1.7 \times 10^{10}$ mol^{-1} sec^{-1}). The low [SF_6] in micellar systems also precludes reaction with the excited singlet states of the aromatic amines which have lifetimes in the region of 10^{-8} sec. It has been suggested that in radiolysis of aqueous SF_6 solutions the reaction of e^-_{aq} with SF_6 produces six fluoride ions via the mechanism:

$$SF_6 + e^-_{aq} \rightarrow SF\cdot_5 + F^- \qquad (5)$$

$$SF_5\cdot + 2H_2O \rightarrow OH + F^- + H_3O^+ + SF_4 \qquad (6)$$

$$SF_4 + 9H_2O \rightarrow SO_3^{2-} + 4F^- + 6H_3O^+ \qquad (7)$$

In the present work where e^-_{aq} is produced photolytically the same mechanism for F^- production is assumed, that one e^-_{aq} produces six F^- ions.

Laser flash photolysis with light of $\lambda = 3471$ Å also was used to identify the ionic products of photolysis in some of the micellar solutions (*13*). In some systems the ionic products are sufficiently stable to be used as direct monitors of I_p following steady state photolysis.

Results

Photoionization in Alkanes. Figure 1 shows typical data for the photoionization of several molecules in aerated tetramethyl silane (TMS), and isooctane (IO). The photocurrent vs. excitation energy is shown and an extrapolation of the current back to the energy axis gives I_p. Deoxygenation of the solutions by bubbling with N_2 only increased the small photocurrents in the tails of the onset wavelengths, presumably because

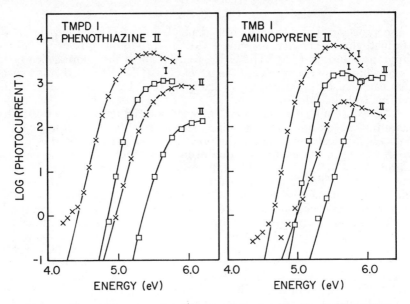

Figure 1. Photoionization of solutes in nonpolar liquids TMPD, TMB, Ph and AP in TMS and isooctane

Table I. Cation Polarization Energies

	I_g (eV)	I_s (eV) (TMS) (10)	P_+ (eV)	P_+ (avg) (eV)	r (Å)	R (Å)
TMPD	6.25	4.27	−1.37	−1.36	2.51	3.73
		4.72	−1.36			
TMB	6.40	4.52	−1.27	−1.32	2.59	4.11
		4.87	−1.36			
Phenothiazine	6.96	4.80	−1.55	~1.57	2.18	3.87
		5.20	−1.59			
1-Aminopyrene	7.02	4.77	−1.64	−1.67	2.05	4.08
		5.15	−1.70			
3-Aminoperylene	6.94	4.75	−1.58	−1.62	2.11	4.32
		5.12	−1.65			
Pyrene	7.55	5.46	−1.91	−1.91	1.66	3.98
Perylene	6.92	4.82	−1.49	−1.52	2.25	4.23
		5.20	−1.55			
Anthracene	7.42	5.60	−1.21	−1.30	2.63	3.84
		5.87	−1.38			
Tetracene	6.88	4.92	−1.33	−1.43	2.39	4.18
		5.19	−1.52			

of biphotonic processes of the triplet state (4). The ionization onset (potential) data for several compounds in TMS and IO are summarized in Table I. The I_p's of TMPD in the two solvents were $I_{TMS} = 4.27$ eV and $I_{IO} = 4.72$ eV, which compare well with the values reported by Holroyd et al. (3) and Lipsky and Wu (9).

The I_s of a molecule in solution is related to that in the gas phase (I_g) by:

$$I_s = I_g + P_+ + V_0$$

where P_+ is the polarization energy of the cation and V_0 the energy state of the electron in the liquid. The V_0 for several liquids have been measured (3), and they enable the calculation of P_+. These data also are shown in Table I. For the most part P_+ is identical in both liquids although the unsubstituted arenes, apart from pyrene, tend to give a slightly larger P_+ in IO compared with TMS.

Photoionization in Alcohols. The effect of $CHCl_3$ on the fluorescence activation spectrum (described in the experimental section) was investigated for several arene and aromatic amines in methanolic and ethanolic solutions. No effect of $CHCl_3$ was observed up to concentrations of 1M and down to wavelengths of 2200 Å. This technique is successful in "alkane" solution for both one- and two-photon photoionization (10, 11) at $CHCl_3$ concentrations of 0.1M. Electrons are not generated on the photolysis of alcoholic solutions of tetramethylbenzidene (TMB), aminopyrene (AP), tetracene, and pyrene at wavelengths greater than 2200 Å.

Earlier work on laser and flash photolysis of TMPD in alcoholic solution with light of $\lambda = 3471$ Å or 2650 Å did not produce observable yields of electrons (*14, 15, 16*). However, a long-lived species with the kinetic and spectral properties of the TMPD cation was observed. Recent data (*5, 6*) indicate that phenothiazine and TMB are photoionized weakly by one photon of 3471 Å light.

Photoionization in Aqueous Micellar Solutions. Laser excitation of molecules in anionic micelles leads to efficient photoionization. In some cases the photoionization is two photon with quanta of wavelength 3471 Å. The excited singlet state produced first is subsequently photoionized by absorption of a second photon (*11, 17*). The molecule is often pumped up to an energy of 7 eV or close to the I_g and photoionization then occurs readily in condensed systems. This type of behavior is exhibited in unsubstituted arenes such as pyrene, tetracene, perylene, and anthracene. However, in some cases such as phenothiazine, (Ph), TMB (*5*), AP, and aminoperylene (*7*), the photoionization is one photon of $\lambda = 3471$ Å. The energy of photoionization at 3.6 eV is much less than the I_g (~ 6.5 eV) for these molecules. Typical data for laser photolysis of ·AP follows.

Laser Photoionization of AP. Figure 2 shows a grouping of spectra produced by the pulse radiolysis or laser photolysis of AP in various solvents. The pulse radiolysis technique has been well studied and can be used to produce triplet or ionic states of a solute under controlled conditions (*18*). For example, in the pulse radiolysis of *p*-xylene only excited singlets and triplets are formed, and at times greater than 30 nsec the excited singlets cross over, yielding triplets. The *p*-xylene triplet state transfers energy to AP, forming the triplet state. This spectrum is shown in Figure 2A, where $\lambda_{max} = 4200$ Å. The radiolysis of methanol produces solvated electrons that have a broad absorption spectrum with $\lambda_{max} = 6750$ Å; free radicals such as CH_2OH are produced also but have low reactivity in most systems. Pulse radiolysis of AP in methanol gives rise to the spectrum of solvated electrons which decay rapidly giving AP⁻. These data are shown in the inset of Figure 2 where the e_s^- at $\lambda = 6000$ Å decays rapidly while the absorption of the AP anion at $\lambda = 5000$ Å grows. The immediate absorption at this latter wavelength is caused by overlap of the e_s^- and AP⁻ spectra, $\lambda_{max} = 4950$ Å.

Laser photolysis of AP gives rise to different transitory species depending on the solvent. In hexane (Figure 2B) the spectrum observed at 1 μsec after the pulse is that of the triplet state, which compares well with that found in the radiolysis of AP in *p*-xylene (Figure 2A). AP fluorescence also is observed as well as a short-lived absorption in the 5250-Å region. The decays of the fluorescence and absorption are identical, which suggests that the short-lived absorption is that of the excited

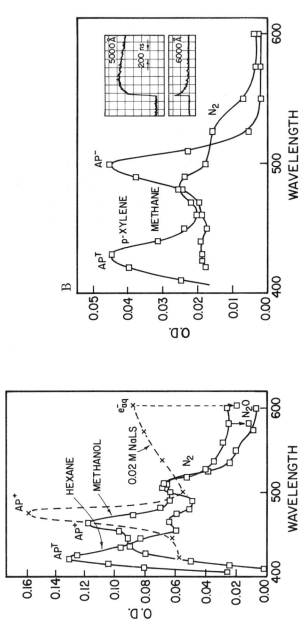

Figure 2. (A) Short-lived intermediates of AP following laser photolysis in hexane, methanol, and 0.02M NaLS. (B) Short-lived intermediates following pulse radiolysis of AP in p-xylene or methanol. Insert. Decay of e_{aq}^- in methanol and rise of AP^-.

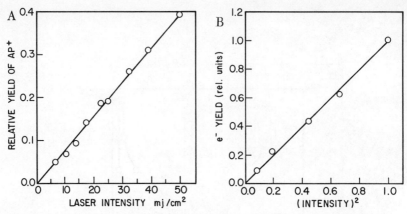

Figure 3. Intensity dependence of cation yield (A) of AP in laser photolysis in NaLS solution and (B) of e_s^- yield in methanol.

singlet state of AP (AP^S). Laser excitation of AP produces AP^S, which decays giving fluorescence and the triplet, AP^T. Figure 2B shows the spectra of transitory species formed in the laser photolysis of AP in methanol. AP fluorescence and absorption attributable to AP^T at 4100 Å are observed, but the main features of the spectrum are the peak at 4600 Å and the broad absorption to the red, which disappears with N_2O. This latter behavior is typical of solvated electrons, which cause the red absorption. Photoionization occurs and it is suggested that the species with $\lambda_{max} = 4700$ Å (the decay of which is independent of $[O_2]$) is the AP cation, AP^+. Laser photolysis in micellar solution accentuates the ionization effect (Figure 2B) where both the cation AP^+ ($\lambda_{max} = 4700$ Å) and hydrated electrons ($\lambda_{max} > 6000$ Å) are observed, removing the latter species by saturating the solution with N_2O.

The data of Figure 2 show the pronounced effect of phase on the nature of the final products (10^{-8} sec) in the photolysis of AP. The ionization process is two photon in methanol and one photon in micelles or in a mixture of 1:1 water–methanol. This is shown in Figure 3 where the yield of photoionization is plotted against laser intensity, I, or laser intensity squared; linearity is obtained with I in micelles and I^2 in methanol. It is also pertinent to note that the amino group in AP changes the photoionization process in micelles from two photon with pyrene to one photon with AP. Similar effects are noted for perylene and aminoperylene (7).

Onset of Photoionization in Micelles. It is desirable to obtain some estimate of the onset of photoionization (I_p) in micellar solution. This was achieved using the SF_6 technique in the presence of O_2. Typical data for Ph, TMB, and AP are shown in Figure 4. It is noted that the I_p occurs close to the onset for absorption of light by these molecules, and at a wavelength close to 4000 Å (3.15 eV).

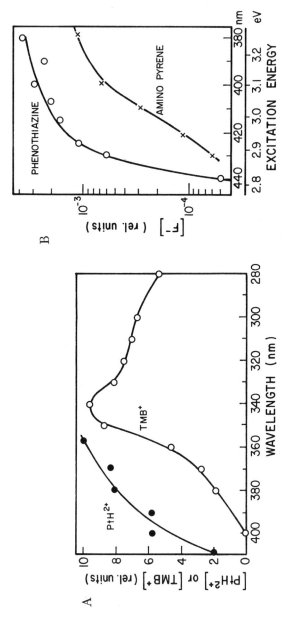

Figure 4. I_p in NaLS solution: (A) yield of Ph di-cation $P + H^{2+}$ and TMB^+, (B) yield of F^- in photolysis of solutions of Ph or AP containing $10^{-4}M$ SF_6 and $10^{-4}M$ O_2. Yields are not corrected for absorption.

Table II. Quantum Yields for Photoionization in Anionic Micelles

Solute	Method	λ (Å)	ϕ
TMB	laser	3471	~ 0.5 (5)
Aminopyrene	laser	3471	$< 4 \times 10^{-2}$
	SF_6/F^-	3500	2×10^{-2}
Phenothiazine	laser	3471	~ 0.5 (5)
	SF_6/F^-	380	0.3
		415	0.2
Aminoperylene	laser	5300	< 0.1
	SF_6/F^-	5300	0.06

Similar onset data also were obtained by observing the final products of photoionization of these molecules, Ph given as the di-cation, and TMB giving the TMB cation. These cations are quite stable in micellar solution, particularly if an electron scavenger such as NO_3^- is present to remove the negative charge (5, 6). The photoproduction of these cations also was confirmed by observing EPR spectra characterization of the phenothiazine di-cation and TMB cation in irradiated micellar solutions of these solutes. Table II shows some data concerning the quantum yields of photoionization of solutes in anionic micelles measured via laser excitation or via the F^- method. The agreement between the two methods is satisfactory.

Photolysis in the Presence of Benzoquinone. Laser photolysis of micellar solutions containing pyrene or AP and benzoquinone (BQ) leads to formation of anions of BQ and cations of the arenes. The fluorescence

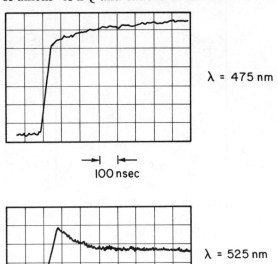

Figure 5. Growth of AP^+ ($\lambda = 4750$ Å) and decay of AP^T ($\lambda = 5250$ Å) in the laser photolysis of AP and BQ in NaLS solutions

of the arenes is diminished in the presence of the quinone, while the triplet state decays rapidly leading to the arene cation and BQ anion. Typical data are shown in Figure 5 for the photolysis of AP and BQ in NaLS micelles where a rapid decay of AP^T is observed at 5250 Å. This leads to a concomitant increase in the AP^+ yield at 4700 Å. In the presence of O_2 the AP^T decays rapidly but no formation of AP^+ is observed, indicating that BQ is essential to promote the electron transfer reaction.

Discussion

The foregoing data in alkanes and aqueous micellar solutions reinforce the concept that the I_p of a molecule in the condensed phase is reduced by several electron volts compared with the gas phase. A convenient way of discussing this effect of phase on the I_p is via the cation polarization energy (P_+) and the energy level of the electron in the medium (V_0), i.e. $I_s = I_g + P_+ + V_0$ where I_s and I_g are the ionization potentials in solution and the gas phase, respectively. The V_0's for many liquids (3) as well as the I_g's for numerous molecules are known, and I_s is measured via experimental data such as those shown in Figure 1. Thus P_+ may be calculated for several molecules of interest, some of which are given in Table I. Two radii values, r and R, are also included, the significance of which is given below.

The P_+ may be expressed by the Born equation:

$$P_+ = -\frac{z^2 e^2}{2r}\left(1 - \frac{1}{\epsilon}\right)$$

where ϵ is the rapid dielectric constant of the medium and is taken to be the square of the refractive index, e is the electronic charge, and z is a radius of the ion in question. A value for r may be obtained by using the radius of a sphere of equal volume to that of the ion (19, 20). This is carried out using the expression:

$$\frac{4}{3}\pi R^3 = \frac{1}{N}\frac{M}{d}$$

where N is Avogadro's number, M the molecular weight, and d is the density of the molecule considered. Table I lists values of r derived from the Born equation and values of R from the sphere model. The calculated radii R are always larger than the Born radius. This might be expected if the positive charge of the cation is not distributed uniformly over the molecule. A region where the positive charge may be high, e.g. in the vicinity of a nitrogen atom, leads to a more compact interaction with the solvent molecules, which is reflected by a smaller radius of interaction.

Along these lines the P_+'s for TMPD and TMB, which are diamines, are lower than P_+ for the monoamines, AP, aminoperylene, and Ph. This may be rationalized as a sharing of the charge among two N atoms in the diamines compared with one in the monoamines, with a correspondingly decreased interaction with the solvent. This is indicated by a closer agreement of r and R for the diamines. It is surprising that P_+ for pyrene in isooctane is much larger than that for the other molecules. The symmetrical nature of pyrene and its cation, however, may lead to a larger interaction with the solvent if the charge is localized in a central region of the molecule. The onsets are rather low and close to the limits of accurate detection; the possibility of impurities in the pyrene producing the cation also increases at these lower wavelengths. Hence the P_+ for pyrene is viewed with caution at this stage.

In the alkanes P_+ plays the major role in decreasing I_p by about 1.5 eV. A discussion of P_+ via the Born equation helps with some qualitative features of the cation solvent interaction. However, it is desirable to have a more detailed description of this energy parameter for future discussion. Perhaps even a discussion of I_s in terms of P_+ and V_0 is inadequate generally, and the data in alcohols seem to indicate this.

The photoejection of an electron from the selected amines and alkanes in alcoholic solution was not observed in the present work. There also is some corroboration of this observation in the literature. To a first approximation the I_s of a molecule in alcoholic solution should be similar to that in an alkane. It has been observed that the absorption spectrum of an electron ejected into alcohol occurs initially in the IR, and is quite similar to that in an alkane (21). The spectrum subsequently relaxes to give the normal e_s^- spectrum with an absorption maximum in the orange or red part of the spectrum. This indicates that V_0 in an alcohol may be similar to that in alkanes. The more polar nature of alcohols could increase P_+, again leading to an I_s in alcohols at longer wavelengths than in alkanes. The lack of any observation of strong polarization in the photolysis of molecules in alcohols except at high energies (> 7.0 eV) is puzzling. The simple picture of photoionization in terms of P_+ and V_0 seems to break down and specific molecular processes leading to alternate products have to be considered.

The most dramatic solvent effects on photoionization appear in micellar solution. The data on AP are typical of what has been observed previously for Ph and TMB in solution. The photoionization with light of $\lambda = 3471$ Å is two photon in alcohols and one photon in anionic micelles. The I_p observed via e_{aq}^- capture by SF_6 or by direct observation of a stable product also indicate I_p's close to the onset of the molecular absorption. The I_s is approximately 3.0 eV and is 3.5–4.0 eV lower than I_g. For aminoperylene, photoionization at 5300 Å has been observed. This corresponds to a decrease in I_s for I_g of 4.6 eV. Electrochemical values for V_0 in water of -1.6 eV have been reported (22). This requires that P_+ is greater than 2.5 eV to account for the lowering of I_p by the micelle system. The values of P_+ for alkanes were all approximately 1.5 eV for the molecules studied (see Table I). However, the environment of the molecules in the region of the sulfate head groups of the micelle could lead to a much larger P_+ in these systems compared with alkanes. The I_p's of anthracene ($I_g = 7.2$ eV) and pyrene ($I_g = 7.5$ eV) are beyond the limit of detection via the SF_6 method, i.e. $\lambda < 2300$ Å (5.5 eV). This indicates that $\Delta I = I_g - I_s < 1.7$ eV, and is similar to that found in a hydrocarbon liquid. Molecules containing polar groups such as AP and TMB have ΔI's of approximately 4.0 eV or greater. If V_0 is -1.6 eV, then P_+ is approximately -2.4 eV, and greater than that in liquid hydrocarbons. Both pyrene and anthracene are located in the micelle interior and not in the head group region. Thus the polarization energy is only that which is found in an alkane, i.e. approximately -1.5 eV; V_0 must also be small. The polar amino compounds are located at the micelle surface leading to an increase in both P_+ and V_0. This illustrates the important role played by the micelle surface in decreasing I_p. The micelle also plays a secondary role in repelling e_{aq}^- from the micelle and into the water bulk, thus decreasing ion neutralization or back reaction of e_{aq}^- and the solute cation.

In micellar systems V_0 can be large and unlike in alkanes it plays an important role in reducing I_p. From an energy storage point of view, using light to produce direct photoionization is not as attractive as photoassisted electron transfer. The electron affinity of the acceptor solute also helps to drive the electron transfer reaction. This has been illustrated previously in the pyrene–dimethylaniline system of cationic micelles. In the present case AP and BQ illustrate the principle; other similar systems using TMB and duroquinone have been reported elsewhere (5). Excitation of AP leads to photoionization and excitation yielding AP^+ e_{aq}^-, AP^S, and AP^T. The excited states AP^S and AP^T then react with BQ giving BQ^- and AP^+. The e_{aq}^- produced via photoionization also reacts with the quinone, forming the anion. The net process of absorption of light is to produce ionization giving AP^+ and an anion. The anionic surface of the micelle prevents back reaction of the products, and leads to efficient separation of the ion pair for subsequent use in the chemistry of the system.

Conclusion

There is a pronounced effect of solvent on the nature of the photoionization process. This is partly understood in alkanes and in micellar systems in terms of P_+ and V_0 for the electron. The situation differs in alcoholic solutions and is not understood at this stage. The most dramatic effects reside in micellar systems where the energy required to eject electrons is low and can be carried out with visible light. Choice of an anionic micelle also leads to effective separation of the ionic products. From the energy storage systems' viewpoint it is more useful to transfer the e⁻ to an acceptor rather than to the water bulk, and several redox systems operating in the visible spectrum then can be used to produce ions. Future work requires that these ions are converted to useful forms of energy, e.g. decomposition of H_2O to H_2 and O_2 or to produce an electrical current. Modification of the micellar system is required to achieve this end. One possibility is to transfer the charges to colloidal catalysts, which then could decompose water (23).

Acknowledgment

The research described herein was supported by the Office of Basic Energy Sciences of the Department of Energy. This is Document No. NDRL-1925 from the Notre Dame Radiation Laboratory.

Glossary of Symbols

I_p = ionization potential
I_g = ionization potential in gas phase
I_s = ionization potential in solution
NaLS = sodium lauryl sulfate
CTAB = cetyltrimethylammonium bromide
TMPD = N,N'-tetramethylparaphenylenediamine
TMS = tetramethylsilane
P_+ = polarization energy
TMB = tetramethylbenzidine
AP = aminopyrene
AP^S = singlet-excited aminopyrene
AP^T = triplet-excited aminopyrene
AP^+ = aminopyrene cation
Ph = phenothiazine
BQ = benzoquinone
BQ^- = benzoquinone anion
IR = infrared

Literature Cited

1. Claesson, S., Engstrom, L., Eds. "Solar Energy and Photochemical Conversion and Storage"; *NE* **1977**, *6*.
2. Lesclaux, R.; Jousset-Dubien, J. In "Organic Molecular Photophysics"; Birds, J. B., Ed.; Wiley: London, 1973; Vol. I.
3. Holroyd, R. A.; Russell, R. L. *J. Phys. Chem.* **1974**, *78*, 2128.

4. Jarnagin, R. C. *Acc. Chem. Res.* **1971**, *4*, 420.
5. Alkaitis, S. A.; Beck, G.; Grätzel, M. *J. Am. Chem. Soc.* **1975**, *97*, 5723.
6. Alkaitis, S. A.; Grätzel, M. *J. Am. Chem. Soc.* **1976**, *98*, 3549.
7. Thomas, J. K.; Piciulo, P. *J. Am. Chem. Soc.* **1978**, *100*, 3239.
8. Razem, B.; Wong, M.; Thomas, J. K. *J. Am. Chem. Soc.*, **1978**, *100*, 1679.
9. Waka, Y.; Hamamoto, K.; Mataga, N. *Chem. Phys. Lett.* **1978**, *53*, 242.
10. Wu, K. C.; Lipsky, S. *J. Chem. Phys.* **1977**, *66*, 5614.
11. Piciulo, P.; Thomas, J. K. *J. Chem. Phys.* **1978**, *68*, 3260.
12. Asmus, K.-D.; Grunbein, W.; Fendler, J. H. *J. Phys. Chem.* **1974**, *96*, 5456.
13. Richards, J. T., West, G.; Thomas, J. K. *J. Phys. Chem.* **1970**, *74*, 4137.
14. Richards, J. T.; Thomas, J. K. *Trans. Faraday Soc.* **1970**, *66*, 621.
15. Ottolenghi, M. O. *Chem. Phys. Lett.* **1971**, *12*, 339.
16. Ottolenghi, M .O. *J. Phys. Chem.* **1969**, *73*, 1912.
17. Wallace, S. C.; Grätzel, M.; Thomas, J. K. *Chem. Phys. Lett.* **1973**, *23*, 349.
18. Thomas, J. K. In "Charged Particles and Tracks in Solids and Liquids"; Boag, J., Ed.; Institute of Science, 1970; p 174.
19. Nakato, Y.; Chiyoda, T.; Subomura, H. T. *Bull. Chem. Soc. Jpn.* **1974**, *47*, 3001.
20. Henglein, A. *Ber. Bunsenges. Phys. Chem.* **1975**, *79*, 129.
21. Richards, J. T.; Thomas, J. K. *J. Chem. Phys.* **1970**, *53*, 218.
22. Barker, G. C.; Bottura, G.; Cloke, G.; Gardner, A. W.; Williams, M. J. *J. Electroanal. Chem. Interfacial Electrochem.* **1974**, *50*, 323.
23. Korvakin, B. V.; Dzhaviev, T. S.; Shilov, A. E. *Dokl. Phys. Chem.* **1977**, *233*, 359.

RECEIVED October 2, 1978.

Interfacial Photoprocesses in Molecular Thin Film Systems

LARRY R. FAULKNER[1], HIROYASU TACHIKAWA, FU-REN FAN, and SHIRLEY G. FISCHER

Department of Chemistry, University of Illinois, Urbana, IL 61801

Thin films (less than 10,000 Å in thickness) in assemblies (several films overlaying one another) allow one to create environments where chemical processes can occur under special conditions. One important problem is the prevention of recombination after light-induced electron transfer. Thin film systems offer some promising possibilities blocking the thermalizing. A related problem concerns the collection of light over a large area and transmission of its energy to an interfacial reaction center. Thin film systems accomplish this in a manner that mimics the antenna function of photosynthetic organisms. New types of electrode surfaces can be fabricated by depositing molecular solids over surfaces such as gold. Photovoltaic effects corresponding to nearly ideal Schottky junctions have been observed in assemblies involving these same materials.

Controlling chemical reactions via the synthesis of special environments will be a general theme in the future of chemical research. The basic ideas certainly are not new. Researchers in electrochemistry and heterogeneous catalysis have discussed them for years. Still, the recent past has seen a considerable broadening of interest beyond the groups traditionally associated with surface chemistry.

There are several trends contributing to this development. Biochemical advances have sensitized many chemists to the enormous impact

[1] To whom correspondence should be addressed.

the details of environmental structure can have on a reaction. There also have been improvements in the methods for synthesizing large-scale structures and for characterizing environments, particularly on surfaces, where reactions might occur. Added motivation has been concern over new techniques for energy conversion, because much of the chemistry of conversion is inherently interfacial (e.g., catalytic conversion, electrochemical storage) or involves intermediates (such as excited states) that have several reaction paths. Only one path may be useful; by controlling the reaction environment it may be taken exclusively. The efficiency of charge separation during the primary events of photosynthesis is a good example of a scheme for energy conversion that relies on environmental structure to enforce a favorable result (1–4). This volume testifies to the developing interest in "microstructural synthesis" and its potential.

In creating a special reaction environment, the usual intent is the introduction of some sort of asymmetry. In a semiconductor photovoltaic cell, for example, absorbed photons produce electron–hole pairs that are born into a region in which there is an electric field (5, 6). This field provides the environmental asymmetry needed to force the carriers apart and prevent the loss of energy by recombination. In chemical systems, the easiest way to introduce anisotropy at reaction sites is to carry out the reactions at interfaces; thus the problem of microstructural synthesis becomes quickly associated with aspects of surface chemistry.

In our laboratory, an emphasis has been placed on chemical processes taking place in assemblies of thin films, because these systems permit a ready experimental stress on interfacial events. We are interested in applications of thin-film technology to problems of environmental synthesis. This technology has been extensively developed by industries dedicated to electronics and optical coatings (7, 8, 9). Techniques and apparatus are readily available to chemists and are applicable to a wide range of interesting chemical problems. Over the past four years, we have explored their prospects in several areas, and in this chapter we offer a summary of our results and a prognosis. In keeping with the theme of this volume, systems involving photoprocesses are emphasized.

Experimental Aspects

In the experiments described below, films of metals and small molecules were deposited by evaporation at 10^{-5}–10^{-7} torr in a Varian NRC Model 3117 bell jar system. The substrates were always glass or quartz plates that had been cleaned, degreased, and treated with a static eliminator. Molecular species were evaporated from crucibles; metals were evaporated from boats or filaments. The details of various operations are available in the literature (*10–15*). Patterns were laid down by depositing the films through machined masks. Thicknesses were monitored and controlled by a Sloan Model DTM-200 digital thickness monitor based on a quartz crystal oscillator. For the metals, published densities were used in computing film thickness; for the molecular solids, a working density of unity was used. In general, the real thicknesses of molecular films were probably smaller than the nominal values quoted below by factors ranging from 1–2. The substrate temperature was usually the ambient value, but for some work a thermoelectric cold stage was used to cool the substrate to approximately 0°C.

Submonolayer coverages of perylene were produced by deposition of a cold vapor. To carry out this operation, perylene was first evaporated from a heated crucible onto the walls of a special chamber inside the bell jar. Then the system was opened and substrates to be coated were placed inside the special chamber. The system was then re-evacuated to operating pressure and held there for lengths of time ranging from 30 min to several hours. Both the chamber and the substrates were maintained at ambient temperatures. Vapor transport of perylene to the substrates was observed under these conditions. The amounts deposited were assayed by the methods described below.

Films of polymers and dispersions of small molecules in polymers were deposited by spin-coating (*9, 16*). In this method, a precleaned glass substrate was spun on an axis normal to its surface at a rate of 1000–10,000 rpm. A solution of the polymer (typically 4–5% polystyrene in xylene) was dropped onto the moving surface, and the wetting phenomenon caused a thin film to be bound to the surface. In most cases the solvent was lost from the film quite rapidly (< 1 min). Deposits ranging from 1000 Å to several μm could be made in this way. Dispersions of small molecules, such as fluoranthene, were made in the polystyrene by dissolving them directly in the coating solution. Such dispersions are assumed to be homogeneous, but more will be said about this point later.

The compositions of various molecular films were assayed by dissolving the films in appropriate solvents and performing conventional analytical procedures on the resulting solutions. Measurements were usually made by fluorometry or by absorption spectroscopy.

The chemicals used in all of the work reported here were generally the best available commercial grades and were used without purification. The phthalocyanines were supplied by Eastman Organic Chemicals; Aldrich furnished fluoranthene and perylene. Polystyrene was in the form of beads sold by Polysciences, Inc. Styrene monomer was removed by repeated dissolution in xylene and reprecipitation with methanol.

Luminescence spectra were taken with an Aminco–Bowman spectrophotofluorometer via procedures described elsewhere (*15, 16*). Absorption measurements were made on a Beckman Model MVI Acta Spectrophotometer.

Electrical properties of thin film sandwich cells and electrochemical characteristics of thin film electrodes were evaluated using a custom-built potentiostatic apparatus or a Princeton Applied Research Model 173 potentiostat operated under computer control. The details of these measurements are available elsewhere (*10–13*).

Exciton Collection from an Antenna System into Interfacial Reaction Centers

One difficulty in doing photochemistry at tailored reaction centers is that the probability of light absorption at any given center, or in fact in a whole monolayer of centers, is not very high. Thus efficient utilization of the incident energy is imperative whenever reaction centers are fabricated at interfaces.

Photosynthetic organisms circumvent the problem by using arrays of chromophores, called "antenna" chromophores, to absorb and transmit photon energy to relatively few reaction centers, where the apparatus for efficient chemical use of the energy is localized (1, 4, 17). This strategy lends a certain simplicity to the whole system because it provides for a

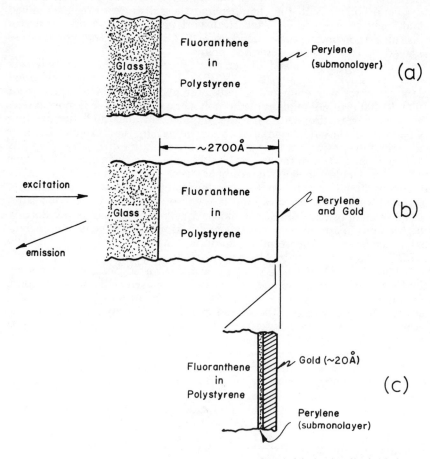

Journal of the American Chemical Society

Figure 1. Schematics of film assemblies. Fluoranthene was contained at 6.5×10^{-9} mol/cm² (13).

high absorption cross-section, but does not require special chemical apparatus at each chromophore. Systems of similar nature could obviously be useful in completely synthetic photochemical devices with tailored reaction centers.

We have recently demonstrated such an antenna effect in thin film systems like that shown in Figure 1a (*13, 15*). The absorbing chromophore was fluoranthene dispersed homogeneously in polystyrene at thicknesses of 1000–3000 Å. On the outer surface of such a film, perylene was deposited at submonolayer coverage from a cold vapor. The luminescence from these systems was studied by front-face optics, as indicated in Figure 1.

Figure 2 shows some of the important results. Frames a and b display emission spectra of fluoranthene in polystyrene and perylene on the polystyrene surface. The former is the usual broad and featureless emission of fluoranthene, while the latter is the characteristic structured distribution of isolated perylene molecules (*16*). Both of these spectra are essentially indistinguishable from corresponding ones taken from molecules dissolved in xylene solution.

Before the surface layer of perylene was deposited, the emission spectrum of the system corresponded to Figure 2a. After deposition, the emission was that shown in Figure 2c. A mixed spectrum was observed, but there was a strong perylene component despite the fact that excitation was carried out at a deep valley in perylene's absorption spectrum. Moreover, fluoranthene molecules outnumbered perylene species in this experiment by a factor of 180. At the excitation wavelength employed, direct excitation of perylene can account for no more than 1% of the observed emission. Figures 3 and 4 demonstrate that the excitation spectrum for total emission at 470 nm is that of fluoranthene absorbance in the 300–380 nm region.

The composite emission spectra were digitized at 1-nm intervals and resolved by a least-squares method into the two pure emission components from perylene and fluoranthene. In this manner, we determined that deposition of the perylene had a quenching effect on the fluoranthene band. The size of the effect was 9–30%, depending on the amount of perylene placed on the surface, the thickness of the film, and the age of the perylene deposit.

These points demonstrate clearly that perylene surface traps are being supplied with energy that is absorbed initially by fluoranthene. The fluoranthene–polystyrene layer has increased the effective absorption cross section for perylene at 337 nm by a factor on the order of 100, hence it acts as an antenna zone for photochemistry at the surface sites. The magnitude of the quenching effect on fluoranthene suggests that chromophores as far as 500 Å from the surface are effective at energy transfer to the surface layer.

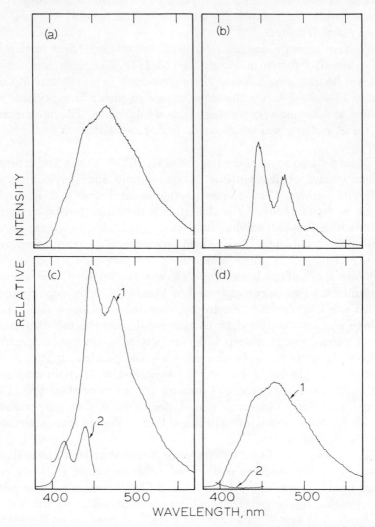

Figure 2. Fluorescence emission spectra: (a) Fluoranthene in polystyrene; excitation at 337 nm. (b) Perylene deposited from a cold vapor on polystyrene. Coverage, 1.6 ng/cm², excitation at 390 nm. (c) Fluoranthene in polystyrene with perylene overlay at 9.1 ng/cm². Curve (1) is emission excited at 337 nm. Curve (2) shows direct excitation of perylene surface layer that emits at 478 nm. (d) System in (c) with 20-Å gold overlay added. Intensity scale is the same as (c). Curves (1) and (2) correspond to those in (c) (13).

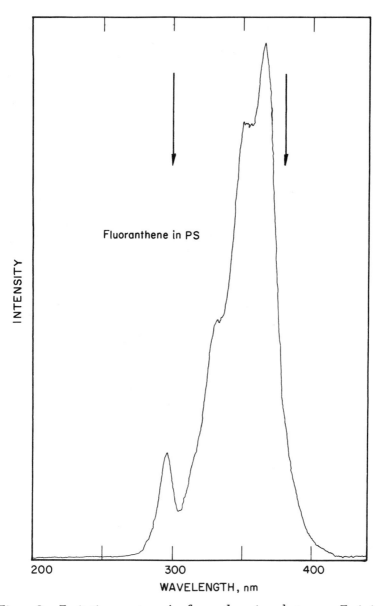

Figure 3. Excitation spectrum for fluoranthene in polystyrene. Emission monitored at 478 nm.

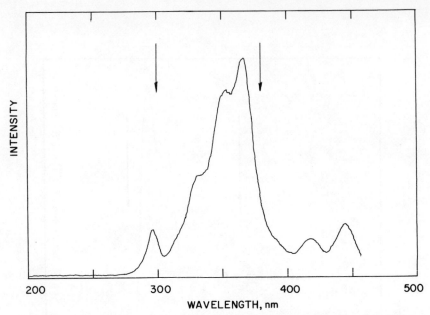

Figure 4. Excitation spectrum for a typical film assembly. Perylene surface layer was ~ 6 ng/cm², emission monitored at 478 nm.

The most straightforward explanation for our observations is to propose that excitons migrate (18, 19, 20) among fluoranthene sites until they approach the surface sufficiently closely to be trapped, probably by Förster's transfer (20, 21), in a perylene site. Since the singlet of perylene lies about 0.3 eV below that of fluoranthene, the trapping effect should be essentially irreversible. Surface trapping effects of this type have been used to evaluate exciton diffusion lengths in single molecular crystals (22, 23), but no reports have described similar effects in systems of randomly-oriented, dispersed chromophores like those involved here.

This interpretation of the antenna effect could be complicated by diffusion of perylene molecules into the bulk polystyrene to produce homogeneous, rather than heterogeneous, trapping centers. This issue was addressed via experiments in which the assemblies of Figure 1a were overlaid by gold at 20-Å nominal thickness, as shown in Figures 1b and 1c. Figures 2c and 2d, both recorded for the same sample, show the effect. Placement of the gold selectively removes the perylene components from the emission and excitation spectra of the assembly. Analysis of the final assemblies by dissolution and fluorometry revealed that about half the perylene was lost during deposition of the gold. Apparently the hot incoming gold atoms imparted enough energy to etch some perylene molecules from the surface. Selective total quenching of the remaining perylene indicates that it lies near the gold surface and is not homogeneously distributed throughout the film. Figure 5 shows that ageing of the

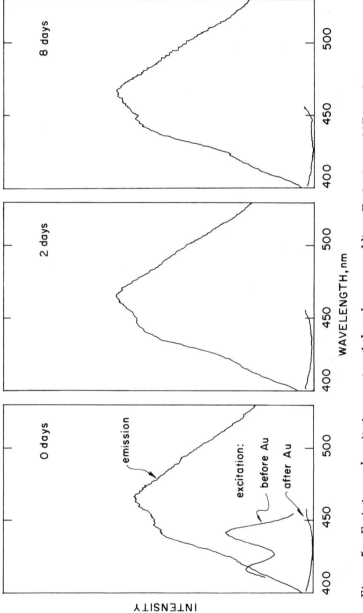

Figure 5. *Emission and excitation spectra of three-layer assemblies. Excitation at 337 nm for emission spectra; emission monitored at 478 nm for excitation spectra. Perylene coverage before deposition of 20-Å gold was ~ 6 ng/cm^2 (15).*

complete assembly over an eight-day period did not result in a reappearance of perylene's luminescence components. The sum of these observations provides strong support for the idea that perylene is indeed a surface trapping center in our experiments and that it remains so for at least several days.

The estimated range of exciton collection cited above rests essentially on an assumption that excitons are initially created uniformly throughout the fluoranthene–polystyrene film. This assumption is not vitiated by optical attenuation of the excitation beam as it passes through the sample, because the film has a low absorbance at 337 nm. Besides, excitation takes place from the side opposite the trapping layer. On the other hand, there is a possibility that the true antenna zone is a surface excess of fluoranthene; about 20% quenching could be effected within a short distance of the surface layer.

At the moment, we are unable to entirely rule out this possibility, but we have conducted some experiments bearing on the issue. In one set of experiments, we cast fluoranthene–polystyrene films of various thicknesses, redissolved them in chloroform, and measured the amounts of both components in the actual films. The measurements were made by absorption spectroscopy at 360 nm, where only fluoranthene absorbs, and at 262 nm, where both absorb. A plot of fluoranthene content vs. polystyrene content (both in moles) was linear and showed a zero intercept within one standard deviation. The virtually zero intercept indicates that there is no intrinsic excess or deficiency created in the overall fluoranthene content during the coating process. The slope of the plot was $(2.1 \pm 0.1) \times 10^{-2}$, which is in good agreement with the mole ratio of 2.1×10^{-2} used in the coating solution. These results show that the overall composition of the films is the same as that of the coating solution. If a surface excess of fluoranthene arises, it is created during the drying process.

Preliminary results are available from another set of experiments in which we measured the quenching of fluoranthene's fluorescence by a 20-Å gold surface layer. Two configurations were examined: (a) glass substrate covered by fluoranthene–polystyrene covered by gold and (b) glass substrate covered by gold covered by fluoranthene–polystyrene. In general, the quenching effect in Configuration b is about one-third smaller than in Configuration a. The effects in both cases are 10–30%, depending on the thickness of the fluoranthene–polystyrene phase. The difference in quenching effect might reflect a surface excess at the outer boundary of fluoranthene–polystyrene. However, it could also arise from differences in the gold layer (e.g., average "island" size) on glass vs. polystyrene or from differences in coating on a glass surface vs. a gold-covered surface. Further study is required to resolve these points. The comparability in the quenching effect for Configurations a and b suggests that the effectiveness of the antenna function in fluoranthene–polystyrene–perylene systems is not wholly based on a compact surface excess of fluoranthene at the outer surface, although some inhomogeneity in concentration may exist.

The molecular requirements for an effective antenna system is a topic of extensive current research (*4, 17, 19, 24*). Emphasis is being placed on photosynthetic systems and on models of photosynthetic apparatus; the points of major concern are the required concentrations of chromophores and the conditions of relative alignment. The exciton collection distance indicated by our work is surprisingly large for a system of dispersed, randomly oriented chromophores. If verified by additional work, it would suggest that effective antenna regions can be synthesized without special alignments among chromophores and without very high chromophore concentrations.

The contribution of thin film techniques to this problem is in the production of accessible trapping centers. One is left here with considerable freedom to assemble additional apparatus to ensure that the trapped energy is used effectively. The technique is fairly general and lends itself to further development.

Photovoltaic Activity from Schottky Barrier Cells Based on Phthalocyanine Films

A general problem of interest to us is the separation of the high-energy intermediates produced in a light-induced electron-transfer reaction. For example, consider an electron acceptor A and a donor species D; a classic case is anthracene as A and dimethylaniline as D (*20, 25, 26*). These species do not react in the ground state, but undergo a ready charge transfer after light absorption to yield the ion pair $A\cdot^-D\cdot^+$,

$$A + D \xrightarrow{h\nu} {}^1A^* + D \rightarrow A\cdot^-D\cdot^+$$

This "carrier pair" is a chemical embodiment of the photon energy, and if the reagents could be separated and stored for later use, such a process could provide important approaches to energy conversion. However, in an isotropic medium, $A\cdot^-D\cdot^+$ will either recombine quickly or dissociate to have the separate carriers recombine with other counter carriers elsewhere (*20, 25, 26*),

$$A\cdot^-D\cdot^+ \rightarrow A + D + h\nu$$

Thus the photon energy is lost to heat. Much of this volume concerns the search for general means to separate photoredox carrier pairs before they can recombine.

The only technologically useful solution at present is found in semiconductor devices, where light produces electron–hole pairs with a similar propensity to recombination (5, 6),

$$\text{Light (photon)} \rightarrow n + p \rightarrow \text{Heat (phonons)}$$

Recombination is successfully prevented by creating the carrier pair in a region immersed in an electric field, where the electron moves in one direction and the hole in another. The fields are generated near interfaces (or junctions) with other phases.

Even in a conventional semiconductor device, such as a silicon solar cell, there is a close analogy between the problem of carrier separation after light absorption and the problem of redox pair separation in a chemical system. The analogy is even closer in a semiconductor device based on a molecular solid, which is bound together by relatively weak van der Waals forces. In these solids, carriers are in many ways similar to radical ions in a concentrated solution.

We have recently completed (11, 12) a study of photovoltaic activity in some devices based on metal-free and zinc phthalocyanines (H_2Pc and $ZnPc$), which are widely studied molecular solids (27–34, 36, 37). They have a natural p-type semiconductivity, which seems to arise largely from the incorporation of oxygen and water from the atmosphere (10, 29, 30, 31). Apparently the oxygen acts as an acceptor impurity and results in mobile holes in the lattice. In our work, the phthalocyanines were studied as evaporated films 100–3000 Å thick. These films are very smooth and comprise very small grains ($\sim$ 200 Å typical dimension) of the α crystal modification (32, 33).

Figure 6 shows a typical device used for studying photovoltaic activity. The substrate was glass and the lower metal contact M_1, gold. The top contact was indium in the work discussed here. The current path through the cell was the circular overlap region in the center.

A metal with a low work function (relatively high electron energy), such as indium, will form a "blocking" contact with a p-type semiconductor (5, 6), because its electrons tend to fill in the holes. Thus there are few carriers in the interfacial region, and conductivity is low. A bias voltage applied externally tends to intensify the effect if the metal is positive, but tends to overcome it with the opposite polarity. Thus current flow becomes markedly nonlinear and rectification effects are observed. The initial interaction between the metal and the semiconductor results in a charge separation at the interface, hence there is an electric field inside the semiconductor at the blocking contact. This field is a manifestation of the barrier to current flow, which is known as a "Schottky barrier." The whole assembly is a "Schottky junction."

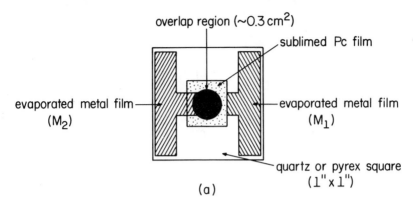

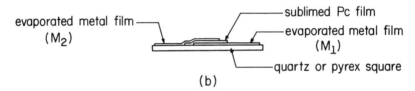

Journal of Chemical Physics

Figure 6. Arrangement of films in metal–phthalocyanine–metal sandwich cells. Substrates are 1 × 1 in. glass plates. (a) Top view. (b) Side view (11).

A metal with high work function (relatively low electron energy), such as gold, will form an "ohmic" contact with a p-type material (5, 6); it tends to extract electrons from the semiconductor and leave an excess of holes at the interface. Plenty of carriers are available, hence there is ready charge transfer at the boundary.

In a device like that shown in Figure 6, the properties of the system are controlled by the Schottky barrier at the indium–phthalocyanine junction (11, 12, 34) because it limits electron flow. Whenever a Schottky junction exists, photovoltaic activity from photons absorbed in the space-charge region or from excitons diffusing into it can be expected (5, 6) because, (a) there are few holes there to annihilate the light-induced electrons and (b) the electric field there tends to pull the photogenerated hole into the bulk semiconductor and the photogenerated electron into

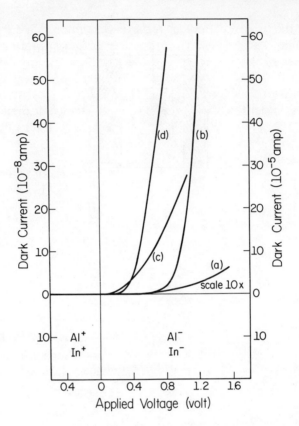

Figure 7. Current–voltage characteristics of sandwich cells in the dark: (a) (Al–H₂Pc–Au) cell; (b) (Al–ZnPc–Au) cell; (c) (In–H₂Pc–Au) cell; (d) (In–ZnPc–Au) cell. Phthalocyanine film thickness, ~ 3000 Å; (a) and (b) use left hand scale; (c) and (d) use right hand scale (11).

the metal. This action effects a safe storage of these excess carriers until they can be recombined in a controlled (and useful) manner via an external circuit.

The actual performance of sandwich cells based on H_2Pc and $ZnPc$ is highlighted in Figures 7–11. The first of these illustrates the rectification effects observed in the cells in the dark. In general, devices with aluminum top layers are defective because an Al_2O_3 layer intervenes between the metal and the phthalocyanine phases (11, 12); they will not be discussed further here. Indium contacts are much closer to ideal.

True Schottky junctions should pass current under forward bias according to a modified Shockley equation (5, 6),

$$J = J_o[\exp\,(eV/mkT) - 1]$$

where J is the current density, J_o is the reverse saturation current density, V is the bias, and m is the modification factor accounting for certain nonidealities in the junction. The other parameters are defined in the usual way. For bias voltages greater than mkT/e (several tens of millivolts), the first term overwhelms the second, and a linear semilogarithmic plot of J vs. V is expected. Figure 8 shows that such behavior is actually observed in the phthalocyanine cells for bias voltages less than 250 mV. At higher voltages, the current falls below the modified Shockley values,

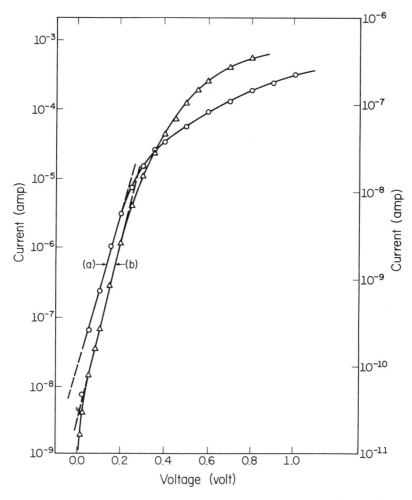

Journal of Chemical Physics

Figure 8. Semilogarithmic plots of the forward-biased dark current–voltage characteristics: (a) (In–H_2Pc–Au) cell, using right hand scale; (b) (In–ZnPc–Au) cell, using left hand scale. Phthalocyanine film thickness, ~ 3000 Å (11).

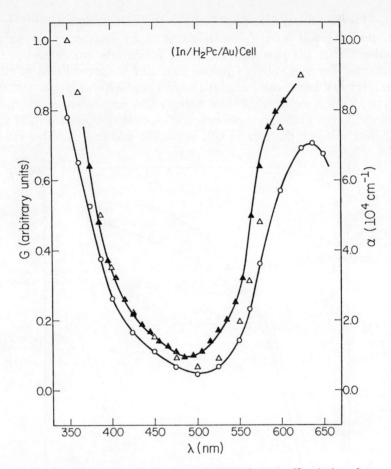

Figure 9. Absorption spectrum of a 3000-Å H_2Pc film (○) and action spectra of the short-circuit photocurrent of an $(In–H_2Pc–Au)$ cell. Illumination at 6328 Å on the In contact; (▲) experimental action spectrum, (△) theoretical points calculated according to Ref. 12.

probably because resistance losses in the bulk phthalocyanines become appreciable (11, 28–31). The slopes of the linear segments provide $m = 1.3$–1.5, which are reasonable values, even for Schottky barrier devices based on conventional materials (6). The reverse saturation current density J_0 yields estimates of the barrier height (6, 35) that agree very well with predictions based on independent measurements of the work function of indium and the ionization potentials of the phthalocyanines (12, 35–38).

Of particular interest here is the photovoltaic activity (12, 34). Light produces a photovoltage at open circuit or under load and the polarity is such that indium is always negative, in accord with expectations for the anticipated Schottky junction. The photocurrent at short circuit is strictly linear with the intensity of irradiation in devices based on ZnPc, and it is nearly linear with intensity in cells involving H_2Pc.

The action spectra of the system provide strong evidence that photovoltaic activity arises from a zone very near the indium contact. Figures 9 and 10 display the results. The ordinate in each case is the ratio of photocurrent to the light intensity actually reaching the phthalocyanine. Corrections were made to compensate for the filtering effects of the metal layers. When illumination was through indium, the action spectrum

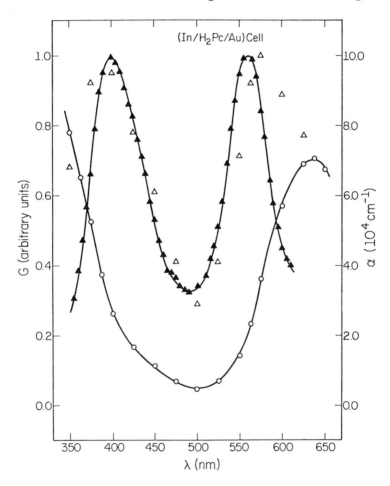

Journal of Chemical Physics

Figure 10. Experimental (▲) and calculated (△) action spectra of the short-circuit photocurrent of an (In–H_2Pc–Au) cell. Illumination at 6328 Å on the Au contact; (○) show the absorption spectrum of an H_2Pc film of 3000-Å thickness (12).

followed the absorption spectrum of the phthalocyanine film. In contrast, illumination through the gold contact yielded an action spectrum with peaks at the inflections of phthalocyanine absorption.

The latter behavior arises from an optical filtering effect of the phthalocyanine phase (*12, 34, 39, 40*). Photons must be absorbed in or near the field-containing region of the Schottky junction to produce a photocurrent. Strongly absorbed light incident on the back surface penetrates only weakly to the junction. Weakly absorbed light penetrates, but is ineffective in producing action. The strongest effect is therefore seen in regions of moderate absorption, because those wavelengths offer the optimal compromise between penetrating power and a capacity for action. When the junction is illuminated directly, there is no filtering effect, hence the action spectrum follows the phthalocyanine's absorptivity.

The expected relationship between the open-circuit photovoltage V_{oc} and the short-circuit photocurrent density J_{sc} is (*6*) as follows:

$$V_{oc} = \frac{mkT}{e} \ln (J_{sc}/J_o + 1)$$

Under ordinary illumination intensities, J_{sc} is much greater than J_o, and J_{sc} is proportional to intensity, hence there should be a semilogarithmic relationship between intensity and V_{oc}. Figure 11 shows that our phthalocyanine sandwich cells do adhere to this expectation. Moreover, the slopes of the plots yield values of m that are in good agreement with values obtained separately from Shockley plots (*11, 12*).

The quantum efficiencies of these cells are surprisingly high. In the case of an In–ZnPc–Au device irradiated at 6328 Å, we find the photons incident on the ZnPc phase to have a 14% efficiency in producing carriers for the external circuit. There is some evidence that the efficiency for photons actually absorbed in the junction region approaches unity (*12*).

This discussion provides strong support for the idea that relatively well-behaved Schottky junctions can be formed at phthalocyanine surfaces. To our knowledge, these data supply the first extensive evidence that even the expected quantitative relationships can be observed in devices fabricated from molecular solids, though blocking metallic contacts to molecular materials has been known for some time (*27, 34, 39, 40, 41*). The phthalocyanines generally seem to be far better semiconductors than one might expect from reports in the literature. Even more ideal solid state properties could probably be obtained for them if more controllable doping techniques could be applied and if larger grain sizes could be produced.

The chief value of this work to chemists is in calling attention to the importance and predictability of solid-state electronic properties in one class of molecular solids. These properties may be very important to

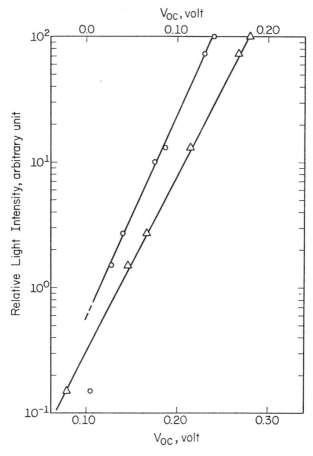

Figure 11. Semilogarithmic plots of the relative light intensity vs. open-circuit photovoltage. (○) *(In–ZnPc–Au), using top scale;* (△) *(In–H_2Pc–Au), using bottom scale. Irradiation at 6328 Å (12).*

chemical processes (e.g., catalysis) taking place on surfaces of materials like the phthalocyanines. In the future, it may also be of interest to exploit the controllable electric fields at molecular semiconductor junctions by constructing tailored chemical reaction centers susceptible to their influence.

Photoelectrochemistry at Phthalocyanine Electrodes

Electrochemistry at phthalocyanine electrodes has received much attention lately because some of these materials catalyze the electroreduction of molecular oxygen (42–46). Electrodes of various types have been employed. Phthalocyanines have been coated thinly onto conventional electrode materials by adsorption from solution (*45*), by chemical vapor deposition (*44*), or by vapor phase transport (*44*); powders have been compressed with carbon or metal powders to give composite electrodes (*42, 44*); and thick phthalocyanine films (> 100 Å) have been deposited by evaporation onto gold substrates or glass covered with gold (*10, 43, 47*). In many of these systems, it is not known whether or not the phthalocyanines are involved in the actual charge transfer events. Studies involving thin films offer greater control of the system, in terms of limiting direct contact between the electrolyte and the conducting underlayer, and they have accordingly provided the most extensive information about faradaic activity directly on phthalocyanine surfaces. It is clear now that even reversible voltammetry can be observed at these interfaces (*10*).

Given the evidence of the previous section, one might expect phthalocyanine electrodes to show electrochemical behavior concordant with its observed *p*-type character (*48, 49, 50*). Extensive studies of H_2Pc and ZnPc in our laboratory have borne out this expectation (*10, 14*).

Our electrodes are fabricated by depositing first a patterned Cr underlayer at 40–70 Å on a cleaned glass substrate, then a coincidently patterned gold contact layer at 200–300 Å thickness over the Cr, which serves only to improve adhesion, and lastly a general phthalocyanine overlayer of 100–3000 Å thickness (*10, 14*). The working electrode is the portion of the phthalocyanine that directly overlies a gold disk of 0.3-cm^2 area.

The starting point for our discussion of results is the dark electrochemistry. Consider Figure 12, which shows cyclic voltammograms (*14*) for six complexes of Fe(II). The couples involved are all virtually reversible at Pt electrodes and have standard potentials ranging from -0.13–0.82 V vs. SCE. At H_2Pc electrodes, there is a systematic improvement in reversibility as $E^{\circ\prime}$ becomes more positive. The most negative couple, Fe(EDTA)$^{-/2-}$, shows essentially complete irreversibility. The Fe(II) species can be oxidized, but the resulting Fe(III) complex is hardly reduced. A steady improvement of kinetic performance is observed for more positive $E^{\circ\prime}$ values, and one sees virtually reversible responses for complexes involving 4,4′-bipyridine and 1,10-phenanthroline. We have been unable to observe faradaic activity at potentials more negative than $+ 0.3$ V vs. SCE, aside from a large current rise yielding catastrophic failure of the electrode at potentials on the order of -1 V vs. SCE.

These voltammetric characteristics are generally concordant with a view of the phthalocyanines as *p*-type semiconductor electrodes having a flatband condition near 0.3 V vs. SCE. At more negative potentials, carriers are removed from the electrode surface and a depletion layer is formed in the material. This condition is a fatal impediment to faradaic activity. At potentials more positive than 0.3 V, carriers accumulate in

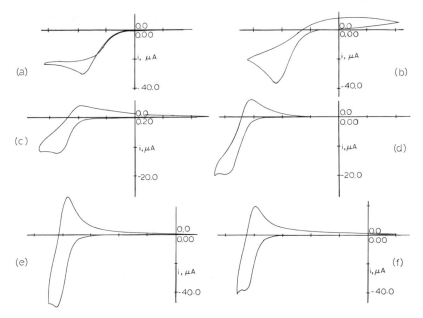

Journal of the American Chemical Society

Figure 12. Cyclic voltammograms of six iron complexes at H_2Pc electrodes: (a) 1.00mM Fe(EDTA)$^{2-}$; (b) 1.00mM Fe(CN)$_6^{4-}$; (c) 0.45mM Fe(3,4,7,8-tetramethyl-1,10-phenanthroline)$_3^{2+}$; (d) 0.55mM Fe(4,7-dimethyl-1,10-phenanthroline)$_3^{2+}$; (e) 1.00mM Fe(2,2'-bipyridine)$_3^{2+}$; (f) 1.00mM Fe(1,10-phenanthroline)$_3^{2+}$. Supporting electrolyte was 1M KNO_3, except for (c) and (d), which involved 0.5M K_2SO_4; scan rate = 100 mv/sec (14).

excess at the interface, and electrochemistry is facile. In this region, the overlap between electronic states in the semiconductor and the solution is also more conducive to charge transfer (14, 48, 49, 50).

Figure 13 shows a plot of differential capacitance supporting this interpretation. At potentials where a depletion region exists, the capacitance is controlled by the relatively small space–charge components in series with the double-layer capacitance (50). At the flatband potential a sharp rise in capacitance is expected because the restrictions imposed by the depletion layer are removed. One sees such a rise near 0.3 V in the figure. The peak at 0.0 V can be ascribed to a band of intermediate levels located in the band gap (14, 50). Their role in the electrochemistry seems substantial (14), but is beyond our scope here.

If the interpretation advanced here is correct, then one ought to be able to see photoelectrochemistry at the electrode whenever a depletion region exists at the electrode–solution interface. The basis is exactly the same as for the Schottky junctions discussed above. Thus, there ought to be a photoreduction of Fe(EDTA)$^-$, even though that reduction does

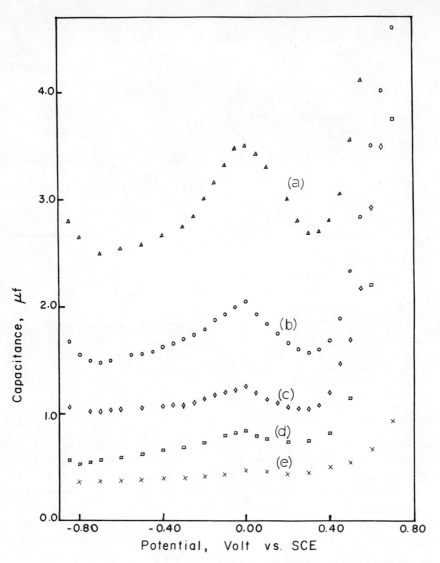

Journal of the American Chemical Society

Figure 13. Differential capacitance of H_2Pc electrodes in 1M KNO_3 at various excitation frequencies: (a) 100 Hz, (b) 350 Hz, (c) 1 kHz, (d) 3.5 kHz, (e) 10 kHz. H_2Pc film thickness was 500 Å (14).

not proceed in the dark. Moreover, the onset of photoreduction should be at the flatband potential of + 0.3 V vs. SCE. Figure 14 shows that the real system actually does meet these expectations. It seems likely that other photoeffects reported for phthalocyanine electrodes are rooted in their properties as semiconductors (10, 46, 47).

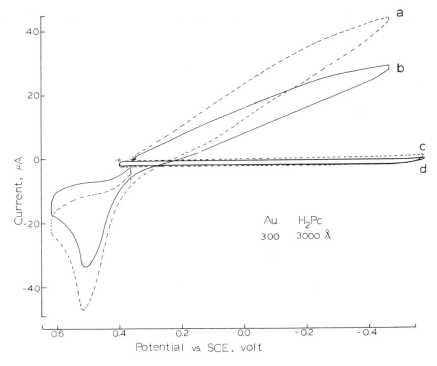

Figure 14. *Photoelectrochemical effect at an H_2Pc electrode. Irradiation at 6328 Å; (———) data taken in the dark, (- - -) data taken under illumination: (a) and (b), 10mM $K_3Fe(CN)_6$ plus 1.0M KNO_3; (c) and (d), 1.0M KNO_3; scan rate = 100 mV/sec; H_2Pc film thickness = 3000 Å (14).*

The message of these results is much the same as that of the preceding section. Molecular solids can behave fairly well as semiconducting materials and their solid state properties may have to be taken into account in studies of their chemical behavior. These results also highlight the point that molecular solids in thin films can provide new electrochemical surfaces with interesting special properties. The ability of CoPc films to reduce oxygen catalytically is an example (44, 51).

Prognosis

We have endeavored here to illustrate the utility of thin-film techniques for creating special reaction environments and for characterizing materials of chemical interest. The technology of thin films is well-developed; yet chemists have exploited it relatively little for any purpose and hardly at all to create complex assemblies in a synthetic manner. A great deal of room for development exists.

Of course, there are drawbacks to this experimental approach. Even though film technology is developed in a general sense and sophisticated apparatus is available and is inexpensive, the literature overwhelmingly stresses conventional semiconductors, metals, and insulators. The fabrication of films of molecular materials is usually required in studies of chemical interst; therefore the experimenter must develop his own technology as he proceeds. Films of most substances are impractical for vacuum processing because they are too volatile. Even perylene, which for most chemical purposes is quite involatile, is difficult to handle in a film processor for this reason. Cast films, like our spin-coated examples, are easy to prepare, but only materials with virtually no tendency to crystallize can be deposited in that manner. Moreover, it remains unclear if homogeneous dispersions are obtained for small molecules included in the casting. Some of these problems are bound to be solved, and others will be circumvented as experience grows; but for the present they are significant obstacles.

Here we advance thin-film methods essentially as architectural tools, but in the interest of balance we must also point out that present procedures are not usually capable of architecture on the molecular level. Isolated species tend to migrate and agglomerate on surfaces (7, 8), hence one is hard pressed to deposit submonolayer to monolayer coverage by conventional vacuum deposition techniques. Architectural control can be maintained only on a 50–100-Å distance scale in most circumstances. Monolayer assembly by the Langmuir–Blodgett method (52–55) or assembly by covalent reactions on surfaces (56, 57) has much finer resolution, but both methods have difficulties in creating a phase of any appreciable thickness. The deposition of thin films under ultrahigh vacuum conditions with molecular beam sources directed at cooled substrates promises to provide a route to architectural control on the molecular level, but developments along this line are still in the future.

Acknowledgment

We are grateful for continuing financial support from the National Science Foundation under Grant CHE-78-00584.

Literature Cited

1. Rabinowitch, E.; Govindjee. "Photosynthesis"; Wiley: New York, 1969.
2. Clayton, R. K. *Annu. Rev. Biophys. Bioeng.* **1973**, *2*, 131.
3. Sauer, K. In "Bioenergetics of Photosynthesis"; Govindjee, Ed.; Academic: New York, 1975.
4. Sauer, K. *Acc. Chem. Res.* **1978**, *11*, 257.
5. Sze, S. M. "Physics of Semiconductor Devices"; Wiley: New York, 1969.

6. Hovel, H. J. "Solar Cells," In "Semiconductors and Semimetals"; Willardson, R. K., Beer, A. K., Eds.; Academic: New York, 1975; Vol. 11.
7. Berry, R. W.; Hall, P. M.; Harris, M. T. "Thin Film Technology"; Van Nostrand Reinhold: New York, 1968.
8. Holland, L. "Vacuum Deposition of Thin Films"; Chapman and Hall: London, 1970.
9. DeForest, W. S. "Photoresist: Materials and Processes"; McGraw-Hill: New York, 1975.
10. Tachikawa, H.; Faulkner, L. R. *J. Am. Chem. Soc.* **1978**, *100*, 4379.
11. Fan, F-R.; Faulkner, L. R. *J. Chem. Phys.* **1978**, *69*, 3334.
12. Ibid., p. 3341.
13. Tachikawa, H.; Faulkner, L. R. *J. Am. Chem. Soc.* **1978**, *100*, 8025.
14. Fan, F-R.; Fauffilkner, L. R. *J. Am. Chem. Soc.* **1979**, *101*, 4779.
15. Tachikawa, H.; Fischer, S. G.; Faulkner, L. R., unpublished data.
16. Tachikawa, H.; Faulkner, L. R. *Chem. Phys. Lett.* **1976**, *39*, 436.
17. Knox, R. S. In "Bioenergetics of Photosynthesis"; Govindjee, Ed.; Academic: New York, 1975.
18. Ferreira, J. A.; Porter, G. *J. Chem. Soc. Faraday Trans. 2* **1977**, *73*, 340.
19. Kopelman, R. *J. Phys. Chem.* **1976**, *80*, 2191.
20. Birks, J. B. "Photophysics of Aromatic Molecules"; Wiley: New York, 1970.
21. Förster, Th. *Discuss. Faraday Soc.* **1959**, *7*, 27.
22. Mulder, B. J. *Philips Res. Rep. Suppl.* **1968**, *4*.
23. Cohen, M. D.; Klein, E.; Ludmer, Z. *Chem. Phys. Lett.* **1976**, *37*, 611.
24. Gantt, E.; Lipschultz, C. A.; Zilinskas, B. *Biochim. Biophys. Acta* **1976**, *430*, 375.
25. Leonhardt, H.; Weller, A. *Ber. Bunsenges. Phys. Chem.* **1963**, *67*, 791.
26. Weller, A. *Pure Appl. Chem.* **1968**, *16*, 115.
27. Gutmann, F.; Lyons, L. E. "Organic Semiconductors"; Wiley: New York, 1968.
28. Sussman, A. *J. Appl. Phys.* **1967**, *38*, 2738.
29. Ibid., p. 2748.
30. Harrison, S. E.; Ludewig, K. H. *J. Chem. Phys.* **1966**, *45*, 343.
31. Sakai, Y.; Sadaoka, Y.; Yokouchi, H. *Bull. Chem. Soc. Jpn.* **1974**, *47*, 1886.
32. Zhdanov, G. S.; Vorona, Yu. M. *Izv. Akad. Nauk SSSR, Ser. Fiz.* **1963**, *27*, 1239.
33. Katz, W.; Evans, C. A., Jr.; Eaton, D. R.; Faulkner, L. R. *J. Vac. Sci. Technol.* **1978**, *15*, 1561.
34. Ghosh, A. K.; Morel, D. L.; Feng, T.; Shaw, R. F.; Rowe, C. A., Jr. *J. Appl. Phys.* **1974**, *45*, 230.
35. Hackham, R. H.; Harrop, P. *Solid State Electron.* **1972**, *15*, 1031.
36. Vilesov, F. I.; Zagrubskii, A. A.; Garbuzov, D. Z. *Sov. Phys., Solid State (Engl. Transl.)* **1964**, *5*, 1460.
37. Pope, M. *J. Chem. Phys.* **1962**, *36*, 2810.
38. Formenko, V. S. In "Handbook of Thermionic Properties," Samsonov, G. V., Ed.; Plenum Press: Data Division, New York, 1966.
39. Tang, C. W.; Albrecht, A. C. *J. Chem. Phys.* **1975**, *63*, 953.
40. Merritt, V. Y.; Hovel, H. J. *Appl. Phys. Lett.* **1976**, *29*, 414.
41. Kampas, F. J.; Gouterman, M. *J. Phys. Chem.* **1977**, *81*, 690.
42. Jasinski, R. *Nature (London)* **1964**, *201*, 1212.
43. Savy, M.; Bernard, C.; Magner, G. *Electrochim. Acta* **1975**, *20*, 383.
44. Appleby, A. J.; Savy, M., presented at the 151st Meeting of the Electrochemical Society, Philadelphia, PA, May, 1977, Paper No. 345, proceedings in press.
45. Zagel, J.; Sen, R. K.; Yeager, E. *J. Electroanal. Chem.* **1977**, *83*, 207.
46. Alferov, G. A.; Sevast'yanov, V. I. *Elektrokhim.* **1975**, *11*, 827.
47. Meshitsuka, S.; Tamaru, K. *J. Chem. Soc., Faraday Trans. 1* **1977**, *73*, 236.

48. Gerischer, H. *Adv. Electrochem. Electrochem. Eng.* **1961**, *1*, 139.
49. Gerischer, H. In "Physical Chemistry: An Advanced Treatise"; Eyring, H., Henderson, D., Jost, W., Eds.; Academic: New York, 1970; Vol. 9A.
50. Myamlin, V. A.; Pleskov, Yu. V. "Electrochemistry of Semiconductors"; Plenum: New York, 1967.
51. Good, N. L.; Maher, R. T., unpublished data.
52. Blodgett, K. B. *J. Am. Chem. Soc.* **1935**, *57*, 1007.
53. Langmuir, I.; Schaefer, V. J. *J. Am. Chem. Soc.* **1937**, *59*, 1406.
54. Bücher, H.; Drexhage, K. H.; Fleck, M.; Kuhn, H.; Möbius, D.; Schaefer, F. P.; Sondermann, J.; Sperling, W.; Tillmann, P.; Wiegand, J.; *Mol. Cryst.* **1967**, *2*, 199.
55. Kuhn, H.; Möbius, D.; Bücher, H. In "Physical Methods of Chemistry"; Weissberger, A., Rossiter, B., Eds.; Wiley: New York, 1972; Vol. I, Part 3B.
56. Holland, L. "The Properties of Glass Surfaces"; Chapman and Hall: London, 1964.
57. Hair, M. L. "Infrared Spectroscopy in Surface Chemistry"; Marcel Dekker: New York, 1967.

RECEIVED October 2, 1978.

8

Photoelectrochemistry and Spectroscopy of Metal Phthalocyanine Films on a Transparent Semiconducting Electrode

COOPER H. LANGFORD, BRYAN R. HOLLEBONE, and THEODORE VANDERNOOT

Metal Ions Group, Chemistry Department, Carleton University, Ottawa, Canada K1S 5B6

The photoelectrochemistry of Mg, Fe, Ni, and Zn phthalocyanine films on transparent n-type tin oxide electrodes is reported. Photocurrents of a type expected from a p-type semiconducting film are observed in all cases. These currents are small; slow electron transfer is indicated. There is some evidence of the retardation of the "dark couple" reaction without inhibition of the "photocouple" at the photoelectrode. The most effective film is one containing the β-crystal modification of zinc phthalocyanine. This result is clarified by examination of magnetic circular dichroism spectra of metal phthalocyanine films. The broad line spectra reveal the solid state character of the films. Requirements for efficient photoelectrochemistry in terms of electronic states are discussed.

Albery and Archer have analyzed the ideal diffusion-controlled photogalvanic cell and indicated that it could achieve an efficiency comparable with that of good photovoltaic devices (1, 2). In it there are two redox couples, A/B and Y/Z, where the excited state of A(A*) is the photoreactive species in the cell. The reactions in the cell can be summarized as follows.

$$A \xrightarrow{h\nu} A^* \tag{1}$$

$$A^* + Z \rightarrow B + Y \text{ occurring near illuminated electrode} \tag{2}$$

$(B + Y \to A + Z)$ energy wasting (3)

$B \pm e^- \to A$ occurring at illuminated electrode (4)

$Y \to Z \pm e^-$ occuring at dark electrode (5)

For the operation of an efficient cell, conditions should include an illuminated electrode that can mediate the A/B couple but that is unreactive to the Y/Z couple paired with a dark electrode that can mediate the Y/Z couple. Additionally, the back reaction (Reaction 3) must be inefficient. Analysis of the kinetics of homogeneous systems aided by the Marcus–Weller theory (3, 4) of electron transfer reaction rates suggests that these criteria are very hard to satisfy. We must look for "non-Marcusian" electron transfer. An additional feature is that it is desirable to have light absorption occur in a limited zone near the illuminated electrode. Together, all of these features suggest a heterogeneous system with the A/B couple incorporated into a surface layer of the illuminated electrode. Such a cell approaches the regenerative semiconductor cells (5) based on CdS or GaP and discussed elsewhere in this volume, but the use of molecular layers as absorbers has distinctive features that lead us to prefer modelling our discussions on the comparison to homogeneous cells. A third model, a conductor–insulator–conductor sandwich has recently been considered by Lyon's group (6).

It turns out that there are related requirements that can be specified for a system that will carry out efficient photochemical synthesis. Efficient synthesis requires concentration of light absorption into a narrow zone to avoid costly catalyst recovery. It also requires means of favoring reversibility. Moreover, electrochemically mediated syntheses may permit a light reaction on a wide area collector to be coupled to a more compact and manageable dark electrode reaction where the desired product is to be collected. (It's attractive to think that the main link to a large solar collector field could be electrolyte and wires only). Again, an attractive response to these requirements is to attempt to build photosynthetic electrochemical cells based on heterogeneous systems.

It is clear in the photogalvanic cell case and probable in the synthesis case, that the requirements specified lead to the investigation of ways of producing a sort of "rectification" at a photoelectrode. The subject of this chapter is our preliminary work on such rectification and its relationship to the electronic properties of "semisolid" materials as revealed by their electronic spectra under magnetic circular dichroism (MCD) investigation. Relations of reactivity to spectroscopic information on electronic structure is always important to effective photochemical advance. These experiments have employed metal phthalocyanine (Pc) dyes.

The choice of Pc's was made for several reasons. Pc's have attracted attention as organic semiconductors from an early date (7, 8) and the rich literature concerning mechanism of charge transport in these films invites attention. Photovoltaic cells based on Pc's have been explored for some time and several mechanisms for photovoltaic action have been proposed. These include models based on Shottky barrier theory for doped insulators (9, 10, 11). However, results have been reported that do not indicate appropriate dependence on work function of electrode metals and this leads to an alternative model based on single carrier theory (6). Our photogalvanic work to date does not lead to a specific connection with photovoltaic experiments although suggestive connections exist. The main factors relevent to our present choice of Pc's for study of heterogeneous photogalvanic and photosynthetic effects are the following. First, these materials are interesting spectroscopically. Second, they bear a relationship to both photosynthetic pigments and the metal diimine complexes, which have attracted much recent attention as photochemical electron transfer agents. Additionally, they are convenient commercial dyes. The results suggest that there remains considerable scope for synthetic investigations. We emphasize that only the barest start has been made in the field of organometallic film photoelectrodes.

Theoretical Remarks

The characteristics of a heterogeneous system meeting the Albery–Archer criteria can be sketched, at least in part. Let's suppose that Reaction 2 is reduction of Z by an excited dye A^* in the cathode. A convenient way to block reoxidation of Y would be for the cathode to be a p-type semiconductor with the band bent to deplete the surface layer of holes. This is just the condition for a p-type semiconductor to act as a photoreductant (5). Semiconductor photoelectrochemistry is promoted when the hole–electron pair created by photon absorption is prevented from recombining by migration of the majority carrier to the interior and the minority carrier (electrons in our case) to the surface to react with the electrolyte. Note also, that photons absorbed at various depths in the space change layer can be harvested and delivered to the reactive site (surface) if this mechanism is sufficiently efficient. There is an element functionally analogous to light harvesting in photosynthesis.

For B to regenerate A, it is necessary to have an electron source underlying the semiconducting dye film. This may be either a metal or an n-type semiconductor. The latter possibility offers the prospect of an n–p junction rectifier underlying the p-electrolyte junction to enhance rectification. The $B + Y$ wasting reaction is also inhibited if B is associated with a hole that migrates toward the semiconductor interior and away from Y.

An ideal cell is conceivable as an illuminated electrode composed of an n-type semiconductor (or perhaps a metal) coated by a p-type dye layer that is photochemically active (the A/B couple) in contact with an electrolyte containing the Y/Z couple. The electrolyte makes contact with a metal electrode at which the Y/Z couple is reversible. An efficient cell based on the diffusion of the Y/Z couple should be thin and the metal electrode is rarely transparent. Therefore, it is useful to consider transparent n-type semiconductors coated on glass as supports for the dye layer and entry of light to the illuminated electrode from the support side rather than the electrolyte side. This has the added advantage of allowing the use of colored electrolytes.

In the experiments described below, the barest skeleton of such an ideal cell is visible. α-Crystal modification metal phthalocyanine dye films have p-type semiconducting properties when equilibrated with air/moisture (*11*). (This is indicated in Figure 1.) When they are sublimed onto transparent n-doped conducting SnO_2 on glass electrodes, a photo-active p-type semiconductor is formed. Contact of this solid and semi-solid (the dye layer) rectifier layer cake with an electrolyte containing a couple such as quinone/hydroquinone (in 1.0M KCl as supporting electrolyte), in turn in contact with Pt, completes a potential realization of the "ideal" cell. A few of the ideal features are realized in this cell but many others are not. Especially, the reaction of reduction remains a slow

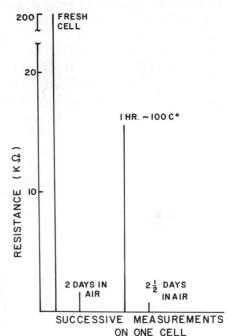

Figure 1. Effect of exposure of an α-crystal modification ZnPc to the atmosphere. (Data from Ref. 6). The resistance reported is for a 3000-Å film sandwiched between two metals.

photoprocess at the electrodes. This cannot be tolerated for an efficient photon energy converter. (It is also possible that the flaw in the ideal analysis as applied to this cell is the use of the model of Ref. 5 rather than the model of Ref. 6.)

Experimental

Materials. Metal Pc's were obtained from Eastman Organic Chemicals and purified by sublimation. The doped SnO_2-coated glass was obtained from O. H. Johns Ltd. The resistance was under 100 Ω m^{-2} and light transmission in the visible region exceeded 80%. Electrolyte components were prepared from reagent grade chemicals without special purification. Water was redistilled from potassium permanganate.

Electrode Preparation. Films of phthalocyanine were deposited on n-doped SnO_2-coated glass (or in one experiment, glass) by sublimation from a quartz vessel in a tube furnace. The vacuum line was evacuated with an oil diffusion pump (minimum pressure = 10^{-6} torr) except in experiments intended to produce the green α modification of ZnPc. The electrode substrate was mounted on a "sled" that could be magnetically pushed past a hinged barrier to the mouth of the tube furnace. At the (undetermined) operating temperature, films of approximately 500-Å thickness were deposited in just over 15 min; 1000-Å films required nearly 30 min. Film thickness was estimated from absorbance at the wavelength of maximum absorbance using extinction coefficient data from Ref. 12 and an assumed film density of 1.5 g cm^{-3}. Thus, the thicknesses cited are quite approximate. This seems appropriate since the films were, unfortunately, recognizably nonuniform. Reproducibility from one phthalocyanine to another is better.

Electrochemical Studies. The cell was fabricated from Teflon with an opening to which the transparent electrode could be held by a brass plate. Electrical contact to the SnO_2 was established by a brass ring of diameter larger than the opening in the cell. A neoprene O-ring and silicone grease ensured a seal from the electrolyte. The cell admitted a nitrogen bubbler placed to sparge the electrode surface; a saturated calomel reference electrode was used in the potentiostatic circuit and a bright, large area, platinum electrode formed the counter electrode. Current–voltage curves were recorded with the aid of a PAR Model 364 polarograph operated at a scan rate of 2 mV/sec.

Small background waves were observed when dye coated electrodes were analyzed in KCl solution. These appeared to be time dependent and led to adoption of a procedure in which the first scan of any electrode–electrolyte combination was ignored.

Photochemical Aspects. The light source irradiating the metal phthalocyanine films through the glass and SnO_2 plate was a 1000-W Hg–Xe arc lamp. Filters were placed in the light path as follows. A 10-cm path, water-filled quartz absorption cell reduced IR and heating effects; and a UV cutoff filter that was tested to establish that it eliminated the photoeffects that arise from SnO_2 band gap absorption. The intensity of the light reaching the film was estimated using broad band Reineckate chemical actinometry (13) as 1.6×10^{-6} einstein sec^{-1}.

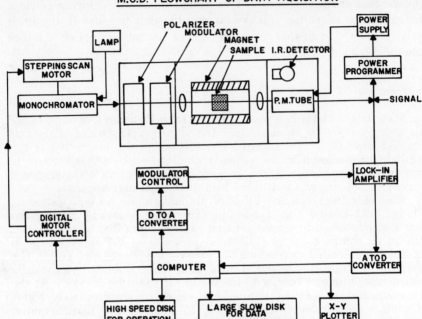

Figure 2. *Block diagram of the MCD spectrometer*

MCD Spectra. MCD spectra were recorded on an improved design of the cryomagnet instrument described in Ref. *12*. Its general features are indicated in Figure 2. The spectra were obtained at a field of 4.7T. Absorption spectra were recorded on a Cary 14 spectrophotometer. MCD spectra were recorded for films on quartz plates. The technique of preparation was similar to that adopted for electrode preparation.

Survey of Photoelectrochemistry

Current–voltage curves have been recorded for n-SnO_2 electrodes coated with α-MgPc, α-FePc, α-NiPc, α-ZnPc, and β-ZnPc. In all cases, the supporting electrolyte was 1.00M KCl and reactive species in the solution included the quinone/hydroquinone couple, the ferricyanide/ferricyanide couple, tris(ethylenediamine)cobalt(III), and N,N' diheptyl-5,5' dipyridine (heptylviologen). The general features are similar on all n-SnO$_2$ electrodes; Figures 3, 4, and 5 provide a summary. A curve on α-MgPc, which is representative for "thin" films of approximately 500 Å, is shown in Figure 3. Figure 4 shows results on NiPc that provide a comparison of films of 500 Å and 1000 Å and Figure 5 displays results for the β-crystal modification of ZnPc. Several general observations may be made.

1. Photocurrents of the sort expected for p-type semiconducting films are seen in all cases. It should be noted especially that this includes the lower conductivity β-ZnPc film.

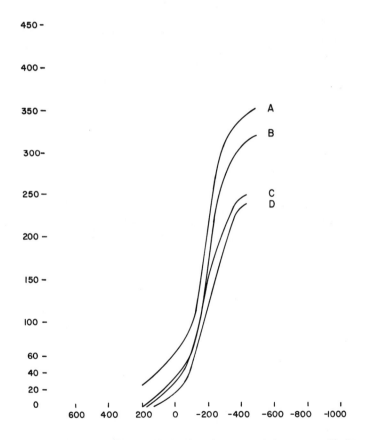

Figure 3. Current–voltage curves for 5.00 × 10^{-4} quinone/hydroquinone solutions in 1.00M KCl at Mg and "green" (α) ZnPc electrodes. Curves are labelled as follows: (A) photo Zn; (B) dark Zn; (C) photo Mg; (D) dark Mg. The film thickness is near 500 Å in both cases; V is in mV, I is in μA.

2. The photocurrents are small and the morphology of the current–voltage curve implies kinetically slow electron transfer in all cases—both photo and dark. This is the limiting feature of the performance of these electrodes.
3. The photoenhancement of currents in the limiting region suggest that the carrier generation process plays a part in establishing the limiting current. This is confirmed by light intensity dependence that is linear with a slope a little less than one in the limiting region.
4. There is very little dependence on the metal in the phthalocyanine over the range Mg, Fe(II), Ni, and Zn. The only unique metal feature observed occurred with Fe(II). In

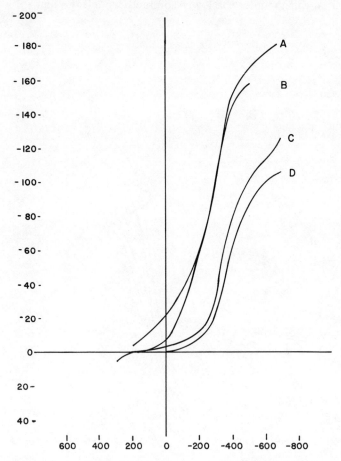

Figure 4. Current (in μA) vs. voltage (in mV) for the quinone/hydroquinone electrolyte described in Figure 3 at NiPc electrodes. Curve (A) photo and (B) dark were recorded on an electrode with a 500-Å film. Curve (C) photo and (D) dark on a 1000-Å film.

an irregular (and undiagnosed) event, illuminated electrodes exhibited a catastrophic current increase accompanied by loss of the illuminated zone of the film. We suspect conversion to Fe(III) accompanied by axial ligation may be involved.

The major variables leading to significant changes of behavior were film thickness and phthalocynaine crystal modification. Increasing film thickness decreases thermal currents as higher resistance might imply. But the decrease of photocurrent is not proportional to the decrease in dark current. Also, larger open circuit photogalvanic potentials were recorded for the 1000-Å films. A change from the α- to β-crystal modifica-

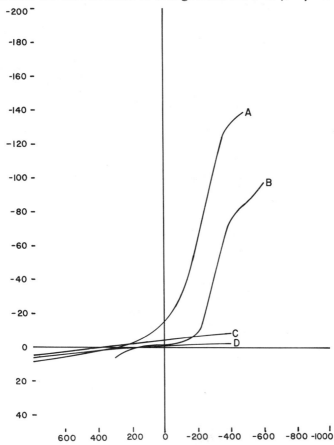

Figure 5. Current (in μA) vs. voltage (in mV) for electrolytes at a "blue" (β) ZnPc electrode. Curves (A) photo and (B) dark are for the quinone/hydroquinone electrolyte described above. Curves (C) photo and (D) dark are for a similar concentration of $K_3Fe(CN)_6K_4Fe(CN)_6$ in KCl. Although the present scale suppresses detail in (C) and (D), limiting regions at cathodic potential were observed.

tion produced similar but emphasized effects. The photogalvanic open circuit voltage varied from 40–80 mV on α films. It reached 180 mV on the β films. This leads to the most interesting issue, what is the advantage of β-crystal modification? We return to this question in the spectroscopic discussion below.

A final experiment is worth recording. To attempt evaluating the role of n-SnO_2 a film of ZnPc was deposited on glass. In this experiment, the electrical contact, which was the brass ring used in other experiments, was at the same front surface of the film that was in contact with the electrolyte. The defect on this experimental arrangement is that the film was polarized radially. Nonetheless, the result is interesting; the poorly conducting (*11*) β-crystal modification acts as a photocatalyst for both oxidation and reduction of the quinone/hydroquinone couple. The open circuit photogalvanic potential is essentially zero. This emphasizes the molecular photochemistry of which this film is capable. Treating organometallic films as semiconductors is only part of an appropriate perspective (*see* Ref. 6). This point is confirmed by the limited (if large width of) spectral band to be discussed below. A suggestion for future development is controlled doping of the films to try to produce both n- and p-semiconducting properties.

The reduction of the "viologen" relative is also significant. Reduced viologens can catalytically (colloidal Pt) generate H_2 (*14*) from water. One electrode for the photoelectrolysis of water is implied.

Spectroscopy and Structure

Comparison of absorption and MCD spectra of phthalocyanines in solution with those in the solid state reveals many substantial differences. Changes occur in band centers, widths, and relative intensities that exceed those associated with a state change of an isolated chromophore. Several different regions of these spectra have been examined in detail and the differences between gas phase or solution spectra on one hand and solid state spectra on the other have been attributed to polymerization in various forms in the solid state (*15*).

These effects are readily observed in the Q band, a series of adsorptions in the red region. In solution spectra a sharp strong transition, often labelled α, at low energy is followed by one or more less intense bands, labelled β, to the high energy side. These are usually assigned as a vibronic series and MCD spectrum shows that the 0–0 transition leads to a degenerate excited state while the 0–1, 0–2, etc. bands yield a nondegenerate system (*16*). From this it is clear that the 0–0 band is the "long axis" transition polarized in the plane, while the 0–2 are "short

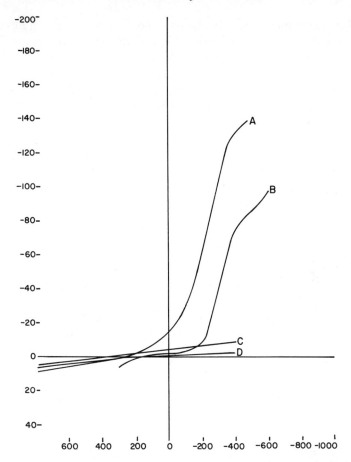

Figure 6. Absorption (upper) and MCD (lower) Q band region of α-NiPc on a glass support. The resolution of the envelope (after Ref. 12) is shown.

axis" or perpendicular transitions. The 0–0 band can thus be assigned as $^1A_{1g} \rightarrow {}^1E_u$ in the D_{4h} point group. This is usually further interpreted, from extended Hückel calculations, as a $(a_{1u})^1 (e_g)^1$ excitation with little contribution from other configurations. (17).

The solid state spectra are quite different; Figure 6 presents the α-NiPc Q band as representative. Films prepared at low temperature and pressure show a single broad adsorption in the Q region with a low energy shoulder. This reversal of relative intensities is accompanied by energy shifts and a considerable broadening from a few hundred to a few thousand wave numbers. Recent MCD results (12) show that two pairs of transitions occur under this envelope. One pair of transitions is represented by A terms implying degenerate excited states while the second pair, displaying B terms, are nondegenerate.

This observation can be interpreted most easily as a strong interaction between molecular pairs on identical symmetry sites in the unit cell. The states of such pairs are readily derived using a Davydov model. The degenerate ^1E_u states of the monomers may be linearly combined to provide two degenerate states in the site group of the unit cell. The nondegenerate states may be recombined similarly, producing two states of the dimer.

The splitting between states in either pair can be parameterized in a point dipole model or treated more completely in a molecular orbital calculation. While the latter is usually a better approximation, the point dipole highlights the role played by relative orientation of the dimer molecules and is useful in predicting structural features of the solid.

In the simplest model (18) using two molecules per unit cell on identical sites, the magnitude of the splitting between pairs of states becomes;

$$V = \frac{\rho_a \rho_b}{d_{ab}^3} [\sin\theta \sin\phi \cos(\theta - \phi) - 2\cos\theta \cos\phi]$$

in which ρ_a and ρ_b are the transition dipole moments for molecules a and b, d_{ab} is the distance between dipole centers, and θ and ϕ are the angles formed by the dipoles on a and b, respectively, with the displacement vector. When $\theta = \phi = 0°$, the dipoles are colinear and when $\theta = \phi = \pi/2$, they are coparallel. At the colinear limit $V_{ab}^{--} = \rho_a\rho_b/d^3$ and the transition dipoles become; $\rho_+ = 0$ and $\rho_- = 2\rho_a$ where ρ_+ is the dipole of the transition to the symmetric combination while ρ_- refers to the antisymmetric state.

In contrast at the coparallel limited $V_{ab}^{=} = \rho_a\rho_b/d^3$ and the transition dipoles become; $\rho_+ = 2\rho_a$ and $\rho_- = 0$.

Thus a nearly coparallel or "sandwich" dimer would display two bands, one strong to high energy of a monomer transition by $V_{ab}^{=}$ and one weak to low energy. In contrast, a nearly colinear dimer would yield two bands one weak to high energy by V_{ab}^{--} and one strong to low energy. Since $V_{ab}^{=} = \frac{1}{2} V_{ab}^{--}$ for identical molecules ($a = b$), the dimer geometries not only yield a reversal of intensities but the interaction energy doubles on changing from the coparallel to colinear limits. Between these limits the angle between dipoles can be estimated from the complete expression for V_{ab}. Estimates of angles from relative intensity measurements are complicated by vibronic enhancement of forbidden dipole terms.

Application of this model to the low temperature crystal form of phthalocyanines, usually labelled α, indicates that it is a coparallel dimer conformation in the unit cell (12). The two in-plane dipoles display a strong high energy and weak low energy spectrum; the two perpendicular

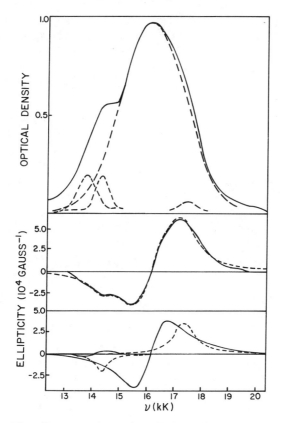

Figure 7. The absorption (upper) and MCD (lower) spectra of β-ZnPc in the Q band region.

bands have reversed intensities and a separation 1.5 times as large as the in-plane transitions.

A second crystal modification grown at higher temperatures and pressures, usually labelled β, displays a very different spectrum. Here the in-plane transition intensities are the reverse of those observed in the α form and the separation is larger (15). This implies a more colinear arrangement of the dimer, which is perhaps derived from the α form by maintaining a constant value of d_{ab} but reducing the angles θ and ϕ. This cannot, incidentally, be confirmed by crystallography since the α phase is always microcrystalline.

In any case the increased interaction energy indicates that the β modification is a more compact lattice than the α and the stacking angle $\theta = \phi = 46°$ has been estimated for the β form (15). This complements the observations of Tachikawa and Faulkner (11) on the high resistance of the β form.

In all present experiments films have been aged by exposure to the atmosphere. This aging increases the conductivity of sublimed films. The mechanism for this increase is not well understood. A simple model would be based on oxidation of some phthalocyanine molecules in the film creating conducting holes in the lattice. Evidence of this is not observed in the MCD spectra. A substantial concentration of such oxidized species would lead to temperature-sensitive molar ellipticities called C terms associated with the spin degenerate ground states. Spectra of aged α-modification films at 6 and 300 K showed no appreciable differences.

However, it is clear from present results that the conductivity of aged films depends on the lattice modification. The conductivity of the β films examined is substantially reduced in comparison with the α films (11). It seems clear that whatever the atmospheric reaction may be, that conductivity depends upon penetration of O_2 and H_2O molecules into the lattice and that the more compact β form is more resistant to such attack.

Spectroscopy and Photochemical Properties

In the absence of the dopants, the MCD spectra of the pure films can give some indication of the type of photoelectric behavior to be expected. The valence band corresponds to the molecular $^1A_{1g}$ state, while promotion to the first conductor band corresponds to excitation to the ^1E_u state in the isolated molecule. The effects of the lattice in this state have just been discussed but it is clear from the distribution of charge in the $0 \rightarrow 0$ excited state of the isolated species that the electrons would remain localized in the solid. The transition dipole resides in the plane of the ring and provides little movement of electron density away from the plane towards the neighboring molecule. This in turn means that the lattice is a good insulator in both the valence and conduction bands. In contrast to this analysis of the 0–0 band the vibronic overtones observed in the MCD are nondegenerate. This must imply an active vibration, or other perturbation, of an e_g symmetry. All nondegenerate transition moments bear some out of plane components and the only way such components can arise is from the coupling, indirect product notation, $^1E_u \times e_g \in A_{1u} + A_{2u} + B_{1u} + B_{2u}$. This out-of-plane transition moment is capable of moving charge towards neighboring molecules and hence reducing film resistance.

Recent theoretical analysis of transition moments in **vibronic transi**tions of simple chromium complexes show that the overall vibronic selection rule should be at least octupolar in character (19). This implies that while the 0–0 band, which is purely electronic, is allowed to be dipolar, the succeeding vibronic progression must display the octupole transition rule. Under this assumption the generative label (*19*) to be used to characterize the complete vibrational basis vector should be at least 2, a conclusion completely consistant with the use of an e_g vibration. This does mean, however, that the proper assignment of the vibrational progression is probably $^1A_{1g} \rightarrow {}^1E_u(0-0)$, $^1E_u(0-2)$, $^1E_u(0-4)$, ... in first order, assuming the 0 → odd transition moments are negligible since they would require g → g transitions. This conclusion provides an answer to a long standing difficulty in the vibronic assignment of phthalocyanines. The observable separation in the progression is 1100 cm^{-1} or more, which is very high for a ring vibration assignment. Clearly this vibration may now be assigned the more reasonable value of approximately 550 cm^{-1}.

If the phthalocyanine film is to be modified to improve the conductivity of the conductance band, two factors emerge from the foregoing analysis. First, the nature of the ^1E_u state could be changed by preparation of heavy metal derivatives. This would provide stronger delocalization of charge in the $e_g{}^*$ orbital of the ring through overlap with filled $d\pi^*$ orbitals of the metal. Secondly, the amplitude of the e_g vibration could be increased by heavy nonmetal substituent groups on the pyrote rings. Each of these substitutions would increase the magnitude of the individual transition moments, which would in turn be reflected in the magnitude of the out-of-plane moment formed as the vector cross product. Such effects would be clearly visible in MCD spectrum of the exciton structure. In particular, the high energy nondegenerate transition would increase in rotational strength because of increased magnetic and electric dipoles in the antisymmetric state.

Acknowledgment

We thank the donors of the petroleum research fund for support.

Literature Cited

1. Albery, W. J.; Archer, M. D. *J. Electrochem. Soc.* **1977**, *124*, 688.
2. Albery, W. J.; Archer, M. D. *Electrochim. Acta* **1976**, *21*, 1155.
3. Marcus, R. A. *Annu. Rev. Phys. Chem.* **1964**, *15*, 155.
4. Rehm, D.; Weller, A. *Ber. Bunsenges. Phys. Chem.* **1969**, *73*, 834.
5. Tributsch, H.; Gerischer, H. *Ber. Bunsenges. Phys. Chem.* **1969**, *73*, 251, 850.
6. Hall, K. J.; Bauham, J. S.; Lyons, L. E. *Austr. J. Chem.* **1976**, *31*, 1661.
7. Gutman, G. F.; Lyons, L. E. "Organic Semiconductors"; Wiley: New York, 1967.
8. Lever, A. B. P. *Adv. Inorg. Chem. Radiochem.* **1967**, *7*, 27.
9. Usov, N. N.; Benderskii, V. A. *Sov. Phys.—Semicond. (Engl. Transl.)* **1968**, *2*, 580.
10. Ghosh, A. K.; Feng. T. *J. Appl. Phys.* **1973**, *44*, 2781.
11. Tachikawa, H.; Faulkner, L. R. *J. Am. Chem. Soc.* **1978**, *100*, 4379.
12. Hollebone, B. R.; Stillman, M. J. *J. Chem. Soc., Faraday Trans. II* **1978**.

13. Wegner, E. E.; Adamson, A. W. *J. Am. Chem. Soc.* **1966**, *88*, 394.
14. Moradpour, A.; Amouyal, E.; Keller, P.; Kagan, H. "Abstracts of Papers," 2nd International Conference on Photochemical Conversion and Storage of Solar Energy, Cambridge, August 1978, p. 31.
15. Sharp, J. H.; Lardon, M. *J. Phys. Chem.* **1968**, *72*, 3230.
16. Stillman, M. J.; Thomas, A. J. *J. Chem. Soc., Faraday Trans. II* **1974**, *70*, 805.
17. Schaffer, A. M.; Gouterman, M. *Theor. Chim. Acta* **1972**, *25*, 62.
18. Davydov, A. S. "Theory of Molecular Excitons"; McGraw Hill: New York, 1962.
19. Hollebone, B. R. *Theor. Chim. Acta*, in press.

RECEIVED October 2, 1978.

9

Charge Transfer at Illuminated Semiconductor–Electrolyte Interfaces

A. J. NOZIK[1], D. S. BOUDREAUX, and R. R. CHANCE

Corporate Research Center, Allied Chemical Corporation, Morristown, NJ 07960

FERD WILLIAMS

Department of Physics, University of Delaware, Newark, DE 19711

> *A new heterojunction model for the semiconductor–electrolyte interface is presented that considers the electrolyte as a doped semiconductor and that predicts that hot photogenerated minority carriers can be injected into the electrolyte. Preliminary calculations are presented in support of the hot carrier injection hypothesis. Recent experimental results on the photoenhanced reduction of N_2 on p-GaP cathodes are discussed and they appear to provide experimental evidence for hot electron injection. The importance of hot carrier injection for photoelectrochemical cells is also discussed.*

In recent years a great deal of interest has developed in the field of photoelectrochemistry based on photoactive semiconducting electrodes, especially in the application of these systems to solar energy conversion and chemical synthesis (*1–9*). In Figure 1, a classification scheme is presented for the various types of photoelectrochemical cells. The first division is into: (a) cells wherein the free energy change in the electrolyte is zero (electrochemical photovoltaic cells), and (b) cells wherein the free energy in the electrolyte is nonzero (photoelectrosynthetic cells). In the former cell, only one effective redox couple is present in the electrolyte—the oxidation and reduction reactions at the anode and cathode are inverse to each other. The net photoeffect is thus the

[1] Present address: Solar Energy Research Institute, Golden, CO 80401.

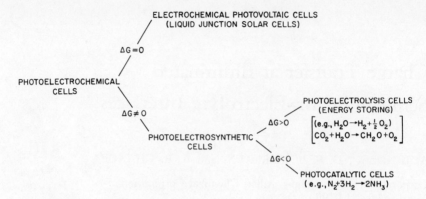

Figure 1. Classification scheme for photoelectrochemical cells

circulation of charge external to the cell, producing an external photovoltage and photocurrent (a liquid junction solar cell); no chemical change occurs in the electrolyte.

In the photoelectrosynthetic cell, two effective redox couples are present in the electrolyte and a net chemical change occurs upon illumination. If the free energy change of the net electrolyte reaction is positive, optical energy is converted into chemical energy and the process is labeled photoelectrolysis. On the other hand, if the net electrolyte reaction has a negative free energy change, optical energy provides the activation energy for the reaction, and the process is labeled photocatalysis. Energy level diagrams for these three types of cells are shown in Figures 2 and 3.

The most important aspects of all photoelectrochemical cells are the nature of the semiconductor–electrolyte junction and the photo-induced charge transfer process across the junction. A new model for

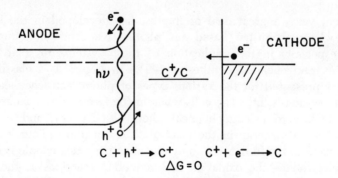

Figure 2. Energy level diagram for electrochemical photovoltaic cells. C^+/C is a redox couple in the electrolyte that produces the indicated anodic and cathodic reactions such that no net chemical change occurs.

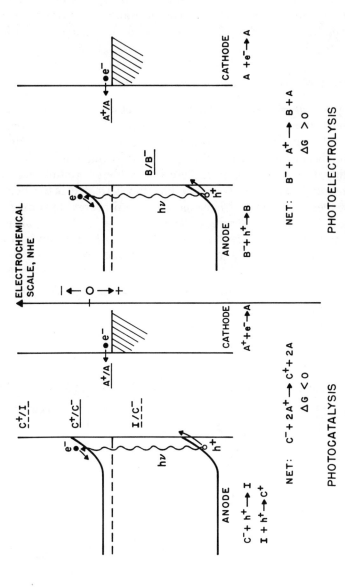

Figure 3. Energy level diagrams for photoelectrolytic and photocatalytic cells. A^+/A, B/B^-, and C^+/C^- are generalized redox couples; I/C^- and C^+/I are redox couples for intermediate steps of the C^+/C^- couple.

this charge transfer process has been suggested (*10*); one feature of this model is the possibility of "hot carrier injection." The new model and the likelihood of hot carrier injection are discussed further here and qualitative experimental evidence is presented in support of hot carrier injection during photoelectrochemical charge transfer. This evidence is based on experimental studies of photoenhanced N_2 reduction on p-GaP cathodes.

Heterojunction Model of Semiconductor–Electrolyte Junctions

In all previous models for the semiconductor–electrolyte junction, the electrolyte is considered to behave like a metal and the electrolyte redox potential is equated to the Fermi level of the electrolyte, which is analogous to the metal work function (*1–9*). In a different model proposed recently (*10*), aqueous electrolyte is considered a semiconductor (*11*) and the semiconductor–electrolyte junction is treated like a heterojunction between two different semiconductors. An energy level diagram for this model is shown in Figure 4. The band gap of H_2O is taken as about 9 eV and its electron affinity is 0.5 eV (*11*). Redox couples in the aqueous electrolyte are treated as extrinsic electronic states.

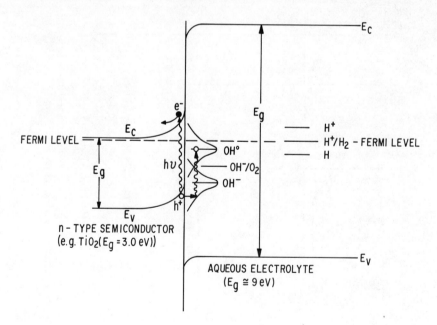

Figure 4. Energy diagram of heterojunction model for semiconductor–electrolyte interface

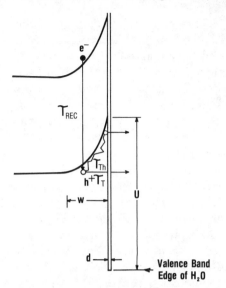

Figure 5. Potential well in depletion layer at semiconductor–electrolyte interface for n-type semiconductors. τ_{REC} is recombination time for electrons and holes. For hot hole injection out of the well, $\tau_T < \tau_{TH}$, τ_{REC}.

In the heterojunction model, photogenerated minority carriers created in the depletion layer ($1/\alpha < w$, where α is the absorption coefficient and w is the depletion layer width) find themselves in an asymmetric potential well (see Figure 5). One side of the well has a parabolic shape with a height equal to the band bending (V_B); the other side is a vertical wall with a height (U) equal to either the difference between the valence-band edge of water and the valence-band edge of the semiconductor (for n-type semiconductors) or the difference between the conduction-band edge of water and the conduction-band edge of the semiconductor (for p-type semiconductors). The thickness of the vertical barrier (d) depends upon the distance from the semiconductor surface of the relevant electrolyte species participating in the charge transfer reaction. For adsorbed species, this distance is 2–3 Å.

In this model, charge transfer of photogenerated minority carriers can be considered to proceed via quantum mechanical tunneling from the potential well of the depletion layer into the electrolyte. To make the model amenable to a quantum mechanical treatment, the parabolic wall is approximated by a linear potential, which makes the well triangular in shape; the effects of the periodic potential in the well are handled via the effective mass approximation. These approximations permit separation of variables and the transformation to a one-dimensional model in which the Schrodinger equation is exactly solvable.

Inside the electrode surface the Schrodinger equation (after a change of variables) takes the form of the Airy equation (12): $d^2\psi/dZ^2 - Z\psi = 0$, where Z, the transformed variable, is a function of the well dimensions. In the barrier, the potential is assumed to be constant at the vacuum level and the wave function has the form of a decaying exponential function. In the electrolyte, localized bound states are assumed.

Since there are no analytic representations of the Airy functions, the connection of the solutions at the boundaries of the triangular well must be done numerically. The result is a set of quantized stationary states in the triangular well. The spacing of the quantized levels is determined by the width of the depletion layer, the amount of band bending in the depletion layer, and the effective mass of the minority carriers.

Tunneling probabilities are calculated from the quantized levels in the depletion layer to isoenergetic electrolyte states. The usual tunneling probility depends upon the barrier height and thickness, and the effective mass of the tunneling particle. A better calculation considers the electrolyte site into which tunneling occurs to be represented by a narrow square well (2 Å in width) with the depth adjusted to put the lowest state at the same energy level as that in the depletion layer. This calculation then takes into account resonance tunneling effects between the depletion layer states and bound states in the narrow well of the electrolyte. Details of these calculations will be published elsewhere (13).

In the following section, the characteristic times to tunnel from the quantized states in the depletion layer into the electrolyte states are calculated and compared with thermalization times to establish the likelihood of a hot carrier injection process across the semiconductor–electrolyte interface.

Hot Carrier Injection Across Semiconductor–Electrolyte Interfaces

Hot carrier injection is defined as a process wherein photogenerated minority carriers are injected into the electrolyte before they undergo complete intraband relaxation (thermalization) in the semiconductor depletion layer. The total time for thermalization (τ_{TH}) is the time required for minority carriers to dissipate all of the band-bending potential energy through carrier–phonon collisions. If the thermalization time is greater than the effective residence time of the minority carriers (τ) in the depletion layer, then hot carrier injection can occur.

The τ is determined by the time required to transfer the photogenerated carrier out of the depletion layer and the relaxation time (τ_R) of the extrinsic electrolyte state in going from an initial state of occupancy to a final state of occupancy such that reverse charge transfer from the electrolyte back into the semiconductor is prevented.

In the quantum mechanical treatment, the time to transfer the photogenerated carriers out of the depletion region and into the electrolyte is the tunneling time (τ_T). Alternatively, this charge transfer time

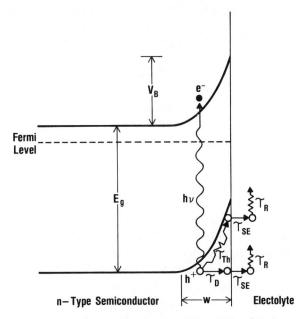

Figure 6. Competing processes for the fate of the photogenerated hole in n-type semiconductor electrodes.

can be calculated classically as the sum of the time required for the carrier to diffuse across the depletion layer (τ_D) plus the time required for the carrier to drift across the semiconductor–electrolyte interface (τ_{SE}). Since the charge transfer process from the depletion layer and the relaxation process in the electrolyte are in series, they both must be faster than the thermalization process in the depletion layer for hot carrier injection to be possible; i.e., for hot carrier injection, $\tau_{TH} > \tau_T$, τ_R or $\tau_{TH} > \tau_D, \tau_{SE}, \tau_R$. A schematic representation of these competing processes is depicted in Figure 6 for *n*-type semiconductors.

***n*-TiO$_2$. HOLE TUNNELING TIMES.** Quantum mechanical tunneling times were calculated from the quasiclassical frequency factor of the quantized states in the depletion layer and their tunneling probabilities (*13*). The calculations depend upon the minority carrier effective mass and the barrier thickness. Detailed calculations have been made for the case of *n*-TiO$_2$ in which d is taken as 10 Å, and values of 0.01 and 0.1 are used for the hole effective mass (m_h^*).

A d of 10 Å represents an upper limit that yields very conservative estimates of the tunneling time; more probable values of d are from 2–4 Å. The correct value of m_h^* for TiO$_2$ is not known, but it is expected to be low because the valence band of TiO$_2$ is a very wide 2p oxygen band; a value of $m_h^* = 10^{-2}$ for TiO$_2$ has been published (*14*). Present calculations use values of 0.1 and 0.01 for m_h^* to test the sensitivity of the results to m_h^*. The dimensions of the potential well for the case of n-TiO_2 in aqueous electrolyte are $U = 3$ eV (10), $V_B = 1$ eV, $w = 200$ Å (10).

The results of calculations for n-TiO_2 that were made by simply multipyling the frequency factor of the quantized state in the depletion layer by the transmission coefficient for the potential barrier show that for the case where $m_h^* = 0.1$, quantized states near the bottom of the well have a tunneling time of about 5×10^{-13} sec; the spacing between levels is about 0.04 eV. Quantized states near the top of the well show tunneling times of about 1×10^{-12} sec, and are spaced about 0.01 eV apart. When the value of m_h^* is reduced to 0.01, only one state appears in the well with an energy level 0.56 eV from the well bottom; for this state $\tau_T = 1 \times 10^{-14}$ sec.

Calculations based on resonance tunneling between the semiconductor and electrolyte potential wells show that for $m_h^* = 0.1$ $\tau_T = 2 \times 10^{-13}$ sec at the bottom of the semiconductor well; for a state of 0.5 eV from the bottom of the semiconductor well $\tau_T = 3 \times 10^{-13}$ sec. Reduction of the hole effective mass to 0.05 reduces the tunneling time for the state near the bottom of the semiconductor well to 1×10^{-13} sec.

Thus, if it is assumed that m_h^* lies between 0.1 and 0.05, then one would expect τ_T's to be of the order of 5×10^{-13} sec for d's of 10 Å, and energy level spacings to be of the order of 0.1 eV.

Values of d less than 10 Å decrease the τ_T's accordingly. For a d of 3 Å and with $m_h^* = 0.1$, $\tau_T = 3 \times 10^{-14}$ sec for simple tunneling, and $\tau_T = 3 \times 10^{-13}$ sec for resonance tunneling for states near the top of the semiconductor well. If m_h^* is reduced to 0.01, then the respective τ_T's for the 3-Å barrier are reduced to 3×10^{-15} sec and 4×10^{-14} sec for the single state in the semiconductor well. Thus, if m_h^* lies between 0.1 and 0.01, then one would expect τ_T's to be of the order of about 5×10^{-14} sec for a d of 3 Å.

HOLE DIFFUSION TIMES. The time for photogenerated minority carriers to diffuse across the depletion layer under the influence of the internal electric field (τ_D), and be transferred to the electrolyte across the semiconductor electrolyte interface (τ_{SE}) can be estimated using a classical model and compared with the quantum mechanical tunneling time. The drift velocity (V_D) in the depletion layer is $V_D = \mu E$, where E is the electric field ($E = V_B/w$) in the depletion layer, and μ is minority carrier mobility. Hence, $\tau_D = w/V_D = w^2/\mu V_B$. The time required to cross the semiconductor–electrolyte interface (τ_{SE}) is d/V_T, where V_T is the thermal velocity of the carrier. If it is assumed that the hole mobility, μ_h, for TiO_2 is at least 100 cm^2/V sec (14), and that $V_B = 1$ eV, $d = 10$ Å, and $w = 200$ Å, then the results yield $\tau_D = 4 \times 10^{-14}$ sec, and

$\tau_{SE} = 3 \times 10^{-15}$ sec. Thus, the classical diffusion times are of the same order of magnitude as the tunneling times for barrier thicknesses of the order of 3–5 Å, and effective hole masses between 0.01 and 0.1.

HOLE THERMALIZATION TIMES. If it is assumed that thermalization of minority carriers in the depletion layer occurs via consecutive single phonon–carrier collisions, then τ_{TH} will be equal to the number of such collisions required to dissipate the band bending potential energy multiplied by the characteristic time between carrier–phonon collisions. The latter time is called the scattering time (τ_{scat}) and by definition, $\tau_{scat} = m_h^* \mu/e$. If the energy loss per single phonon–carrier collision is taken to be 0.06 eV (15), then τ_{TH} via single phonon interactions is $\tau_{TH} = V_B m_h^* \mu/0.06e$. For n-TiO$_2$ with $V_B = 1$ eV, $m_h^* = 0.1$, and $\mu_h = 100$ cm^2/V sec, this yields $\tau_{TH} = 1 \times 10^{-13}$ sec.

Consideration of the quantization of energy levels in the depletion layer leads to longer τ_{TH}'s. This occurs when the spacing between quantized levels is greater than the phonon energy, such that multiple phonon–carrier interactions are required to dissipate energy. The probability of multiple phonon–carrier collisions is much smaller than that of single phonon–carrier collisions, and the corresponding characteristic time constants are longer. Initial estimates indicate that quantized level spacings of about 0.1 eV produce τ_{TH}'s of the order of 10^{-11} sec.

Thus, for n-TiO$_2$ in aqueous electrolyte, the τ_{TH}'s in the semiconductor, involving either single phonon–carrier collisions ($\tau_{TH} \sim 10^{-13}$ sec) or multiple phonon–carrier collisions ($\tau_{TH} \sim 10^{-11}$ sec), are expected to be longer than either the expected tunneling time ($\tau_T \sim 10^{-14}$) or the classical diffusion time across the depletion layer ($\tau_D \sim 4 \times 10^{-14}$ sec). This means that hot photogenerated holes can be expected to reach the TiO$_2$ surface. Their injection into the electrolyte as hot holes will finally depend upon τ_R being faster than τ_{TH}.

RELAXATION TIMES IN ELECTROLYTE. For complete dipolar relaxation in aqueous electrolyte, the characteristic time constant is 10^{-11} to 10^{-12} sec (16, 17); this is the time required for H$_2$O molecules to reorient themselves into a new equilibrium solvation structure around a donor or acceptor species after the species has participated in a charge transfer process. However, complete relaxation is not required to prevent reverse charge transfer from the electrolyte to the semiconductor. The relaxation need only be sufficient to produce misalignment with the quantized energy levels in the depletion layer, or at most to bring the electrolyte energy level outside the energy range of the depletion layer.

For the case of n-TiO$_2$ with large separation between the quantized levels in the depletion layer ($\tau_{TH} \sim 10^{-11}$ sec), the effective τ_R is expected to be faster (13) than τ_{TH}, so hot carrier injection is feasible. For the case where thermalization occurs via single phonon–carrier collisions ($\tau_{TH} \sim 10^{-13}$ sec), it has not yet been established if τ_R is faster than τ_{TH}; calculations on this problem are in progress (13).

Hot hole injection from metal electrodes into liquid electrolyte has been experimentally observed (18). This supports the idea that effective relaxation processes in liquid electrolyte can be fast compared to electronic relaxation processes in solids.

p-GaP. The occurrence of hot electron injection from *p*-GaP photocathodes into vacuum is well known in solid state physics (19, 20, 21). This effect is produced in semiconductors that have a surface layer with a small work function (e.g., Cs or Cs_2O) such that a large degree of band bending is induced in the semiconductor; such systems are labeled "negative electron affinity photocathodes" (19, 20, 21). In Figure 7, an energy level diagram is shown for a *p*-GaP photocathode with a Cs_2O surface layer. The large band bending produced by the Cs_2O layer places the energy of the conduction band (E_c) of bulk *p*-GaP above the vacuum level (this creates a condition of negative electron affinity). Photogenerated electrons subsequently injected into vacuum suffer only a few electron–phonon collisions and are emitted with a hot energy distribution as shown in Figure 7.

In the next section, experimental results are reported for a *p*-GaP electrode that provide qualitative support for a photoelectrochemical hot electron injection process. In this experiment, the carrier density of the *p*-GaP is 5×10^{17} cm^{-3} and $V_B \cong 2.5$ eV. Therefore, the following calculations on diffusion and thermalization times are based on these parameters. Tunneling times have not yet been calculated for the *p*-GaP case.

ELECTRON DIFFUSION TIMES. For *p*-GaP, $m_e^* = 0.5$ and $\mu = 100$ cm^2/V sec (22); with $N = 5 \times 10^{17}$ cm^{-3} and $V_B = 2.5$ eV, $w = 700$ Å. Hence, $\tau_D = w^2/\mu V_B = 2 \times 10^{-13}$ sec. Also, $\tau_{SE} = d/V_T = 7 \times 10^{-15}$ sec for $d = 10$ Å.

ELECTRON THERMALIZATION TIMES. For consecutive single phonon–electron collisions, the energy loss per collision is 0.05 eV (20). Hence, $\tau_{TH} = (V_B/0.05) \tau_{scat} = V_B \mu m_e^*/0.05e = 2 \times 10^{-12}$ sec; $\tau_{scat} = 3 \times 10^{-14}$ sec.

Thus, for the *p*-GaP case, τ_{TH} is longer than the τ_D, and hot electrons arrive at the semiconductor–electrolyte interface. For this case, the actual number of single phonon–electron collisions resulting from diffusion across the depletion layer is equal to $\tau_D/\tau_{scat} = 7$ collisions; hence only about 0.35 eV of the 2.5 eV available from the band-bending potential is dissipated.

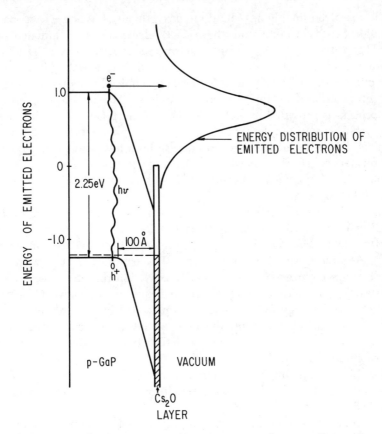

Figure 7. Hot electron injection from a negative electron affinity photocathode (Cs_2O on p-GaP) (19)

The τ_{TH} for single phonon–electron collisions (2×10^{-12} sec) is slower than that for the n-TiO_2 case, and quantization effects could slow it down even further as previously discussed. Therefore, τ_R in the electrolyte can be faster than the τ_{TH} in the semiconductor for the reasons discussed earlier. This means that hot electron injection from p-GaP electrodes into liquid electrolyte should be possible.

Photoenhanced Reduction of N_2 on p-GaP Cathodes

Recent experiments (23) on the photoenhanced reduction of N_2 in a photoelectrochemical cell appear to provide qualitative evidence for a hot electron injection process. The system studied is a photoelectrochemical cell which contains a p-GaP cathode and an Al-metal anode immersed in a nonaqueous electrolyte of titanium tetraisopropoxide and

AlCl$_3$ dissolved in glyme (1,2-dimethoxyethane). When N$_2$ is passed through the electrolyte and the p-GaP electrode is illuminated with band gap light, the N$_2$ is reduced and is recovered as NH$_3$; Al is consumed in the process and acts as the reducing agent. Although the reduction of N$_2$ to NH$_3$ with Al is thermodynamically favored ($\Delta G < 0$), the cell reaction does not proceed in the dark. The activation energy for the process is provided by light absorbed in the p-GaP electrode; hence, this system is an example of photocatalysis in a photoelectrochemical cell. The cell has been successfully operated in both flow and static modes; in the former, N$_2$ is continuously bubbled through the electrolyte. Experiments using ^{15}N$_2$ have also been carried out and ^{15}NH$_3$ has been identified from FTIR spectra.

The cell and electrolyte used in this work are closely related to those used by Van Tamelen and co-workers (24, 25, 26) to demonstrate normal electrolytic fixation of N$_2$. In those previous experiments, an external voltage source was used with either two Pt electrodes (24) or with an Al anode and a nichrome cathode (25) to fix N$_2$. In the photoelectrochemical system, no external voltage source is required to achieve N$_2$ fixation; the activation energy for the reaction is provided by light alone.

In typical flow runs, the reduced nitrogen yields, expressed as moles NH$_4^+$ per mole of Ti ion, varied between about 2%–5%; this corresponded to reaction rates of about 10^{-4} mol NH$_3$/hr/cm^2 electrode (23). Blank runs, in which either N$_2$ was replaced by Ar, p-GaP was replaced by Pt, or no light was used, produced insignificant yields of NH$_4^+$ (20).

The chemical processes occurring in the cell are equivalent to those for the pure electrolytic case as described by Van Tamelen et al. (24, 25, 26). Titanium (IV) isopropoxide is first reduced to a state wherein molecular N$_2$ can be bound; this is evidenced by the development of an intense blue–black color which is attributed to a Ti(II) complex. The reduced titanium–molecular nitrogen complex is then reduced further to produce a reduced nitrogen–reduced titanium complex. Finally, NH$_3$ is produced through protonation of the reduced nitrogen–reduced titanium complex. Alternatively, Van Tamelen had suggested that reduced nitrogen in the reduced complex is transferred to an Al(III) complex, then NH$_3$ is produced via protonation of the reduced nitrogen–aluminum(III) complex (24, 25, 26).

The overall reaction in the cell can be represented as:

$$N_2 + 2Al + 6H^+ \xrightarrow{h\nu} 2NH_3 + 2Al^{3+}, \Delta G/e^- = -1.72 \text{ eV} \qquad (1)$$

Although Reaction 1 is favored in the dark, it does not proceed because of the high activation energy of intermediate steps. The occurrence of

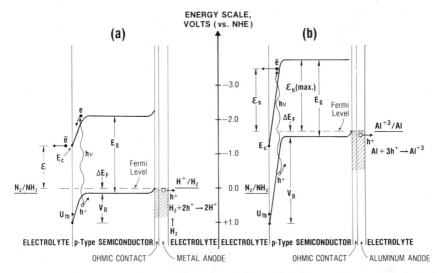

Figure 8. Energy diagram for photoenhanced reduction on p-GaP electrodes. (a) Anode reaction is oxidation of H_2 to H^+ and thermalized electron is injected at the E_c; (b) anode reaction is the oxidation of Al to Al^{+3} and a hot electron is injected above E_c. Redox potential for the reduction of N_2 is below H^+/H_2 ($\Delta G < O$); system is therefore catalytic and light provides the activation energy for the reaction. ε is the photoenhanced reduction potential available for thermalized electrons; ε_h is that for hot electrons. Thermalized electrons are available for reduction at E_c, independent of the anode reaction. Dependence of the cathode reaction on the nature of the anode reaction indicates that a hot electron process is involved.

Reaction 1 under illumination is an example of a photocatalyzed reduction process.

The energetics of general photoenhanced reduction on p-type semiconductors is represented in Figure 8, and is based on the usual principles of photoelectrochemistry with semiconductor electrodes (2). In Figure 8a, the energy level scheme is shown for the case where the reducing agent (e.g., H_2) has a redox potential that lies within the band gap of the semiconductor (e.g., p-GaP). Illumination of the semiconductor with band gap light creates electrons in the semiconductor that are available for reduction at a redox potential more negative than the standard redox potential of the reducing agent oxidized at the anode. Thus, the system produces a photoenhanced reduction effect.

If the standard redox potential of the cathodic reaction is more positive than that of the anodic reaction, then the overall reaction is thermodynamically favored in the dark and the light-driven reaction is photocatalytic (this is the situation with the present cell driving Reaction 1). On the other hand, if the standard redox potential of the cathodic reaction is more negative than that of the anodic reaction, then the overall photoenhanced reduction reaction would be endoergic and light would be converted into stored chemical free energy (photoelectrolysis). This would be the case, for example, with the following reaction:

$$N_2 + 3H_2O \xrightarrow{h\nu} 2NH_3 + \tfrac{3}{2} O_2 \qquad \Delta G/e^- = +1.17 \text{ eV} \qquad (2)$$

The important experimental result that provides evidence for hot electron injection from the p-GaP cathode is that substitution of a H_2/Pt anode for the Al anode did not result in N_2 reduction at the p-GaP cathode (23). As seen in Figure 8, for thermalized injection the energy of the injected electrons (and, hence, the nature of the allowed cathodic reaction) is independent of the nature of the anodic reaction (and, hence, independent of V_B) since the energy level of E_c is pinned at the semiconductor–electrolyte interface. That is, the cathodic reaction occurring at the p-GaP electrode should be independent of the identity of the anode if the photogenerated electrons in the p-GaP cathode were thermalized in the depletion layer before being injected into the electrolyte. Therefore, the observation that N_2 is only reduced with an Al anode and not with a H_2/Pt anode indicates that the cathodic reaction is dependent upon V_B, and that a hot electron injection process is probably involved.

Although Figure 8 is based on aqueous electrolyte and the experiment was performed in nonaqueous electrolyte, the above arguments are nevertheless valid since the relative differences between the energetics for the H_2/Pt anode and the Al anode would be about the same for the two electrolytes. That is, the V_B for p-GaP using the H_2/Pt anode will be much smaller than that produced using the Al anode in both aqueous and nonaqueous electrolyte. However, further experimental work is required to establish the existence of hot carrier injection with more certainty.

Summary and Conclusions

The heterojunction model for the semiconductor–aqueous electrolyte interface considers aqueous electrolyte as a doped semiconductor with a band gap of 9 eV and an electron affinity of 0.5 eV. The resulting heterojunction creates an asymmetric potential well for minority carriers. One side of the well has a parabolic shape with a height equal to the band bending in the semiconductor; the other side is a vertical wall with a height equal to the absolute difference between the minority band edge of the semiconductor and the corresponding band edge of water.

In the model, charge transfer of photogenerated minority carriers is considered to occur via tunneling from the potential well in the semiconductor depletion layer into the electrolyte. Preliminary estimates of the effective minority carrier residence time in the depletion layer, the thermalization time in the depletion layer, and the relaxation time in the electrolyte indicate that hot minority carrier injection into the electrolyte is feasible. This is a charge transfer process wherein the photogenerated minority carriers are injected into the electrolyte before they undergo complete intraband relaxation (thermalization) in the semiconductor depletion layer. The calculations, made for n-TiO_2 anodes and p-GaP cathodes, contain many simplifying assumptions and uncertainties; more refined calculations are required to establish the theoretical foundation for hot carrier injection into electrolytes with more certainty. The effect is well known for photoemission into vacuum from certain p-type semiconductor photocathodes (known as negative electron affinity photocathodes), but analogous effects in electrolyte require relaxation processes in the electrolyte that are fast compared with thermalization in the semiconductor. Effects of quantization in the potential well in the depletion layer will tend to favor the latter requirement; further theoretical development of this particular problem is required and is in progress.

Experiments on the photoenhanced reduction of N_2 in a photoelectrochemical cell provide qualitative evidence for hot electron injection. These experiments showed that the cathodic reaction on p-GaP depends upon the amount of band bending in the depletion layer; this is the expected behavior for hot carrier injection. For thermalized injection, the energy of the injected electron, (and, hence, the nature of the cathodic reaction) is independent of the band bending since the energy level of the conduction-band edge is pinned at semiconductor–electrolyte interface.

The occurrence of hot carrier injection in photoelectrochemical reactions would be very significant for the following reasons: (1) the nature of the permitted photoinduced reactions at semiconductor electrodes could be controlled either by the nature of the anode reaction or by an external bias; (2) the photogenerated carriers would not be in thermal equilibrium in their respective bands so that quasithermodynamic arguments, such as the use of the quasiFermi level to describe the energetics of photoelectrochemical reactions (3), would not be valid; (3) the influence of surface states would be restricted to the class of states originating from the chemical interaction of the electrolyte with the semiconductor surface; and (4) the maximum theoretical conversion efficiency for photoelectrochemical energy conversion would be different compared with the case of thermalized injection. Further experimental and theoretical work is required to firmly establish the importance of hot carrier injection in photoelectrochemical systems.

Glossary of Symbols

d = thickness of potential barrier, Å
e = electron charge, colombs
E = electric field, eV/cm
ΔG = free energy change, eV
m_e^* = electron effective mass
m_h^* = hole effective mass
N = carrier density, cm^{-3}
U = potential barrier to tunneling, eV
V_B = band bending in depletion layer, eV
V_D = drift velocity, cm/sec
V_T = thermal velocity of minority carrier, cm/sec
w = depletion layer width, Å

α = optical absorption coefficient, cm^{-1}
μ = mobility of minority carriers, cm^2/V sec
τ = effective residence time of the minority carriers in depletion layer, sec
τ_D = time for minority carriers to drift across depletion layer, sec
τ_R = relaxation time of extrinsic electrolyte state required to prevent reverse charge transfer from electrolyte to semiconductor, sec
τ_{REC} = recombination time of electrons and holes, sec
τ_{scat} = scattering time of minority carrier, sec
τ_{SE} = time for minority carriers to cross semiconductor–electrolyte interface, sec
τ_T = tunneling time from depletion layer to electrolyte, sec
τ_{TH} = thermalization time of minority carriers in depletion layer, sec
ψ = wave function

Literature Cited

1. Fujishima, A.; Honda, K. *Nature* **1972**, *238*, 37.
2. Nozik, A. J. *Annu. Rev. Phys. Chem.* **1978**, *29*, 189.
3. Gerischer, H. In "Solar Power and Fuels"; Bolton, J. R., Ed.; Academic: New York, 1977; pp. 77–119.
4. Bard, A. J. *J. Photochem.* **1979**, *10*, 1.
5. Krauetler, B.; Bard, A. J. *J. Am. Chem. Soc.* **1977**, *99*, 7729.
6. Wrighton, M. S.; Ginley, D. S.; Wolczanski, P. T.; Ellis, A. B.; Morse, D. L.; Linz, A . *Proc. Nat. Acad. Sci. U.S.A.* **1975**, *72*, 1518.
7. Wrighton, M. S. *Technol. Rev.* **1977**, *79*, 30.
8. Harris, L. A.; Wilson, R. H. *Annu. Rev. Mater. Sci.* **1978**, *8*, 99.
9. Maruska, H. P.; Ghosh, A. K. *Sol. Energy* **1978**, *20*, 443.
10. Williams, F.; Nozik, A. J. *Nature* **1978**, *271*, 137.
11. Williams, F.; Varma, S.; Hillenius, S. *J. Chem. Phys.* **1976**, *64*, 1549.

12. Abramowitz, M.; Stegun, I. A. "Handbook of Mathematical Functions"; National Bureau of Standards, Applied Mathematics Series No. 55; p. 446.
13. Nozik, A. J.; Boudreaux, D.; Williams, F., to be published.
14. Yahia, J. *Phys. Rev.* **1963**, *130*, 1711.
15. Frederikse, H. P. In "American Institute of Physics Handbook"; McGraw-Hill: New York, 1972; pp. 9–73.
16. Eisenberg, D.; Kauzmann, W. In "Structure and Properties of Water"; Oxford University Press: Oxford, 1969; p. 207.
17. Sass, J. K.; Gerischer, H. In "Photoemission and the Electronic Properties of Surfaces"; Feuerbacher, B., Fitton, B., Willis, R. F., Eds.; Wiley: New York, 1978.
18. Gerischer, H.; Meyer, E.; Sass, J. K. *Ber. Bunsenges. Phys. Chem.* **1972**, *76*, 1191.
19. Pankove, J. I. In "Optical Processes in Semiconductors"; Dover Publications: New York, 1975; pp. 294–300.
20. Moss, T. S.; Burrell, G. J.; Ellis, B. In "Semiconductor Opto-Electronics"; Butterworth: London, 1973, pp. 161–164.
21. Spicer, W. E. *Appl. Phys.* **1977**, *12*, 115.
22. Sze, S. M. In "Physics of Semiconductor Devices"; Wiley–Interscience: New York, 1969; p. 20.
23. Dickson, C. R.; Nozik, A. J. *J. Am. Chem. Soc.* **1978**, *100*, 8007.
24. Van Tamelen, E. E.; Akermark, B. *J. Am. Chem. Soc.* **1968**, *90*, 4492.
25. Van Tamelen, E. E.; Seeley, D. A. *J. Am. Chem. Soc.* **1969**, *91*, 5194.
26. Van Tamelen, E. E. *Acc. Chem. Res.* **1970**, *3*, 363.

RECEIVED October 2, 1978.

10

Short-Lived Radicals at Photoactive Surfaces

Spin Trapping and Mechanistic Consequences

M. L. HAIR and J. R. HARBOUR

Xerox Research Centre of Canada, 2480 Dunwin Drive,
Mississauga, Ontario, Canada L5L 1J9

> *The technique of spin trapping has been successfully to the study of radicals produced when photoactive particles are suspended in either aqueous or insulating fluids and irradiated in the presence of O_2. This trapping technique is reviewed with particular emphasis on the detection and identification of the superoxide anion and hydroxyl radicals. Results are interpreted within the framework of a simple band model for semiconductors. The effects of both anionic and cationic surfactants on the photoprocess are described. The addition of electron-donating molecules to the suspension results in a reaction (in the fluid) that is "pumped" by application of band gap radiation to the solid particle. The radicals that have been identified on irradiating several different photoactive particles are described. The ability to identify these radical intermediates is important in determining the exact reaction mechanism, as exemplified by a discussion of the photosynthesis of H_2O_2 on zinc oxide.*

Many solar energy conversion devices based upon the interaction of light with semiconductors have been proposed. These include photovoltaic devices, photoelectrochemical cells that can directly generate electricity or produce a fuel, and pigment dispersions, which also can produce a fuel (1) or photodecompose a pollutant (2).

In systems where the semiconductor interfaces with a solution, the proposed photochemical mechanisms generally involve radical intermediates. However, there is very little evidence of radical participation or identification. We have therefore begun an experimental program aimed at identifying the radicals photoproduced as a result of irradiation of various pigment dispersions (3). A better understanding of these intermediates and the factors that influence their production and destruction should contribute to the development of these types of solar energy converters. These factors include a detailed knowledge of the role that the surfactant plays. The surfactant is added to pigment dispersions to prevent flocculation of the particles. This is important since such flocculation causes a reduction in surface area as well as an increase in the rate of settling. However, in all these photoactive systems the surfactant plays a dual role and always affects the surface charge as it stabilizes the system.

It is speculated that the photochemistry at an interface occurs through radical intermediates (4). In principle, electron spin resonance (ESR) spectroscopy would be the ideal method for examining this type of interfacial photochemistry. Unfortunately, the direct detection and identification of radicals by this technique is possible only if the radicals are produced in relatively high concentrations in the ESR cavity and are sufficiently long-lived to be detected. In most systems of practical interest there will be relatively large concentrations of both O_2 and H_2O. Therefore, there is a high probability that superoxide (O_2^-) or hydroxyl ($\cdot OH$) radicals will be formed under normal ambient conditions. The half-lives of these radicals (or their spin lattice relaxation times T_1) are sufficiently short that direct detection of them is not always possible. In order to circumvent this problem we have successfully applied the technique of spin trapping to photoactive particulate dispersions.

The use of a radical addition reaction to detect short-lived radicals was first proposed by Janzen (5) in 1965. Early work on this technique centered on the interactions of nitrones with radicals and the consequent production of stable nitroxides. The reader is referred to a review by Janzen (6) which covers the development of the spin-trapping reactions prior to 1971. A major advance in the utility of this technique came in 1973 when Janzen and Liu (7) described the use of a five-membered ring nitrone 5,5-dimethyl-1-pyroline-1-oxide (DMPO). This acted as a spin trap in the following manner:

$$\text{Me}_2\text{C}_4\text{H}_6\text{N}^+\text{O}^- + R\cdot \rightarrow \text{Me}_2\text{C}_4\text{H}_7(R)\text{NO} \quad (1)$$

The spin adduct of DMPO has the advantage that the hyperfine splitting constants are strongly dependent upon the nature of the complexed radical and are sufficiently separated that ready identification of the radical is generally possible. Harbour and Bolton (8, 9) have applied this spin-trapping technique to in vivo studies of chloroplasts and chromatophores. They found that when these systems were illuminated both O_2^- and ·OH could be identified from the spectra of the radical adducts. Application to particulate dispersions of photoconducting particles was first reported in 1977 by Harbour and Hair (3) who showed that when aqueous suspensions of cadmium sulfide were irradiated in the presence of DMPO the O_2^- spin adduct was readily observed.

Experimental

Three types of photoconducting particles have been used in this work. Cadmium sulfide, an n-type semiconductor, was obtained from Fisher and used without further treatment (3). It consisted of particles approximately 0.5 μm in diameter with a BET (N_2) surface area of 10 m^2/g. Metal-free phthalocyanine, an organic photoconducting pigment that is often taken as an analog of chlorophyll, was in the x-crystalline form. This insoluble powder consisted of particles less than 1 μm in diameter with a BET surface area of 70 m^2/g. Distilled water was redistilled from an all-glass apparatus. The spin trap DMPO was synthesized and purified prior to use by bulb-to-bulb distillation on a vacuum system and added directly to the dispersion (~ $0.1M$).

In all cases the pigment suspensions were prepared by ultrasonic dispersion. The samples were illuminated in situ with a tungsten-quartz-iodide lamp described elsewhere (10) or with a Hanovia Model 997B-1KW Hg–Xe lamp in a Schoeffel Model LH151N lamp housing with appropriate filters. The ESR spectra were obtained on a Varian E12 ESR spectrometer. In certain cases either a cationic surfactant, cetyltrimethylammonium bromide (CTAB) from Sigma, or an anionic surfactant, Aerosol OT (AOT) from American Cyanamid, was added to aid dispersion and/or to observe the effect of the adsorbed molecules on photochemistry.

Results and Discussion

·OH Adduct. The formation and identification of the ·OH adduct of DMPO was first reported by Harbour, Chow, and Bolton in 1974 (11). These authors prepared the ·OH radical by UV photolysis of dilute aqueous H_2O_2 solution.

$$2H_2O_2 \xrightarrow{h\nu} 2 \cdot OH \quad (2)$$

In the presence of DMPO the signal shown in Figure 1 was recorded. The signal was characterized by $g = 2.0060 \pm 0.0002$ and $a^N = a_\beta^H = 14.9$ G. The accidental equality of a^N and a_β^H gives rise to the 1:2:2:1 quartet. This assignment has been confirmed by Sargent and Gardy (12)

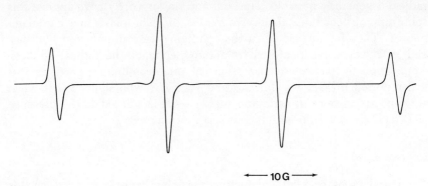

Figure 1. The ESR spectrum of the ·OH adduct of DMPO in water at 25°C

who prepared ·OH by radiolysis of deoxygenated water using 3 MeV electrons. They obtained an identical ESR spectrum using the spin trap DMPO.

As will be discussed in more detail later, band gap irradiation of zinc oxide suspended in water gives rise to a signal identical to that shown in Figure 1.

O_2^- Adduct. The O_2^- adduct was the major product observed by Harbour, Chow, and Bolton (11) when concentrated aqueous solutions of H_2O_2-containing DMPO were photolyzed.

$$·OH + H_2O_2 \rightarrow H_2O + HO_2 \qquad (3)$$

The DMPO adduct in water gives an ESR spectrum with $g = 2.0061$, $a^N = 14.1$ G, $a_\beta^H = 11.3$ G and a_γ^H 1.25 G.

In aqueous systems the O_2^- radical is in equilibrium with the HO_2 radical.

$$HO_2 \rightleftharpoons H^+ + O_2^- \qquad (4)$$

The pK_a for this equilibrium is 4.4 ± 0.4 (13). However, the pK_a of the spin adduct ionization is not known. Thus it is not possible to distinguish between O_2^- and its protonated form when the radicals are incorporated into the DMPO adduct.

Further proof of the correct identification of this spin adduct also has been obtained by independently generating O_2^- by solubilizing potassium superoxide with the K^+-selective 18-crown-6-ether (CE) (14) (*see* Figure 2).

$$KO_2 + CE \rightleftharpoons CEK^+ + O_2^- \qquad (5)$$

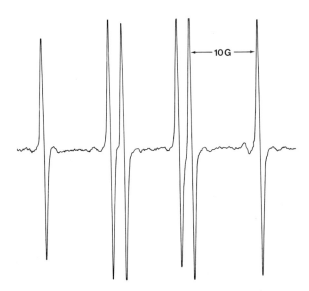

Figure 2. The ESR spectrum of O_2^- adduct in propylene carbonate at 25°C

This technique has been extended to other solvents and the O_2^- adduct has been identified in a series of solvents with polarities ranging from that of water to that of benzene. The nitrogen and β-hydrogen splittings have been determined as a function of solvent polarity. This enables the extension of the spin-trapping technique to any solvent system. Further evidence that O_2^- was actually present in those systems was obtained by rapid freezing experiments in the absence of the spin trap. An ESR signal consistent with the production of an axially symmetric radical was obtained ($g_{\parallel} = 2.08$ and $g_{\perp} = 2.00$).

Application to Dispersions. Aqueous dispersions of either cadmium sulfide or x-phthalocyanine gave no ESR signal upon addition of DMPO to the system. However, illumination with band gap irradiation yielded a small ESR signal. This was similar to the spectrum shown in Figure 2 and is readily identified as being that of the DMPO/O_2^- spin adduct. The spin adduct does not form in the absence of O_2 and the amount of product is dependent upon the O_2 partial pressure. Its formation is consistent with a one-electron transfer from the irradiated solid to the dissolved O_2. In the absence of surfactants or other additives, the intensity of the ESR signal is never very great and thus the apparent efficiency of the photogeneration is quite low.

During illumination in oxygenated non-aqueous suspensions, the signal shows a continual narrowing of the spectral lines. This is consistent with the removal of O_2 from the solution through reduction of O_2 to O_2^- and subsequent trapping by the DMPO. Also during illumination, however, the increase in signal intensity is followed by a slow decay. After turning off the light the signal shows a further decay. This indicates that the O_2^- spin adduct is somewhat unstable in these systems, preventing an accurate quantitative determination of the O_2^- adduct concentration. The ·OH adduct is much more stable in both light and dark conditions. Initial experiments aimed at quantifying radical formation in the aqueous zinc oxide system are described elsewhere (15).

The results obtained above are consistent with a mechanism such as in Figure 3 (16). The incident photon creates an electron–hole pair and, if an acceptor state lies below the conduction band, electron transfer to the acceptor level is thermodynamically favorable. The redox level for the O_2/O_2^- couple is below the conduction band for both cadmium sulfide and x-phthalocyanine, so formation of O_2^- is not unexpected.

The direct detection of an electron transfer reaction in the CdS–H_2O system can be achieved by replacing molecular oxygen with a molecule whose reduced form is relatively stable. Methyl viologen (MV^{+2}) is such a compound. It is water soluble, exists as a colorless cation which has a redox potential of -0.44 (vs. NHE) and can be reduced to a stable blue cation radical provided O_2 is not present. Addition of MV^{+2} to a cadmium sulfide dispersion under N_2 purging and illumination did indeed give rise to the MV^+ signal.

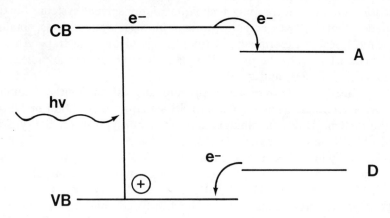

Figure 3. A photon (hv) of light causes the excitation of an electron from the valence band (VB) to the conduction band (CB). If the energy level of the acceptor (A) is below that of the CB then electron transfer can occur as indicated by the arrow. Similarly, if the donor (D) state is above that of the VB electron transfer to the hole can occur.

A direct consequence of the above model is that when light is absorbed a hole also must be created. Initially the hole may be trapped in the photoconducting particle. In this case the particle may assume a reversed charge and this type of charge reversal (in nonaqueous systems and with large applied electric field) forms the basis of a novel type of electrophotographic imaging (*17*). Over a period of time, however, the cadmium sulfide will undergo self-oxidation (*18*) or the hole will react with an electron donor in the surrounding solution. It is well established that EDTA is photooxidized very efficiently (*19*). Addition of EDTA to the CdS–DMPO suspension greatly increases the ESR signal of the O_2^- radical adduct. The cadmium sulfide particles are thus acting as a photopump and driving electrons from EDTA to O_2. A similar large increase in intensity of the signal due to the MV^+ radical cation was observed when EDTA was added to the illuminated MV^{+2}–CdS suspension, thus providing further credibility to the mechanism.

Effect of Surfactants. For any practical system involving photoactive pigment dispersions it is almost certain that a surfactant would be required to prevent flocculation of the particles. Many of the surfactants used in aqueous systems are ionic and can charge the particles either positively or negatively. The effect of altering the surface charge can be predicted from Figure 3. The degree of band bending at the interface defines the space charge region within the photoconductor. When the hole and electron are created by absorption of a photon, the holes and electrons migrate oppositely under the influence of the field. If a positive surface charge exists, electrons will migrate more readily towards the interface and will act more efficiently as reducing agents for the molecular oxygen. However, if the surface is negatively charged, the opposite effect would be anticipated and electron transfer impeded. To test these predictions spin-trapping experiments have been performed using phthalocyanine suspensions which have been dispersed by adsorbed monolayers of either CTAB (which adsorbs via the bulky hydrocarbon moiety to give a positive surface) or AOT (which adsorbs to give a negative surface) (*20*). On irradiation, the yield of the O_2^- adduct is greatly increased for the system that has the increased positive charge and is significantly decreased when the surface assumes a negative charge. The monolayer of surfactant does not prevent the electron transfer from occurring and the effect of the surface charge is more noticeable in the case of the phthalocyanine than the CdS. Thus, in these cases, the surface charge does not significantly alter the primary photochemistry. However, the role of surfactant is crucial and, as will be described elsewhere (*21*), a combination of "right" properties of surfactant and electron donor can be used to achieve photosynthesis of H_2O_2 on a phthalocyanine surface.

Application to Photosynthesis of H_2O_2. When zinc oxide powder is suspended in H_2O and irradiated with light of wavelength less than 380 nm in the presence of DMPO, a large ESR signal is observed. This is identical to that shown in Figure 1 and can therefore be identified as the adduct of DMPO and an ·OH. H_2O_2 is photogenerated under these experimental conditions and the efficiency of the photoreaction is increased by the addition of compounds such as formate and oxalate (4). Many mechanisms have been proposed to account for this synthesis of H_2O_2, and although radical intermediates are often proposed, these spin-trapping studies provide the first direct evidence for their presence. We have recently concluded a study on the photosynthesis of hydrogen peroxide on zinc oxide combining spin-trapping experiments, quantitative measurement of oxygen uptake studies, and peroxide formation in an attempt to define the reaction path. Full details are published elsewhere (15) but a summary is pertinent because the spin-trapping experiments reveal the major effect of the carboxylate-type additives on the system. The salient points are as follows:

(A) A quantitative comparison between product formation and radical concentration demonstrated that the radicals were major participants in the reaction mechanism.

(B) When zinc oxide suspensions are illuminated in the absence of additives only the ·OH radical adduct is observed. The O_2^- adduct is never observed in these experiments and therefore does not exist as a free entity in the external solution. (Although this does not rule out its presence as a surface species.) The time dependence of the radical adduct formation is shown in Figure 4. The intensity peaks with time, probably caused by photoinduced destruction of the radical adduct since the intensity levels off when illumination is blocked.

(C) Despite the fact that O_2^- is never observed in free solution the kinetic curves show that the rate of ·OH production is dependent upon the O_2 concentration in the solution. Moreover, previous tracer studies show that the oxygen that is incorporated into H_2O_2 comes from the molecular oxygen and not water (22).

(D) When formate is added to the aqueous zinc oxide system and irradiated in the presence of DMPO, the ·OH adduct is no longer observed, but is replaced by the large signal shown in Figure 5. This new signal has $g = 2.006$, $a^N = 15.6$ G, and $a_\beta{}^H = 18.7$ G. By a series of experiments analogous to those described earlier for O_2^- and ·OH this signal can be identified as being caused by the DMPO/·CO_2^- radical adduct. The limiting concentration of H_2O_2 formed increases from $1 \times 10^{-4} M$ to $8 \times 10^{-4} M$.

10. HAIR AND HARBOUR *Radicals at Photoactive Surfaces* 181

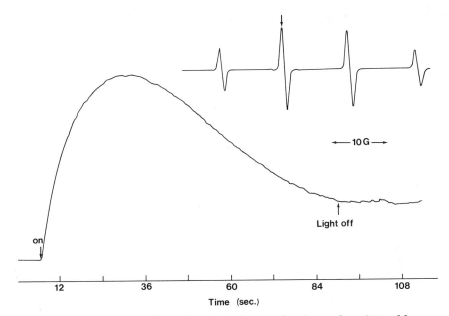

Figure 4. The time dependence of the amplitude of the ·OH adduct signal as a function of illumination. The field is fixed at the point indicated by the arrow in the upper-right portion of the figure.

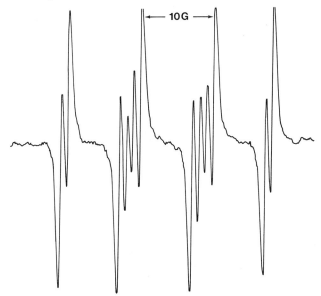

Figure 5. The ESR spectrum of the $·CO_2^-$ adduct of DMPO in water at 25°C

These observations are consistent with the following mechanism.

$$\text{ZnO} + h\nu \rightleftharpoons \underset{\text{hole}}{\oplus} + \underset{\text{electron}}{e^-} \qquad (6)$$

The initial step is a photoreduction of molecular oxygen, where (s) denotes a surface species.

$$O_{2(s)} + e^- \rightarrow O_2^{-}{}_{(s)} \overset{H^+}{\rightleftharpoons} HO_{2(s)} \qquad (7)$$

These must be surface species because they are never detected by the spin trap. This proposal is supported by the separate observation that the rate of O_2 uptake is a function of the square root of O_2 concentration, suggesting a surface effect.

The $HO_{2(s)}$ species can then be reduced by a second photoproduced electron to generate H_2O_2.

$$HO_{2(s)} + e^- \rightarrow HO_2^- \overset{H^+}{\rightleftharpoons} H_2O_2 \qquad (8)$$

Alternatively, since HO_2 is known to dismutate in solution to give H_2O_2, the following reaction could also occur:

$$2\,HO_{2(s)} \rightarrow H_2O_2 + O_2 \qquad (9)$$

For the concurrent oxidation it is clear that $\cdot OH$ radicals must be produced, and this can occur most easily by either of the following reactions:

$$(H_2O)OH^- + \oplus \rightarrow \cdot OH \qquad (10)$$

or

$$Zn\text{—}OH_{(s)} + \oplus \rightarrow Zn^{2+} + \cdot OH \qquad (11)$$

If Equation (10) is occurring, the reaction converts solar energy to chemical free energy since the reaction $H_2O + O_2 \rightarrow H_2O_2$, has a positive free energy ($+$ 25 kcal) (*1*). The oxidation of zinc oxide itself (Equation 11) would be a photocorrosion reaction as discussed by Dixon and Healy (*23*). The limiting concentration of H_2O_2 is then proposed to follow from the reaction

$$H_2O_{2(s)} + \cdot OH \rightarrow H_2O + HO_{2(s)} \qquad (12)$$

The formate is proposed to function, at least in part, as a getter for the $\cdot OH$ radical since no $\cdot OH$ adduct can be observed. It also can function as a reductant.

$$\cdot OH + HCO_2^- \rightarrow \cdot CO_2^- + H_2O \qquad (13)$$

or

$$HCO_2^- + \oplus \rightarrow \cdot CO_2^- + H^+ \qquad (14)$$

The experiments described here clearly demonstrate the utility of the spin-trapping technique as a method for identifying radical intermediates in photoactive systems. The study of the photosynthesis of H_2O_2 on zinc oxide reveals the distinctly different mechanistic pathway which occurs in the presence of additives. We have applied the spin-trapping method to simple photoproduced reactions occurring across the solid–liquid interface in photoactive pigment dispersions. The method is also clearly applicable to photoelectrochemical cells (18, 24) and any other heterogeneous system involving interfacial charge transfer.

Literature Cited

1. Rubin, T. R.; Calvert, J. G.; Rankin, G. T.; MacNevin, W. J. Am. Chem. Soc. **1953**, 75, 2850.
2. Frank, S. N.; Bard, A. J. J. Am. Chem. Soc. **1977**, 99, 303.
3. Harbour, J. R.; Hair, M. L. J. Phys. Chem. **1977**, 81, 1791.
4. Freund, T.; Gomes, W. P. Catal. Rev. **1969**, 3, 1.
5. Janzen, E. G. Chem. Eng. News **1965**, 43, 50.
6. Janzen, E. G. Acc. Chem. Res. **1971**, 4, 31.
7. Janzen, E. G., Liu, J. I-Ping. J. Mag. Reson. **1973**, 9, 510.
8. Harbour, J. R.; Bolton, J. R. Biochem. Biophys. Res. Commun. **1974**, 64, 803.
9. Harbour, J. R.; Bolton, J. R. Photochem. Bhotobiol., in press.
10. Warden, J. T.; Bolton, J. R. J. Am. Chem. Soc. **1973**, 95, 6435.
11. Harbour, J. R.; Chen, V.; Bolton, J. R. Can. J. Chem. **1974**, 52, 3549.
12. Sargent, F. P.; Gardy, E. M. Can. J. Chem. **1976**, 54, 275.
13. Czapski, G.; Bielski, B. H. J. Phys. Chem. **1963**, 67, 2180.
14. Harbour, J. R.; Hair, M. L. J. Phys. Chem. **1978**, 82, 1397.
15. Harbour, J. R.; Hair, M. L. J. Phys. Chem., in press.
16. Gerischer, H. J. Electroanal. Chem. Interfacial Electrochem. **1975**, 58, 263.
17. Weigl, J. W. Angew. Chem. Int. Ed. Engl. **1977**, 16, 374.
18. Gerischer, H. In "Solar Power and Fuels"; Bolton, J. R., Ed.; Academic: New York, 1977; p. 77.
19. Markiewicz, S.; Chan, M. S.; Sparks, R. H.; Evans, C. A.; Bolton, J. R. International Conference on the Photochemical Conversion and Storage of Solar Energy, London, Ontario, Canada, 1976.
20. Harbour, J. R.; Hair, M. L. Photochem. Photobiol. **1978**, 28, 721.
21. Harbour, J. R.; Hair , M. L.; Tromp, J., unpublished data.
22. Calvert, J. G.; Theurer, K.; Rankin, G. T.; MacNevin, W. M. J. Am. Chem. Soc. **1954**, 76, 2575.
23. Dixon, D. R.; Healy, T. W. Aust. J. Chem. **1971**, 24, 1193.
24. Ellis, A. B.; Kaiser, S. W.; Wrighton, M. S. J. Am. Chem. Soc. **1976**, 98, 6855.

RECEIVED October 2, 1978.

11

Luminescent Properties of Semiconductor Photoelectrodes

ARTHUR B. ELLIS[1] and BRADLEY R. KARAS

Department of Chemistry, University of Wisconsin, Madison, WI 53706

> *The use of luminescent, n-type 5–1000-ppm CdS:Te and 10 ppm-CdS:Ag polycrystalline photoelectrodes as probes of recombination in photoelectrochemical cells is reported. Except for intensity, the emission spectra (λ_{max}, 600–700 nm) are insensitive to the presence of S^{2-}/S_n^{2-} electrolyte and to the excitation wavelengths and electrode potentials employed. With ultraband gap irradiation ($\lambda \lesssim 500$ nm) and aqueous S^{2-}/S_n^{2-} or Te^{2-}/Te_2^{2-} electrolytes, optical energy is converted to electricity at 0.1–5% efficiency and to luminescence at 0.01–1.0% efficiency; the effects of surface preparation and grain boundaries in determining efficiency are discussed. Increasingly negative bias applied to CdS:Te and CdS:Ag photoanodes increases emission intensity by 15–100% while the photocurrent simultaneously declines to zero. Band gap edge 514.5-nm excitation yields smaller photocurrents and larger but much less potential dependent emission intensity. These results are consistent with the band bending model presently used to describe photoelectrochemical phenomena.*

The desire to convert optical energy directly into fuels or electricity has led to the rapid development of photoelectrochemical cells (PECs). A typical PEC consists simply of a semiconductor electrode, a counter-electrode, and an electrolyte. The semiconductor is the key element of the PEC, since it serves in the dual capacity of photoreceptor and electrode. Light absorbed by the semiconductor can be channeled into

[1] Author to whom inquiries should be addressed.

electrochemical processes leading to the aforementioned energy conversions. Although the physics governing photoelectrochemical phenomena has been elegantly reviewed (1, 2), a brief description is in order.

Photoelectrochemical events are initiated by ultraband gap photons that, when absorbed by the semiconductor, produce a conduction band electron and valence band hole. The difference between the dark and illuminated electrodes is really the difference between ground and excited states, respectively. Figure 1 illustrates this distinction for an n-type semiconductor. The semiconductor bands are bent in parallel; this is a consequence of the mismatch in chemical potentials between the electrolyte (redox potential) and semiconductor (Fermi level). Band bending occurs over a short distance ($\sim 1~\mu$) from the semiconductor–electrolyte interface into the semiconductor bulk and equilibrates the chemical potentials of the two phases. The distance over which band bending occurs is termed the depletion or space-charge region.

Once the semiconductor excited state has been populated, band bending exerts considerable influence over the attendant deactivation processes. In particular, the potential gradient inhibits the recombination of electron–hole pairs and promotes their separation. The conduction band electron migrates to the counterelectrode where it reduces an electroactive electrolyte species, and the valence band hole migrates to the semiconductor–electrolyte interface where it accepts an electron from an electroactive species, thereby oxidizing it. n-Type semiconductors, the most commonly used photoelectrodes, are thus photoanodes and dark cathodes.

A major obstacle to the practical utilization of these concepts is the undesirable oxidation of the n-type semiconductor electrode itself. Typical is the case of CdS, which undergoes photoanodic decomposition via Equation 1 (3). The problem is minimized by choosing electroactive

$$CdS \xrightarrow{h\nu} Cd^{+2} + S + 2e^- \quad (1)$$

electrolyte species whose oxidation is kinetically rapid enough to quench Reaction 1. For example, sulfide (S^{2-}) or polysulfide (S_n^{2-}) electrolytes greatly inhibit the photoanodic dissolution of CdS (4–8). Polysulfide species can be oxidized at the photoanode and simultaneously reduced at the counterelectrode to yield a PEC that exhibits little change in electrolyte or electrode composition, thus permitting the sustained conversion of optical energy to electricity. This concept has been used to construct PECs employing a variety of photoanodes and electrolytes (9–28).

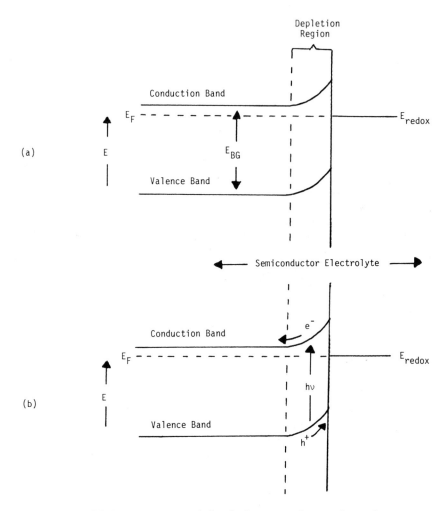

Figure 1. (a) Representation of the dark semiconductor electrode corresponding to the ground state; (b) irradiation of the electrode produces the excited state, which is deactivated here by redox reactions. E_F and E_{BG} are the Fermi level and semiconductor band gap, respectively. E_{redox} is the electrolyte redox potential. Band bending characteristic of the depletion region formed at the n-type semiconductor–electrolyte interface is also shown.

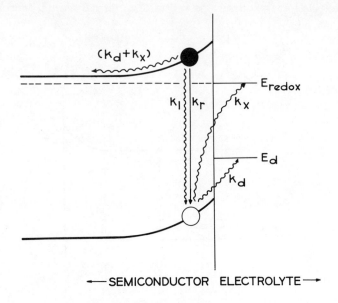

Figure 2. Excited state deactivation pathways of the semiconductor electrode. Wavy arrows signify nonradiative decay routes: k_l, k_d, and k_x correspond to electron–hole recombination leading to heat, electron–hole separation leading to photoanodic decomposition, and electron–hole separation leading to electrolyte redox reactions, respectively. The straight arrow and k_r correspond to radiative recombination, the source of luminescence. E_d is the thermodynamic potential for anodic decomposition. Intraband gap states and defects that might play a role in the various deactivation routes have been omitted for simplicity.

A major thrust of current PEC research is the improvement of energy conversion efficiency. Central to this goal is an understanding of the semiconductor excited state, particularly the extent to which its deactivation routes might be amenable to experimental control. The semiconductor excited state partitions input optical energy into several pathways, as pictured in Figure 2. A broad division into nonradiative and radiative relaxation routes is especially convenient.

At least three nonradiative mechanisms for deactivation are known: heat (lattice vibrations), electrode decomposition, and electrolyte redox reactions with corresponding rate constants k_l, k_d, and k_x, respectively. Heat results from the nonradiative recombination of photogenerated electron–hole pairs and its role in PECs has been explored by photothermal spectroscopy (67). Electrode decomposition and electrolyte redox reactions are also nonradiative but result from separation of electron–hole pairs, as described above. The relationship of k_d to k_x is strongly influenced by the choice of electrolyte (29–32). The sum (k_d +

k_x) is reflected in the current passed in the external circuit, but chemical methods of analysis are required to differentiate between the two current sources. For CdS-based PECs k_d dominates k_x in OH⁻ electrolyte, whereas the opposite is true in polysulfide media. There are thermodynamic potentials (E_d and E_{redox}) associated with these reactions (29, 33, 34); however, the significance of kinetics is underscored by the observation that diffusion-dependent electrolyte redox processes can compete with electrode localized decomposition. Other experimental factors that can affect k_d and k_x are excitation intensity and electrode potential (1, 5, 6, 33, 34).

Against this background we introduce k_r, which represents radiative deactivation resulting from electron–hole pair recombination processes. Luminescence is a powerful tool for characterizing excited states, be they organic, organometallic, or solid state in nature. Emissive properties including spectral distribution, lifetime, and quantum yield permit the calculation of rate constants and the assessment of whether a given reaction is possible during the excited state lifetime. Although a vast literature exists for luminescent semiconductors (36, 37), very little is known about radiative decay in the context of a PEC. Studies that have been carried out focus on electroluminescence resulting from injection processes at extreme potentials or in strongly oxidizing or reducing media (38–44). These are frequently transient effects which are not appropriate for sustained optical energy conversion. Photoluminescence studies that predate our work use n- and p-type GaP (44), n-type ZnO, and Cu-doped ZnO (ZnO:Cu) photoelectrodes (68). Although only p-GaP was photoinert, these systems provide important comparisons with our studies, as will be discussed later.

What we had hoped to find are electrodes that emit while mimicking the essential features of electrodes used in operating PECs. As shown in Figure 3, CdS doped with either Te or Ag (CdS:Te, CdS:Ag) acts as just such a dual electrode–emitter. We find that CdS:Te and CdS:Ag are similar to undoped CdS electrodes in their ability to oxidize aqueous polysulfide and ditelluride species as part of degenerate electrolysis schemes leading to sustained conversion of optical energy to electricity.

Emission from CdS:Te and CdS:Ag involves intraband gap electronic states introduced by the dopant. Tellurium is thought to substitute at S sites and to yield states approximately 0.2 eV above the valence band (45–52, 69). As an isoelectronic dopant, Te is not expected to alter the electrical properties of CdS appreciably. Because it has a smaller electron affinity than S, Te serves as a trap for holes that then can bind coulombically an electron in or near the conduction band, thus forming an exciton. The exciton's binding energy is about 0.22 eV; appreciable exciton concentrations will exist at room temperature. At higher Te doping levels

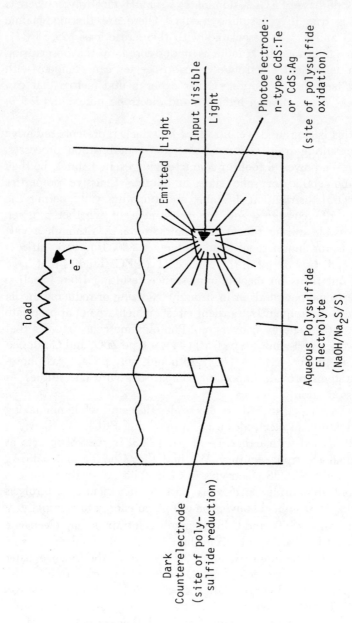

Figure 3. A PEC employing n-type CdS:Te or CdS:Ag photoelectrodes luminesces as it converts optical energy to electricity via a degenerate electrolysis reaction. Polychalcogenide electrolytes have been chosen to minimize photoanodic decomposition of the electrode.

the exciton is thought to be localized over several nearest neighbor Te atoms with a higher binding energy (47, 48). Radiative collapse of the exciton produces the observed luminescence. We have used melt-grown, polycrystalline material that is nominally 5–1000-ppm Te.

The mechanism by which CdS:Ag emits is more complex and depends both on the presence of additional impurities and on whether Ag substitutes at Cd sites or interstitially (53–59). Substitution for Cd would make Ag an acceptor and thus partially compensate the material. We have used melt-grown, polycrystalline 10-ppm CdS:Ag, and the resistivity of approximately 10^3 Ω-cm compared with approximately 1 Ω-cm for undoped CdS (and CdS:Te) is consistent with this role for Ag.

To exploit the emissive properties of CdS:Te- and CdS:Ag-based PECs, the cell is assembled in the emission compartment of a spectrophotofluorometer. Inclining the photoelectrode at about 45° to both a laser excitation source and the emission detection optics permits the sampling of front surface emission during the course of photoelectrochemical events. Thus, changes in the emission spectrum (5-nm resolution) and intensity can be monitored in situ. The excitation source is the continuous output of an Ar ion laser. Generally, the incident power was 1–15 mW, which in the approximately 3-mm diameter beam corresponds to intensities of 14–212 mW/cm^2. All data were obtained with polycrystalline CdS:Te and CdS:Ag samples from Eagle–Picher Industries, Miami, Oklahoma. The grain sizes in the polycrystalline samples are estimated to be 3–8 mm. Preparation and handling of the polysulfide and ditelluride electrolytes has been described previously and differs in the use of N_2 rather than Ar for purging (6).

As will be described below, the key finding that we have made is that the emission from CdS:Te and CdS:Ag photoelectrodes is a very sensitive probe of recombination processes within the depletion region and hence of the semiconductor excited state. Our results thus far are consistent with the aforementioned band bending arguments used to interpret photoelectrochemical phenomena. Importantly, we present evidence that experimental parameters such as electrode potential, electrolyte, and excitation wavelength may be used to manipulate the semiconductor excited state processes and thus influence the course of optical energy conversion.

Results and Discussion

Stability. As a first step in characterizing CdS:Te- and CdS:Ag-based PECs, we wished to determine the extent to which their electrochemistry resembles that of undoped CdS. All three electrode materials undergo photoanodic dissolution in OH$^-$ electrolyte according to Equation 1. Since CdS photoelectrodes are stabilized by polysulfide and ditelluride electrolytes, we investigated whether or not Cds:Te and CdS:Ag are rendered stable by these media using variations in electrode weight, surface quality, photocurrent, and luminescence as criteria. In the kinetic scheme of Figure 2, we are seeking evidence that k_x is much greater than k_d.

The first criterion of stability is met if there is no appreciable weight loss after sufficient current has passed in the external circuit to decompose part or all of the electrode. Data in Table I indicate that this is the case for CdS:Te and CdS:Ag electrodes in S^{2-}/S_n^{2-} and Te^{2-}/Te_2^{2-} electrolytes. The minimal weight losses observed result primarily from chipping of the electrode when it is demounted. All of the experiments listed in Table I were conducted with the electrode at zero or negative bias; under these conditions optical energy is converted to electricity, if the only electrochemistry corresponds to offsetting oxidation and reduction of polychalcogenide species (Figure 3).

A discussion of surface quality should be prefaced with a description of sample preparation. The "as received" polycrystalline samples of CdS:Te and CdS:Ag used in this study were etched in concentrated HCl prior to being used as electrodes. This generally has the effect of increasing emission intensity while leaving the spectral distribution intact.

Depending upon the PEC conditions and the sample employed, we see variable degrees of surface damage. At sufficiently high light intensities (≥ 50 mW/cm^2) and positive voltages (≥ -0.3 V vs. SCE) in polysulfide electrolytes, we occasionally encounter darkening of CdS:Ag and

Table I. Stability of *n*-Type CdS:Te and CdS:Ag

Electrode[b]		Electrolyte[a]	Electrode (mol $\times$ 10^4)[c]	
			Before	After
CdS:Te	1000 ppm	S_n^{2-}	0.847	0.820
CdS:Ag	10 ppm	S_n^{2-}	1.87	1.70
CdS:Te	100 ppm	S_n^{2-}	136.7	136.7
CdS:Ag	10 ppm	Te_n^{2-}	1.27	1.26
CdS:Te	100 ppm	Te_n^{2-}	9.36	8.75[e]

[a] Photoelectrochemical cell with the indicated electrode as photoanode. For the experiments in S_n^{2-} electrolyte, a power supply was used as the load (cf. Figure 3). For CdS:Ag in Te_n^{2-} electrolyte, a third, reference electrode was also used in conjunction with a potentiostat. Pt foil served as the counterelectrode in all cases.
[b] Electrodes were etched in concentrated HCl prior to use. They are melt-grown, polycrystalline samples from Eagle–Picher Industries. The CdS:Te and CdS:Ag specimens have resistivities of ~1 and 2000 Ω-cm, respectively, and were used in irregular shapes.
[c] Electrolyte is 1M OH$^-$/1M S^{2-}/0.2M S (S$_n^{2-}$) or 5M KOH/0.05M Te^{2-}/0.01M Te$_2^{2-}$ (Te$_n^{2-}$). The polysulfide electrolyte was purged with N$_2$ and the electrode was held at ~ −0.05 V vs. the counterelectrode; the negative lead of the power supply was connected to the photoelectrode. Several of the experiments in S$_n^{2-}$ were peri-

CdS:Te surfaces. Even avoiding these conditions, though, there is likely some surface reorganization and we sometimes see discolored electrode surfaces. Recent studies of CdSe electrodes in polysulfide electrolyte indicate that substitution of surface Se sites by S occurs (19, 60). A similar exchange involving Te is plausible as is another surface reorganization mechanism based on the propensity of CdTe to undergo photoanodic dissolution in polysulfide media via Equation 2 (6, 9). It is

$$CdTe \xrightarrow{h\nu} Cd^{+2} + Te + 2e^- \qquad (2)$$

possible that the HCl etch leaves a Te-rich surface, which could then undergo exchange and/or photoanodic decomposition. Consistent with the latter mechanism is the relative lack of surface damage we observe in ditelluride media. Both CdS and CdTe are known to be stabilized by Te^{2-} and Te^{2-}/Te_2^{2-} electrolytes (6). We find, however, that CdS:Ag and CdS:Te exhibit similar properties with regard to surface stability so that the role of Te is not resolved.

The variation in surface stability is repeated in the temporal characteristics of photocurrent and luminescence. In polysulfide electrolyte nominally identical samples of CdS:Te or CdS:Ag have displayed both stable and unstable photocurrents and emissive properties. Declining photocurrents and emission intensity are usually accompanied by the aforementioned surface deterioration. Certainly one of the disadvantages of the polycrystalline material used in this study is its nonuniformity. In particular, grain boundaries are notorious sources of discontinuous

Photoelectrodes in Aqueous Polychalcogenide Electrolytes[a]

Electrons $(mol \times 10^4)$[g]	Average i^h (mA)	Time (hr)	Source[d]
1.73	0.069	67.2	Hg
5.14	0.117	117.8	Hg
5.17	0.066	210.0	Hg
1.31	0.086	40.8	Xe
3.96	0.259	41.0	Xe

odically interrupted to renew the electrolyte, since even with N_2 purging there is slow decomposition from impurities. For the CdS:Ag experiment in Te_n^{2-}, the photoelectrode was held at −1.04 V vs. SCE, the value measured for E_{redox}; the CdS:Te experiment in Te_n^{2-} was run at 0.00 V vs. the counterelectrode.

[d] Hg is a UV-filtered 200-W, super high pressure Hg arc lamp; Xe is an unfiltered 150-W Xe lamp.

[e] Crystal chipped badly upon demounting and not all of it could be recovered.

[f] Moles of crystal determined by weight before and after the experiment.

[g] Moles of electrons passed during the experiment as determined by integrating photocurrent vs. time plots.

[h] Average current during experiment; current densities in mA/cm^2 are roughly a factor of 1–15 larger.

behavior (*61*). Since we do not know the surface composition of the etched CdS:Te and CdS:Ag electrodes, we are reluctant to make a definitive statement regarding their stability in polysulfide media. It can certainly be argued that the intraband gap states might influence the k_x–k_d competition. We suspect that the samples that yield the most stable properties may have surfaces much like undoped CdS. Doped materials with these surfaces could still emit, since the bulk of the excitation beam is absorbed beneath the surface. The studies reported here for polysulfide electrolytes were conducted with samples exhibiting stable, reproducible, photocurrent and emissive properties.

There is less ambiguity in ditelluride electrolyte. We generally see stable photocurrents, emission, and surface properties, although at higher intensities there are often slow monotonic declines in photocurrent and luminescence. There is obvious evidence for the competitive oxidation of colorless Te^{2-} to purple Te_2^{2-} or of either Te^{2-} or Te_2^{2-} to Te. At high light intensities the emission is masked by a layer of Te and/or Te_2^{2-} that can be swept away by greater stirring rates. The visual evidence for oxidation of S_n^{2-} is obscured because the emitting electrode and electrolyte are of similar color. However, we have observed yellowing of the initially colorless S^{2-} solutions as S_n^{2-} forms under PEC conditions. Taken as a unit, the data argue strongly for stability of CdS:Te and CdS:Ag in ditelluride electrolytes; for polysulfide electrolytes sustained photocurrents and minimal weight loss with these electrodes may be obtained, but there is strong evidence that surface reorganization processes are involved. Experiments designed to clarify the complications observed in polysulfide media are in progress.

Optical Properties. As described in the introduction, the semiconductor excited state is reached by absorption of ultraband gap photons. Undoped CdS has a band gap of approximately 2.4 eV and hence a fundamental absorption edge at about 520 nm (*62*). The absorption onset is very sharp because CdS is a direct band gap material. Absorption spectra of single crystal CdS:Te samples have their onset red-shifted by an absorption shoulder; the effect of increasing Te concentration is to extend this shoulder deeper into the red (*46, 47, 48, 62, 69*). Spectra that we have obtained for 100-and 1000-ppm CdS:Te are presented in Figure 4. Although these spectra were obtained from polycrystalline samples, they are at least qualitatively in agreement with the reported single crystal data in shape and color; there is an obvious visual difference, since the yellow undoped CdS becomes orange at 5–100-ppm CdS:Te and red at 1000-ppm CdS:Te. Similarly, 10-ppm CdS:Ag is red and 100-ppm is brown-black. In effect the shoulder masks the band edge and the band gap energy (E_{BG}).

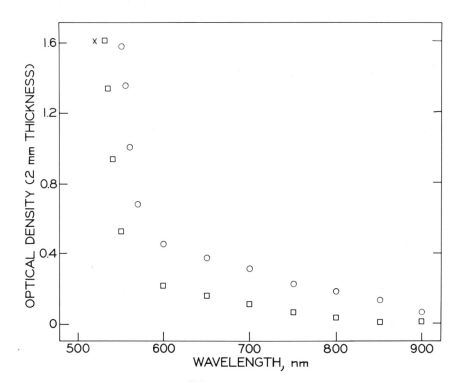

Figure 4. Optical density of polycrystalline 100-ppm CdS:Te (□) and 1000-ppm CdS:Te (○). Thicknesses are 2.0 and 2.2 mm, respectively, and samples have been polished with 1-μ alumina. The x is a literature value optical density of a 2-mm thick, undoped, polished CdS single crystal (5).

What primarily distinguishes CdS:Te and CdS:Ag from undoped CdS is their ability to emit while they serve as electrodes. In Figure 5 we present emission spectra taken at 293 K for 5-, 100-, and 1000-ppm polycrystalline CdS:Te and for 10-ppm polycrystalline CdS:Ag. A systematic study of single crystal CdS:Te emission spectra has shown that peak position and half width may be correlated with doping levels in these samples (48). Our results, though uncorrected for detector response, are qualitatively in agreement: the 5- and 100-ppm emission spectra are very similar, with a peak maximum at around 600 nm, but there is a definite red shift of the emission maximum to around 650 nm for 1000-ppm CdS:Te. The emission itself varies from yellow-orange to red-orange in passing from the 5-, 100-ppm to the 1000-ppm CdS:Te. For CdS:Ag we observe reddish emission and the band maximum appears near 690 nm.

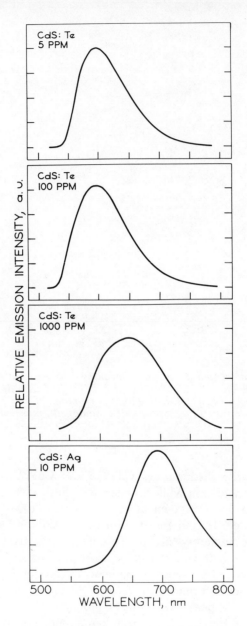

Figure 5. Typical 295 K emission spectra of 5-, 100-, 1000-ppm CdS:Te and 10-ppm CdS:Ag. The CdS:Te samples were excited at 488.0 nm and the CdS:Ag sample at 514.5 nm. Spectra are uncorrected.

We were particularly interested in determining if the PEC environment perturbed the emission spectra of the doped CdS samples. Typical results are given in Figure 6, which shows a sequence of emission spectra for 5-ppm CdS:Te taken without electrolyte, with polysulfide electrolyte (1M OH$^-$/1M S^{2-}/1M S) but out-of-circuit, and in-circuit in the same electrolyte at -0.74 V vs. SCE. Except for the change in intensity, the emission spectra are essentially identical. We interpret this to mean that the intraband gap states responsible for luminescence are influenced in the same manner as the conduction and valence bands and thus would undergo parallel band bending.

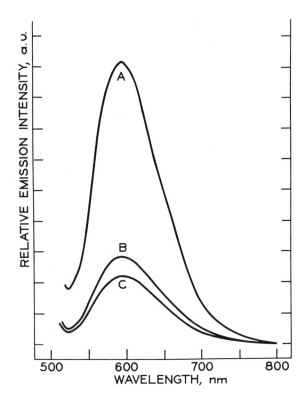

Figure 6. Uncorrected emission spectra of 5-ppm CdS:Te in various environments but in a fixed geometry relative to the 488.0-nm laser excitation source and emission detection optics. For Curve A no electrolyte was present; Curves B and C were both taken with the electrode immersed in 1M OH$^-$/1M S^{2-}/1M S polysulfide electrolyte but out-of-circuit and in-circuit at -0.74 V vs. SCE, respectively. The sharp intensity drop from A to B and C is the result of electrolyte absorption; baseline is not preserved at the high energy end of the emission spectrum due to overlap with the tail of the excitation line.

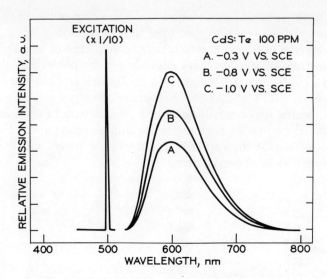

Figure 7. Uncorrected emission spectra of a 100-ppm CdS:Te electrode at several potentials in 1M OH^-/1M S^{2-}/0.2M S electrolyte. Curves A, B, and C correspond to identical experimental conditions except for electrode potentials of -0.3, -0.8, and -1.0 V vs. SCE, respectively. The excitation wavelength is 496.5 nm.

Additional evidence supporting this interpretation is provided by Figure 7. Here the emission of a 100-ppm CdS:Te photoelectrode is recorded at -0.3, -0.8, and -1.0 V vs. SCE in polysulfide electrolyte. Again, although the emission intensity changes, the spectrum does not. Application of negative bias to an n-type semiconductor diminishes the amount of band bending; conversely, positive bias increases band bending (1). Invariance of the emission spectrum under conditions where band bending has been altered is thus consistent with parallel behavior between dopant states and the valence and conduction bands. The spectral distribution of emission appears to be independent of excitation wavelength for the laser lines we normally employ: 488.0, 496.5, 501.7, and 514.5 nm.

We have used these wavelengths as the basis of photoaction and excitation spectra (Table II). There is generally a marked decline in photocurrent in passing from what are certainly ultraband gap energies (488.0, 496.5, 501.7 nm) to 514.5 nm. The data in Table II were obtained in optically transparent sulfide electrolyte at -0.4 V vs. SCE at roughly equivalent intensities for each of the four wavelengths. Photocurrents from 501.7-nm excitation are about 4.5 to 10 times those resulting from 514.5-nm excitation (*see* Table II). For undoped single crystal CdS the corresponding experiment was reported to yield a similar photocurrent ratio of 7–13 for excitation at these same two wavelengths (5). Therefore, although the absorption spectra of doped and undoped CdS differ, the fundamental absorption edge may have the same wavelength dependence; that is, the band gaps might be identical. Complete photoaction and excitation spectra must be obtained to clarify this point, however, and such studies are in progress.

By measuring the emission intensity and photocurrent simultaneously, their relationship as a function of wavelength can be examined. Alongside the photocurrent data in Table II are the corresponding emission intensities measured both in-and out-of-circuit. Two general trends are discernible. One is that the emission intensity increases with wavelength and the other is that the in-circuit values are substantially lower than out-of-circuit values, except at 514.5 nm where the difference is small.

Table II. Wavelength Dependence of Emission and Photocurrent[a]

Electrode[b]	λ (nm)	Relative Emission Intensity (arbitrary units)[c]		Photocurrent (μA)[d]
		Out of Circuit	In Circuit	
1. CdS:Te 5 ppm	514.5	235	232	12
	501.7	187	147	85
	496.5	147	106	84
	488.0	116	88	72
2. CdS:Te 1000 ppm	514.5	300	295	5
	501.7	310	281	22
	496.5	236	204	28
	488.0	187	158	34
3. CdS:Ag 10 ppm	514.5	1623	1612	38
	501.7	355	305	395
	496.5	227	170	347
	488.0	158	115	287

[a] The experiments were conducted in optically transparent $1M$ $OH^-/1M$ S^{2-} electrolyte with the electrodes at -0.4 V vs. SCE. At this potential the photocurrent is saturated with respect to potential. Electrodes were excited by an ~ 3-mm diameter Ar ion laser beam of ~ 3 mW.

[b] Electrodes are irregularly shaped. The surface area exposed to the electrolyte is 0.15, 0.075, and 0.18 cm^2 for Electrodes 1, 2, and 3, respectively. The fraction of the electrode illuminated can be estimated by dividing these numbers into 0.071 cm^2, the laser beam area.

[c] Relative emission intensity was measured with a flat wavelength response radiometer that was filtered to eliminate the exciting wavelengths and positioned to sample front surface electrode emission. A switch on the PAR potentiostat permitted the PEC to be brought in and out of circuit without disturbing the geometry. For a given electrode experimental conditions are identical except for the exciting wavelength. The emission values have been corrected to the same number of einsteins/sec at each wavelength; however, values between electrodes cannot be compared because different geometries and light intensities were employed.

[d] Photocurrents have been corrected for each electrode to equivalent einsteins/sec. For a given electrode experimental conditions are identical except for the exciting wavelength. Values between electrodes are not comparable (cf. Footnote c).

In the region of what appears to be the band gap, then, there is an inverse relationship between photocurrent and luminescence intensity. The penetration depth of the excitation beam is a significant factor in explaining all of the phenomena we observe. Literature values for undoped CdS and CdS:Te indicate that the 300 K absorptivity (α) at 514 nm is about 10^3 cm^{-1} vs. 10^4–10^5 cm^{-1} at 502 nm and shorter wavelengths (46, 47, 62, 69). Electroabsorption measurements on undoped CdS suggest that there should be little variation of α with electrode potential for most of the excitation wavelengths used (63). The excited state deactivation rate constants pictured in Figure 2 should be very dependent on position, since band bending decreases with distance from the semiconductor–electrolyte interface. We would a priori predict that electron–hole pairs produced outside the depletion region are more likely to radiatively recombine than those produced within the depletion region, which can more readily separate to produce photocurrent. Since a greater fraction of 514.5-nm than ultraband gap light will be absorbed outside the depletion region, the low energy photons should be more effective at yielding emission and less effective at producing photocurrent than ultraband gap photons.

The in- and out-of-circuit trends may also be interpreted in this manner. Taking the electrode out of circuit removes deactivation routes corresponding to electron–hole separation and thus increases the probability for deactivation via the remaining pathways. Since photocurrent is least important as a decay route for 514.5-nm excitation, the difference between in- and out-of-circuit emission should be smallest, as is observed. Although this logic intimates at a direct trade-off between emission and photocurrent, the data in Table II are evidence that this need not be so. The declines in emission intensity with wavelength may well be due to surface traps that promote nonradiative electron–hole pair recombination (64). Surface quality can be expected to play a more significant role as the penetration depth decreases.

One further point concerns the long wavelength extreme. We have actually observed emission wavelengths as long as 540 nm beyond which emission intensity becomes limited by the optical density of the sample. This is possible evidence that the absorption shoulder corresponds to the emitting excited state. If this were the case, then an alternate explanation for the decline of emission intensity with wavelength is simply that the absorption band leading to emission has peaked and fallen. A complete excitation spectrum could reveal the shape of this band, since most of it is buried beneath the direct band gap transition.

Current–Voltage–Emission Properties. Undoubtedly the most significant feature of the luminescent photoelectrodes is the opportunity they afford to examine the interplay of radiative and nonradiative excited state deactivation routes. As described in the introduction, the PEC can be assembled inside the sample chamber of an emission spectrophotofluorometer so that front surface emission may be monitored during PEC operation. A standard three-electrode geometry was used in conjunction with a PAR potentiostat that regulates the photoelectrode potential vs. an SCE. A 1.5 × 0.9-cm Pt foil served as the counterelectrode. The photoelectrodes were irregularly shaped but often small enough to be completely illuminated by the 3-mm diameter laser beam.

Electrodes larger than the excitation beam were also used. In this case excitation at a spot on the surface results in emission over the entire sample surface. There are at least three explanations for this. The first is that excitons may migrate and radiatively recombine throughout the sample. We expect exciton diffusion lengths to be small at room temperature, but to our knowledge they have not been measured for CdS:Te and CdS:Ag. We cannot, therefore, rule this out. A second possibility involves reabsorption and reemission of emitted light enough times to transport it throughout the sample. Since most of the emitted light is not appreciably absorbed by the sample, we think that this, too, is unlikely but still possible. The most plausible explanation, we feel, is simply that the emitted light is scattered to the extent that it emerges throughout the sample. Some role may be played by grain boundaries in this process. In several samples where obvious grain boundaries exist, we find luminescence only within the irradiated grain; emission ceases abruptly at the boundary. In this sense the boundary may be acting as a site of nonradiative recombination if migration processes are involved or as a reflecting or absorbing surface if scattering is the dominant mechanism. In either case, grain boundaries are not a requirement for emission; we have recently obtained single crystals of 100-ppm CdS:Te (vide infra) that exhibit the same emission spectrum and this same phenomenon of global emission from local excitation.

The crucial finding with regard to PECs is that electrode potential influences both photocurrent and luminescence efficiency. A typical set of results is presented in Figure 8 for a 100-ppm CdS:Te photoelectrode and $1M$ $OH^-/1M$ $S^{2-}/0.2M$ S electrolyte. With 496.5-nm ultraband gap excitation the photocurrent declines with increasingly negative potential, eventually reaching zero around -1.1 V vs. SCE. Over the same potential range the emission intensity more than doubles. We have chosen the expedient of monitoring emission intensity at the band maximum of 600 nm, since the spectrum is independent of potential (vide supra).

The emission data presented in Figure 8 are similar to the results presented in the preceding section. Out-of-circuit emission values such as those given in Table II should correspond to the emission intensity at zero photocurrent, the point where the current–voltage curve intercepts

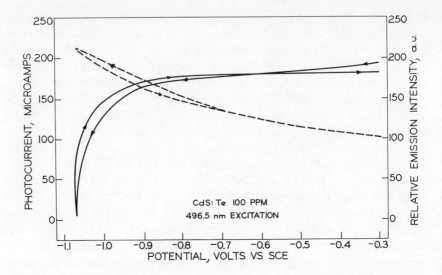

Figure 8. Dependence of photocurrent (———, left hand scale) and relative emission intensity (– – –, right hand scale; monitored at 600 nm) on electrode potential for a 100-ppm CdS:Te-based PEC using 1M OH$^-$/1M S^{2-}/0.2M S electrolyte. Both measurements were made simultaneously at a sweep rate of 13 mV/sec starting at −0.3 V vs. SCE. Electrolyte redox potential is −0.70 V vs. SCE. The 3-mm diameter laser beam only filled part of the irregularly shaped 0.54-cm^2 electrode surface. Because of electrolyte absorption, only an upper limit of approximately 4 mW for the incident 496.5-nm excitation can be given.

the voltage axis. Recall that the out-of-circuit and in-circuit values were most similar for 514.5-nm excitation and quite disparate for ultraband gap irradiation. Complete current–voltage–emission data should reflect this relationship.

The direct comparison is shown in Figure 9, which is based on a 10-ppm CdS:Ag photoelectrode in 1M OH$^-$/1M S^{2-}/0.2M S electrolyte. Comparable intensities of 496.5- and 514.5-nm excitation result in dramatically different photocurrent–luminescence properties. With respect to photocurrent, ultraband gap 496.5-nm excitation yields substantially larger photocurrents than band gap edge 514.5-nm light. The maximum output voltage for a given intensity (E_v), defined as the difference between the voltage at which zero photocurrent obtains and the value of E_{redox}, is about 500 mV for 496.5-nm light and 320 mV for 514.5-nm excitation (from Figure 9). The wavelength dependence of photocurrent and output voltage has been reported for several n-type semiconductor photoelectrodes; larger values for both are obtained with ultraband gap photons than with band gap edge photons because of the different fractions of light absorbed within the depletion region (5, 6, 10).

As the photocurrent in Figure 9 declines to zero in passing to increasingly negative potentials, the emission intensity increases by up to 50% with 496.5-nm excitation, and drops slightly (~6%) with 514.5-nm excitation. The emission intensities at the two wavelengths of Figure 9 are about the same because the 496.5-nm excitation is somewhat

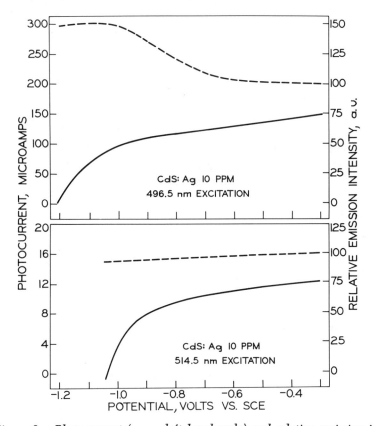

Figure 9. Photocurrent (———, left hand scale) and relative emission intensity (– – –, right hand scale) as a function of electrode potential for a 10-ppm CdS:Ag-based PEC in 1M OH$^-$/1M S^{2-}/0.2M S electrolyte. Experimental conditions in the top and bottom figures are identical except that ~15 mW of 496.5-nm light was used in the former and ~11 mW of 514.5-nm light in the latter. Intensities are upper limits since they are not corrected for electrolyte absorption. Emission intensity in both cases was monitored at 670 nm; the points labelled "100" on the relative emission intensity scales are about the same absolute intensity. Photocurrent and emission intensity measurements were made simultaneously at a sweep rate of 13 mV/sec starting at −0.3 V vs. SCE. The electrolyte redox potential is −0.70 V vs. SCE. Electrode surface area exposed to the electrolyte is ~0.21 cm^2 and only partially filled by the 3-mm diameter laser beam.

more intense. For identical excitation intensities we normally see less luminescence intensity at 496.5 nm than at 514.5 nm (Table II). The dramatic difference in luminescence intensity with electrode potential at these two wavelengths has been observed by us with all of the electrode materials used and in both sulfide and polysulfide electrolytes. Generally, the percentage increase in luminescence in passing to the negative extreme of the current–voltage curve is 15–100% with ultraband gap excitation wavelengths of 488.0, 496.5, and 501.7 nm. Excitation with 514.5 nm yields less than 6% changes and these are often declines. This trend persists even at 514.5-nm light intensities that are sufficiently high to produce photocurrents comparable to those achieved with lower intensity ultraband gap excitation. Again, we ascribe this difference to the smaller fraction of incident light absorbed within the depletion region. Essentially, a greater fraction of light is then absorbed in a region of less band bending, a region where the excited state deactivation rate constants should be more insensitive to changes in electrode potential.

To further probe the generality of this phenomenon, a similar experiment was conducted in ditelluride electrolyte (Figure 10). The electrode is 5-ppm CdS:Te. At comparable incident intensities of 496.5- and 514.5-nm light, approximately 20% and 3% increases in emission intensity occur over the range of $-0.7-- 1.2$ V vs. SCE, respectively. It would be difficult to predict a priori how the luminescence in ditelluride would differ from that observed in polysulfide. If all of the rate constants of Figure 2 were identical for the two electrolytes, then there should be an exact correspondence between the current–voltage–emission curves. This assertion assumes that CdS:Te and CdS:Ag behave as undoped CdS, for which the conduction and valence band energies are relatively independent of polychalcogenide electrolyte (6). Of course, the approximately 0.4 V difference in E_{redox} energies means that output voltages in Te^{2-}/Te_2^{2-} electrolyte will be considerably reduced relative to polysulfide media. Our data thus far indicate similar features for the two electrolytes, but considerably more data need to be collected before comparisons can be comfortably made.

Two features of the range of conditions used are particularly noteworthy. First, the kinds of percentage increases in emission that we observe have been obtained at various sweep rates and with point-by-point equilibration. The difference in emission at maximum and near-zero photocurrents is easily observed visually by pulsing the electrode between the appropriate potentials. We also have confidence that the effect is real because it is reproducible either in point-by-point or in the reverse sweeping of potential. In some cases we have swept through the potential range many times in succession and still see the same variation in emission and photocurrent.

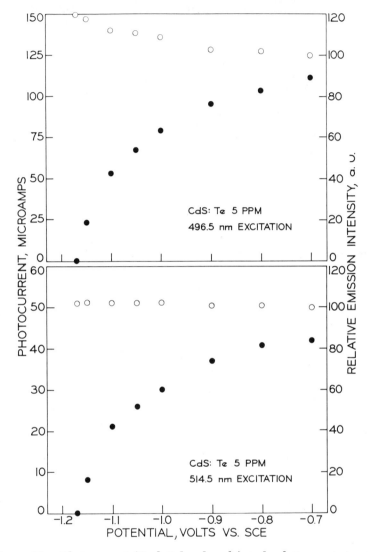

Figure 10. Photocurrent (●, left hand scale) and relative emission intensity (○, right hand scale) as a function of electrode potential for a 5-ppm CdS:Te electrode in 5M KOH/0.05M Te^{2-}/0.01M Te_2^{2-} electrolyte. Emission intensity was monitored with a flat response radiometer suitably filtered to eliminate the excitation lines of 496.5-nm (top figure) and 514.5-nm (bottom figure). The emission intensity point labelled "100" on the top figure is approximately ½ the value of "100" on the bottom figure. Excitation intensities are comparable at ~10 mW, uncorrected for electrolyte absorption. The electrolyte redox potential is −1.04 V vs. SCE. Electrode surface area exposed to the elctrolyte is ~0.12 cm^2 and about half filled by the 2-mm diameter laser beam.

The other feature of interest is that the different doping levels of CdS:Te and 10-ppm CdS:Ag all show similar effects. We believe that the mechanism by which luminescence occurs is likely different for the two dopants, and it is therefore significant that they exhibit the same kinds of potential and wavelength dependence of emission intensity. This observation leads us to believe that luminescence will be a very general probe of recombination processes in PECs.

Efficiency. A complete characterization of the semiconductor excited state requires a detailed energy balance. Although the input optical energy is easily measured with a radiometer, each of the dissipation pathways shown in Figure 2 involves a different technique for its determination. At this point we are able to accurately determine the efficiency of electrochemical redox processes and to estimate the efficiency of luminescence, processes governed by k_x and k_r, respectively. In principle, the efficiency of nonradiative recombination could be estimated by difference, assuming that k_1, k_x, and k_r represent the only significant deactivation routes.

For the degenerate electrolysis pictured in Figure 3, the current–voltage curve permits calculation of the efficiency at which optical energy is converted directly to electricity. Since the Pt counterelectrode is at E_{redox}, passage of photoanodic current at potentials equal to or negative of this value represents optical energy conversion (4). The maximum product of photocurrent and output voltage is the maximum power output, and division by input optical power yields the efficiency (η).

In Table III we have compiled typical efficiencies for CdS:Te- and CdS:Ag-based PECs in polysulfide media. In general they do not exceed 1% and are therefore significantly lower than the approximately 5% efficiency observed for 500-nm monochromatic light with single crystal undoped CdS electrodes (5). We attribute this to both the polycrystallinity of the materials and the surface quality. On occasion we have seen efficiencies that are comparable with the undoped single crystal values; they have been more the exception than the rule, however.

The efficiency is given by Equation 3 where Φ_x is the quantum effi-

$$\eta = \frac{\Phi_x E_V}{E_{BG}} \quad (3)$$

ciency for electron flow in the external circuit. The maximum value of Φ_x is 1.0; for E_V it is the band gap value, E_{BG}. Table III reveals that both Φ_x and E_V are at best one tenth of the maximum values and are thus jointly responsible for the poor efficiencies. Much lower efficiency in Te^{2-}/Te_2^{2-} relative to S^{2-}/S_n^{2-} is expected on the basis of less band bending, a result of the more negative value of E_{redox} (−1.1 vs. −0.7 V vs. SCE, respectively).

Table III. Energy Conversion Characteristics[a]

Property	CdS:Te-, CdS:Ag-based PEC[b]	Undoped CdS-based PEC[b]
η_{max}[c]	0.1–1%	3–7 %
E_v at η_{max}[d]	0.2–0.3 V	0.3–0.4 V
Φ_x at η_{max}[e]	0.01–0.1	0.3–0.5
Φ_x (max)[f]	0.1–0.3	0.8–1.0
Lum. η[g]	0.01–1%	—
Φ_r (max)[h]	0.01–0.1	—

[a] Measures of efficiency for the conversion of $\leq$10 mW/cm², ~ 500-nm monochromatic input optical energy to electricity and/or luminescence in aqueous polysulfide media. Listed values are meant to be representative.

[b] The indicated electrodes serve as the photoanodes of a PEC like that shown in Figure 3. Data were also obtained with a third, reference electrode and a potentiostat. Undoped CdS data were taken from Refs. 5 and 6. All measurements were made with samples etched in concentrated HCl.

[c] Maximum efficiency for the conversion of optical energy to electricity. Obtained from current–voltage curves like those shown in Figures 8–10 by maximizing the product of output voltage (cf. text and Footnote d) and photocurrent, then dividing by input optical power.

[d] Output voltage at maximum efficiency. The output voltage is the absolute value of the difference between the electrode potential on the current–voltage curve and E_{redox}. For the two-electrode cell shown in Figure 3, E_v is the electrode potential, since 0.0 V is E_{redox} (4).

[e] Φ_x is the quantum yield for electron flow in the external circuit, measured here at the potential corresponding to maximum efficiency.

[f] The maximum value of Φ_x, usually measured at up to 0.5 V positive of E_{redox}, where the photocurrent is saturated with respect to potential.

[g] Efficiency for the conversion of optical energy to luminescence, defined here as (energy emitted)/(energy absorbed). A flat wavelength response radiometer is used to estimate the energy emitted (cf. text).

[h] Estimated maximum emission quantum efficiency. Low temperature emission spectra of unmounted CdS:Te and CdS:Ag samples were used for this estimate (see Figure 11 and text).

Determining the luminescence efficiency is difficult because of its spatially diffuse nature. We have used two techniques for estimating its magnitude. The first method provides an upper limit of emission efficiency by finding physical conditions that produce more intense emission from the same excitation intensity. In agreement with the literature (46, 47, 48), we have found that simply lowering the temperature to 77 K generally produces such changes. Figure 11 depicts the emission spectral changes resulting from decreasing the temperature. Although the peak position of the 50-ppm CdS:Te sample does not change very much, the peak sharpens and increases in intensity dramatically. Integration of the peak areas indicates an approximately 40-fold increase in intensity upon cooling and thus sets an upper limit of 0.025 for emission quantum efficiency at room temperature. Data in Table III show that for the samples examined, maximum values of Φ_r determined in this manner are 0.01–0.1. This should be regarded as a crude estimate in that no effort was made to compensate for the differences in optical penetration depth (α) at the two temperatures (62, 69).

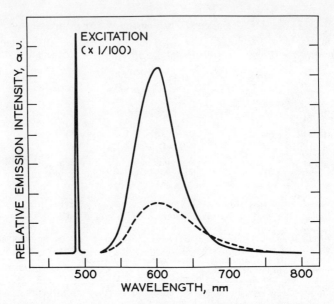

Figure 11. Uncorrected emission spectra of 50-ppm CdS:Te at 77 K (———) and 295K (– – –; 10× scale expansion). The sample was excited with identical intensities of 488.0-nm light at the two temperatures.

We have effectively defined the emission quantum yield as (photons emitted)/(photons absorbed). However, in the context of a PEC an equally satisfying measure would be (energy emitted)/(energy absorbed). The direct interconversion of definitions necessitates integration over the spectral distribution of emitted light. We have sidestepped this problem by using a radiometer with flat wavelength response to sample the emitted light from the back side of a flat electrode. Correcting for the fraction of emitted light sampled and assuming isotropic luminescence, we then typically find values of 0.1–1% efficiency for the conversion of input optical energy to emission in electrolytes of polysulfide. This range is certainly consistent with the upper limits set forth in Table III and reflects dependence on sample, excitation wavelength, and potential. Though it is not the dominant pathway, radiative deactivation can be a significant route for energy dissipation for the CdS:Te and CdS:Ag excited states.

The dominant deactivation mechanism is nonradiative electron–hole recombination based on the low efficiencies of emission and photocurrent. As was mentioned before, it is this feature which masks the extent to which energy trade-offs exist between luminescence and photocurrent. Figures 8–10 confirm the data of Table II, which show that an inverse potential dependence between emission and photocurrent does not always exist. At high excitation intensities we have seen occasionally the emission intensity plateau at negative potentials (*see* Figure 9) or peak and then decline at still more negative potentials. It is intriguing, however, that there is a large potential and hence band bending range over which both photocurrent and emission intensity are relatively constant and that both deviate from this plateau at about the same potential. A complete interpretation of these observations awaits the evaluation of all the rate constants of Figure 2.

Summary and Perspective

We believe that the luminescent properties of semiconductor photoelectrodes represent a powerful technique for probing recombination processes in operating PECs. The key finding is that both emission intensity and photocurrent of n-type CdS:Te and CdS:Ag photoelectrodes are influenced by experimental PEC parameters in a manner consistent with band bending arguments commonly used to describe photoelectrochemical phenomena. Increases in emission intensity of about 15–100% with ultraband gap ($\lambda \lesssim 500$ nm) irradiation are observed with increasingly negative potentials; simultaneously, over the same potential range the photocurrents in aqueous S^{2-}/S_n^{2-} or Te^{2-}/Te_2^{2-} electrolytes decline to zero and underscore the significance of the depletion region in determining the relative rates of electron–hole separation and recombination. Simply put, we find that conditions minimizing band bending in CdS:Te and CdS:Ag photoelectrodes lead to increased emission intensity and reduced photocurrents. Ultraband gap photons are more effective at producing photocurrent and less effective at yielding emission than band gap edge 514.5-nm light because a greater fraction is absorbed within the depletion region. It is gratifying to us that two different materials, CdS:Te and CdS:Ag, with different emission spectra and presumably different emissive mechanisms, both give emissive properties corresponding to this simple model.

The polycrystallinity of the melt-grown CdS:Te and CdS:Ag samples and lack of knowledge of their surface composition precludes definitive conclusions regarding their PEC properties relative to those of undoped CdS. However, we have recently made several observations that cast some light in this direction. First, we have found that Br_2/MeOH etching leads to vastly superior PEC properties relative to HCl etching (65, 66). With respect to Table III, the Br_2/MeOH etchant leads to 3–5% for η_{max} with approximately 500-nm input light. Output voltages of 0.45 V and quantum efficiencies for electron flow of 0.25 (at η_{max}) and 0.35 maximum are typical. These values are almost as good as those previously reported for undoped single crystal CdS samples (5, 6). Additionally, the Br_2/MeOH treatment leads to greater photocurrent stability and far less surface degradation under comparable PEC conditions (*see* Stability section).

These observations lead us to believe that the surface produced by HCl etching is the principal source of poor PEC properties. Chemically, Te and Ag are relatively inert to HCl compared with Br_2/MeOH and the former etchant may not produce as uniform a surface. The differences in etchants seem to be real and reproducible—a sample etched with HCl and exhibiting poor current–voltage properties can be etched with Br_2/MeOH to yield good properties and subsequently re-etched with HCl to regenerate the inferior current–voltage curves.

A second observation we have made concerns experiments with single crystal 100-ppm CdS:Te. As described in the section on Optical Properties, the emission spectrum of the single crystal material matches that of the polycrystalline 100-ppm samples and, in fact, the luminescence efficiencies are very similar. The PEC output parameters of the single crystals (Br_2/MeOH etch) are, however, markedly better. For example, we have observed efficiencies of 7.5% (η_{max}), output voltages at η_{max} of 0.60 V, and quantum efficiencies (Φ_x) at η_{max} of 0.45. These values indicate that grain boundaries may exert some influence on output parameters. A more dramatic illustration is our observation that the 15–100% increases in emission intensity with potential seen with polycrystalline 100-ppm CdS:Te become up to 500% increases with single crystal samples. Again, this effect is observed with ultraband gap wavelengths; at 514.5 nm we see the same independence of emission intensity on potential observed with polycrystalline samples.

We would predict that the effect of grain boundaries on PEC properties would be most pronounced if they occur within the depletion or optical penetration regions of the semiconductor. The grain size of 3–8 mm in our polycrystalline samples is a visual surface measurement and though this dimension is far larger than the 10^{-4}–10^{-2}-mm depths relevant to band bending and optical absorption, we have no way of knowing whether grain boundaries exist beneath the surface at these depths. Even beyond these depths, the boundaries may serve as nonradiative recombination sites and quench emission and/or photocurrent. Further studies on the role of grain boundaries are in progress but at this point our observations serve to emphasize the extent to which current–voltage–luminescence properties are independent of sample preparation.

Two previous efforts were made to observe the luminescent phenomena described in this chapter, and they illustrate the potential generality of the technique. Memming and Beckmann studied photocathodic evolution of H_2 from p-GaP electrodes in acid medium (44). They sought evidence for the quenching of photoluminescence by the passage of photocurrent but found differences in emission intensity of only a few percent over the potential range examined. A more sensitive differential luminescence technique confirmed their observation. n-Type ZnO and ZnO:Cu exhibited potential-dependent emission intensity, which under some conditions bore a nearly mirror-image relationship to the photocurrent–voltage curve; a derivation presented to account for this may be applicable to other systems at high values of Φ_x (68). The electrochemistry in the ZnO study is photocorrosion leading to Zn^{+2} and O_2 (68). These examples illustrate the range of materials and electrode processes with which luminescent PECs can be constructed. In fact, many of the materials now commonly used in PEC studies are either luminescent at room temperature or may be doped to induce emission, as has been done with CdS.

Care must be exercised in extending results for doped semiconductor photoelectrodes to undoped systems. We are encouraged by the insensitivity of the emission spectra of CdS:Te and CdS:Ag to a variety of experimental conditions including the excitation wavelength, the electrode potential, and the presence of S^{2-}/S_n^{2-} electrolyte. We feel that this is strong evidence that the dopant induced states and valence and conduction bands all undergo parallel band bending. It is possible that the dopant states can influence interfacial electron transfer processes, but more systems need to be examined to determine how important this effect will be. We feel, however, that at this point there are enough correlations between the doped and undoped CdS-based PECs to make the pursuit of these systems worthwhile.

Acknowledgment

It is a pleasure to thank the Office of Naval Research, the Research Corporation, and the University of Wisconsin–Madison Graduate School Research Committee for financial support of this work. We also wish to acknowledge many helpful discussions with John Wiley of the UW–Madison Electrical Engineering Department.

Literature Cited

1. Gerischer, H. *J. Electroanal. Chem.* **1975,** 58, 263.
2. Gerischer, H. "Physical Chemistry: An Advanced Treatise"; Eyring, H., Henderson, D., Jost, W., Eds.; Academic: New York, 1970; Vol. 9A, Chapter 5.
3. Williams, R. *J. Chem. Phys.* **1960,** 32, 1505.
4. Ellis, A. B.; Kaiser, S. W.; Wrighton, M. S. *J. Am. Chem. Soc.* **1976,** 98, 1635.
5. Ibid., p. 6855.
6. Ellis, A. B.; Kaiser, S. W.; Bolts, J. M.; Wrighton, M. S. *J. Am. Chem. Soc.* **1977,** 99, 2839.
7. Hodes, G.; Manassen, J.; Cahen, D. *Nature (London)* **1976,** 261, 403.
8. Miller, B.; Heller, A. *Nature (London)* **1976,** 262, 680.
9. Ellis, A. B.; Kaiser, S. W.; Wrighton, M. S. *J. Am. Chem. Soc.* **1976,** 98, 6418.

10. Ellis, A. B.; Bolts, J. M.; Kaiser, S. W.; Wrighton, M. S. *J. Am. Chem. Soc.* **1977**, *99*, 2848.
11. Heller, A.; Chang, K. C.; Miller, B. *J. Electrochem. Soc.* **1977**, *124*, 697.
12. Chang, K. C.; Heller, A.; Schwartz, B.; Menezes, S.; Miller, B. *Science* **1977**, *196*, 1097.
13. Ellis, A. B.; Bolts, J. M.; Wrighton, M. S. *J. Electrochem. Soc.* **1977**, *124*, 1603.
14. Legg, K. D.; Ellis, A. B.; Bolts, J. M.; Wrighton, M. S. *Proc. Nat. Acad. Sci. U.S.A.* **1977**, *74*, 4116.
15. Bolts, J. M.; Ellis, A. B.; Legg, K. D.; Wrighton, M. S. *J. Am. Chem. Soc.* **1977**, *99*, 4826.
16. Miller, B.; Heller, A.; Robbins, M.; Menezes, S.; Chang, K. C.; Thomson, J., Jr. *J. Electrochem. Soc.* **1977**, *124*, 1019.
17. Manassen, J.; Hodes, G.; Cahen, D. *J. Electrochem. Soc.* **1977**, *124*, 532.
18. Noufi, R. N.; Kohl, P. A.; Bard, A. J. *J. Electrochem. Soc.* **1978**, *125*, 375.
19. Heller, A.; Schwartz, G. P.; Vadimsky, R. G.; Menezes, S.; Miller, B. *J. Electrochem. Soc.* **1978**, *125*, 1156.
20. Hodes, G.; Manassen, J.; Cahen, D. *J. Appl. Electrochem.* **1977**, *7*, 181.
21. Tsuiki, M.; Ueno, Y.; Nakamura, T.; Minoura, H. *Chem. Lett.* **1978**, 289.
22. Minoura, H.; Nakamura, T.; Ueno, Y.; Tsuiki, M. *Chem. Lett.* **1977**, 913.
23. Minoura, H.; Tsuiki, M.; Oki, T. *Ber. Bunsenges. Phys. Chem.* **1977**, *81*, 588.
24. Owen, J. R. *Nature (London)* **1977**, *267*, 504.
25. Tributsch, H. *Ber. Bunsenges. Phys. Chem.* **1977**, *81*, 361.
26. Ibid. **1978**, *82*, 169.
27. Tributsch, H. *J. Electrochem. Soc.* **1978**, *125*, 1086.
28. Tributsch, H.; Bennett, J. C. *J. Electroanal. Chem.* **1977**, *81*, 97.
29. Memming, R. *Ber. Bunsenges. Phys. Chem.* **1977**, *81*, 732.
30. Fujishima, A.; Inoue, T.; Watanabe, T.; Honda, K. *Chem. Lett.* **1978**, 357.
31. Inoue, T.; Watanabe, T.; Fujishima, A.; Honda, K.; Kohayakawa, K. *J. Electrochem. Soc.* **1977**, *124*, 719.
32. Madou, M.; Cardon, F.; Gomes, W. P. *Ber. Bunsenges. Phys. Chem.* **1977**, *81*, 1186.
33. Bard, A. J.; Wrighton, M. S. *J. Electrochem. Soc.* **1977**, *124*, 1706.
34. Gerischer, H. *J. Electroanal. Chem.* **1977**, *82*, 133.
35. Fujishima, A.; Inoue, T.; Watanabe, T.; Honda, K. *Chem. Lett.* **1978**, 357.
36. Leverenz, H. W. "An Introduction to Luminescence of Solids"; Wiley: New York, 1950.
37. Aven, M.; Prener, J. S., Eds. "Physics and Chemistry of II-VI Compounds"; North Holland Publishing Co.: Amsterdam, 1967.
38. Gerischer, H. *J. Electrochem. Soc.* **1978**, *125*, 218C.
39. Pettinger, B.; Schöppel, H.-R.; Gerischer, H. *Ber. Bunsenges. Phys. Chem.* **1976**, *80*, 845.
40. Van Ruyven, L. J.; Williams, F. E. *Phys. Rev. Lett.* **1966**, *16*, 889.
41. Pettinger, B.; Schöppel, H.-R.; Yokoyama, T.; Gerischer, H. *Ber. Bunsenges. Phys. Chem.* **1974**, *78*, 1024.
42. Noufi, R. N.; Kohl, P. A.; Frank, S. N.; Bard, A. J. *J. Electrochem. Soc.* **1978**, *125*, 246.
43. Memming, R. *J. Electrochem. Soc.* **1969**, *116*, 786.
44. Beckmann, K. H.; Memming, R. *J. Electrochem. Soc.* **1969**, *116*, 368.
45. Aten, A. C.; Haanstra, J. H. *Phys. Lett.* **1964**, *11*, 97.
46. Aten, A. C.; Haanstra, J. H.; deVries, H. *Philips Res. Rep.* **1965**, *20*, 395.
47. Cuthbert, J. D.; Thomas, D. G. *J. Appl. Phys.* **1968**, *39*, 1573.
48. Roessler, D. M. *J. Appl. Phys.* **1970**, *41*, 4589.
49. Cruceanu, E.; Dimitrov, A. *Sov. Phys.—Solid State (Engl. Transl.)* **1969**, *11*, 1389.
50. Goede, O.; Nebauer, E. *Phys. Status Solidi A.* **1971**, *7*, K85.

51. Nebauer, E.; Lautenbach, J. *Phys. Status Solidi B.* **1971**, *48*, 657.
52. Cuthbert, J. D. *J. Appl. Phys.* **1971**, *42*, 739.
53. Colbow, K.; Yuen, K. *Can. J. Phys.* **1972**, *50*, 1518.
54. Brown, M. R.; Cox, A. F. J.; Shand, W. A.; Williams, J. M. *J. Lumin.* **1970**, *3*, 96.
55. Woodbury, H. H. *J. Appl. Phys.* **1965**, *36*, 2287.
56. Klick, C. C. *J. Opt. Soc. Am.* **1951**, *41*, 816.
57. Lambe, J.; Klick, C. C. *Phys. Rev.* **1955**, *98*, 909.
58. Vydyanath, H. R.; Kröger, F. A. *J. Phys. Chem. Solids* **1975**, *36*, 509.
59. Kröger, F. A.; Vink, J. H.; van den Boomgaard, J. *Z. Phys. Chem.* **1969**, *203*, 1.
60. Gerischer, H.; Gobrecht, J. *Ber. Bunsenges. Phys. Chem.* **1978**, *82*, 520.
61. Tanenbaum, M. In "Semiconductors"; Hannay, N. B., Ed.; Reinhold: New York, 1959; ACS Monograph Ser. No. *140*, Chapter 3.
62. Dutton, D. *Phys. Rev.* **1958**, *112*, 785.
63. Blossey, D. F.; Handler, P. In "Semiconductors and Semimetals"; Willardson, R. K., Beer, A. C., Eds.; Academic: New York, 1972; Vol. 9, Chapter 3.
64. Bube, R. H. In "Photoconductivity of Solids"; Wiley: New York, 1960; pp. 273–324.
65. Ellis, A. B.; Karas, B. R. *J. Am. Chem. Soc.* **1979**, *101*, 236.
66. Ellis, A. B.; Karas, B. R., submitted for publication.
67. Fujishima, A.; Brilmyer, G. H.; Bard, A. J. In "Semiconductor Liquid Junction Solar Cells"; Heller, A., Ed.; Proc. Vol. 77-3: Electrochemical Society, Inc.: Princeton, NJ, 1977; pp. 172–177.
68. Petermann, G.; Tributsch, H.; Bogomolni, R. *J. Phys. Chem.* **1972**, *57*, 1026.
69. Moulton, P. F., Ph.D. Dissertation, Massachusetts Institute of Technology, 1975.

RECEIVED October 2, 1978.

12

Photoelectrochemical Solar Cells

Chemistry of the Semiconductor–Liquid Junction

A. HELLER and B. MILLER

Bell Laboratories, Murray Hill, NJ 07974

The energy conversion efficiency and stability of semiconductor–liquid junction solar cells are critically dependent on the chemistry at the photoactive junction. Improvement of the GaAs | selenide–polyselenide | C cell to 12% solar efficiency and stabilization of the CdS | sulfide–polysulfide | C cells will be discussed in this context. For the GaAs cell, adsorption of ruthenium on the surface has considerably improved the fill factor. Combined with an etching procedure leading to a matte, nonreflective surface, this treatment is responsible for the output enhancement. In the CdSe cell, stability is materially improved by addition of selenium to the redox electrolyte to suppress deleterious ion exchange processes otherwise occurring under high light intensity photoanodic operation.

Junctions between semiconductors and liquids form spontaneously; semiconductor–liquid junction solar cells are simpler to make than other types of solar cells. The quid pro quo is exposure of the junction to chemical processes, such as corrosion, ion exchange, and adsorption of impurities, which may alter the junction and affect the life and output of the cells. However, deliberate modification of the surface to improve performance is possible. We summarize here studies of semiconductor–redox electrolyte combinations in which life has been extended substantially and output importantly improved by analysis of the junction chemistry.

The most efficient semiconductor–liquid junction solar cells made in our laboratory now convert solar to electrical power with a 12% efficiency (1,2), which brings them into the range of solid-state junction devices. This efficiency was reached by modifying both the surface chemistry and the surface topography of gallium arsenide (GaAs). By controlling the topography (3), it was possible to reduce reflection losses (1, 2, 8) at the semiconductor-liquid interface. These aspects of surface modification will be discussed in this chapter.

Another critical factor we wish to discuss in the context of our work on n-cadmium selenide–sulfide–polysulfide cells is the ion exchange (11–14) process at the semiconductor–electrolyte junction and the suppression of life-limiting reactions by changing electrolyte chemistry. We cite some pertinent aspects of the physical chemical and structural properties of junctions between n-type semiconductors and redox couple solutions to establish the background of these developments.

Efficiency and Stability to Dissolution

The theoretical limiting open-circuit photovoltage achievable in semiconductor–liquid junction solar cells under intense illumination is approximately the difference between the fermi level of the semiconductor and the potential of the redox couple in solution (4, 5), as long as this potential is within the semiconductor's band gap. If the redox potential becomes oxidizing with respect to the valence band, the semiconductor is irreversibly oxidized even in the dark. If the potential is too reducing, the redox potential will approach the fermi level of the n-type semiconductor, which is located within approximately 0.2 eV of the conduction band in our materials. The photovoltage will then be small and the cell will be inefficient.

Illumination of the n-type semiconductor causes holes to move to the interface (Figure 1). The holes oxidize the semiconductor unless such oxidation is thermodynamically prohibited or unless the holes selectively react with the adsorbed, reducing member of the redox couple. For semiconductors with band gaps near the optimum for one-step solar energy conversion (1.4 ± 0.4 eV) there are no thermodynamically stable systems with sufficiently oxidizing redox couples that allow photovoltages in excess of about half the band gap. As a result, the simultaneous achievement of solar-to-electrical conversion and stability to oxidation requires control of the kinetics to achieve a situation for which the rate of oxidation of the adsorbed member of a redox couple greatly exceeds the oxidation of the illuminated semiconductor surface.

An example of controlled kinetics is the case of the n-GaAs | K_2Se–K_2Se_2–KOH | C cell (6). GaAs photocorrodes in basic aqueous solutions under illumination by Reaction 1 (7) (where h^+ = hole). In the n-GaAs |

$$GaAs + 6h^+ + 8OH^- \rightarrow Ga(OH)_4^- + AsO_2^- + 2H_2O \qquad (1)$$

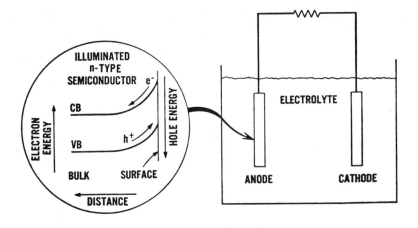

$$2h^+ + 2Se^{2-} \rightarrow Se_2^{2-} \quad \text{ANODE}$$
$$2e^- + Se_2^{2-} \rightarrow 2Se^{2-} \quad \text{CATHODE}$$

Figure 1. Schematic of a semiconductor–liquid junction solar cell. The cell, with the illuminated anode, is shown on the right. Electrode reactions for the n-GaAs|K_2Se–K_2Se_2–KOH|C cell are shown at the bottom. The diagram on the left shows the bending of the bands at the interface, which results in an electric field near the junction. Absorbed photons produce electron–hole pairs, which are separated in the field. The holes move to the liquid interface and oxidize the redox couple. The electrons move through the semiconductor, the back metal contact, and the external load to the carbon cathode, where they reduce the redox couple.

K_2Se–K_2Se_2–KOH | C cells (13) the selenide ion combines with the holes at a rate that greatly exceeds the photocorrosion reaction when the selenide concentration is sufficiently high, and the desired anodic reaction (Reaction 2) dominates. The Se^{2-}/Se_2^{2-} couple is still sufficiently oxidiz-

$$2Se^{2-} + 2h^+ \rightarrow Se_2^{2-} \quad (2)$$

ing relative to the flat band in this electrolyte to allow a photovoltage above 0.7 V.

The kinetics of electrolyte oxidation and semiconductor photocorrosion are different for each semiconductor redox couple system. Generalized kinetic models to predict illuminated semiconductor stability in different redox couples and at different redox couple concentrations do not exist as yet.

Prevention of reflection losses is essential to obtain a high overall energy conversion efficiency. Etching of a semiconductor surface to produce hillocks of micron or submicron size (1, 2) (Figure 2) increases

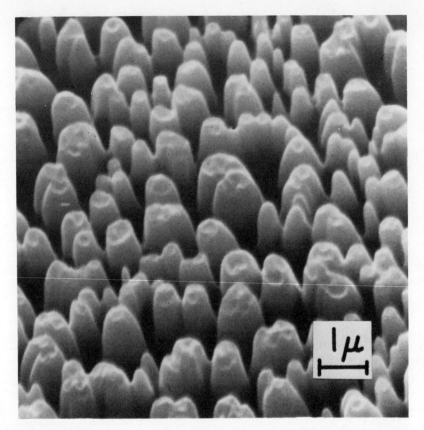

Figure 2. Scanning electron micrograph of a GaAs(100) face etched by a stationary film of a 1:1 30% H_2O_2–H_2SO_4 solution. Because of the submicron size structure ("hillocks"), the reflection of light is substantially reduced. Introduction of the hillocks produces an effect similar to the conversion of "shiny" platinum to platinum "black" upon particle size reduction.

the current efficiency. This increases the overall conversion efficiency, since photons, which would have been reflected because of the difference in index of refraction between the semiconductor and the solution, are trapped between the hillocks. The resulting increase in current efficiency is shown in Figure 3 (8).

Stability to Ion Exchange

Operation of a semiconductor–liquid junction solar cell, or even mere immersion of a semiconductor in a redox couple solution, may lead to ion exchange between the semiconductor surface and the solution. If the exchange results in an epitaxial layer of a new compound, the thick-

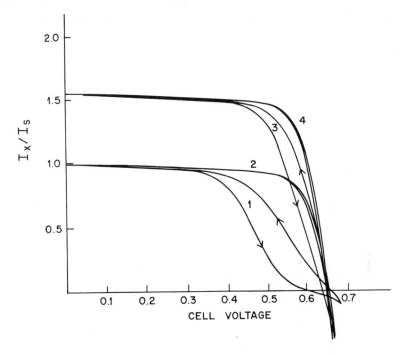

The Electrochemical Society, Inc.

Figure 3. Effects of introduction of surface structure and of Ru(III) treatment in the n-GaAs photoanode on the current–voltage characteristics of the n-GaAs|K_2Se–K_2Se_2–KOH|C cell. Curves 1 and 2, "shiny" smooth electrodes; Curves 3 and 4, "matte" electrodes with hillocks. Curves 1 and 3, untreated electrodes; Curves 2 and 4, Ru(III) treated electrodes. Current (I_x) is normalized to the maximum value (I_s) for the "shiny" electrodes (8).

ness of which exceeds the electron tunneling thickness (30–50 Å), an added junction will exist in the semiconductor. Such a junction may block the flow of holes to the liquid interface, as shown in Figure 4 for the n-CdSe | $1MS^{-2}$–$1MS°$–$1M$NaOH | C cell (8, 9) in which ion exchange results in a CdS layer on the photoanode (10–13). In the resulting structure the valence band of CdS is about 0.5 eV below that of CdSe (10).

The 0.5-eV barrier to hole flow produces a drop in photocurrent (Figure 5) when a CdS layer exceeding tunneling thickness results from surface ion exchange. The ion exchange process is promoted by light and circumstances wherein a photocorrosion step (Reaction 3) can take place. This step is followed by Se° dissolution (Reaction 4) and CdS reprecipitation (Reaction 5).

$$CdSe + 2h^+ \rightarrow Cd^{2+} + Se° \qquad (3)$$

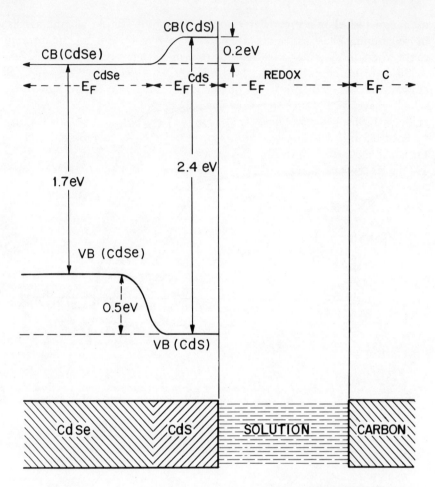

The Electrochemical Society, Inc.

Figure 4. Schematic showing a barrier for hole transport from an n-CdSe photoanode to a redox couple solution created by an n-CdS surface layer under short-circuit conditions. For simplicity, band bending near the liquid interface is not shown. CB, VB, and E_F are, respectively, conduction bands, valence bands, and fermi levels (11).

$$Se^° + S_x^{2-} \rightarrow SeS_x^{2-} \quad (4)$$

$$Cd^{2+} + S^{2-} \rightarrow CdS \quad (5)$$

At low light levels all holes are effectively captured by the redox couple, oxidation of the semiconductor by Reaction 3 does not take place and there is little deterioration in photocurrent. If the intensity is increased to a point where charge transport to the redox couple no longer copes with all the holes arriving at the interface, Reaction 3 and thus ion exchange and photocurrent deterioration are rapid. Rapid stirring assists in maintaining the surface concentration of active species for the desired route of solution oxidation; thus the rate of deterioration is diminished. Addition of elemental selenium to the solution represses the CdSe → CdS conversion. A possible mode of this action is the formation of SeS^{2-} ions, which react with dissolving Cd^{2+} ions to form a $CdSe_{1-x}S_x$ layer. At sufficiently low values of x such a layer has a band gap close to that of CdSe itself and does not block the transport of holes to the liquid interface. At $Se°$ concentrations above $0.07M$ the n-CdSe | $1M Na_2S$–$1M S°$–$1M NaOH$ | C cell is stable to deterioration upon passage of 20,000 C/cm^2

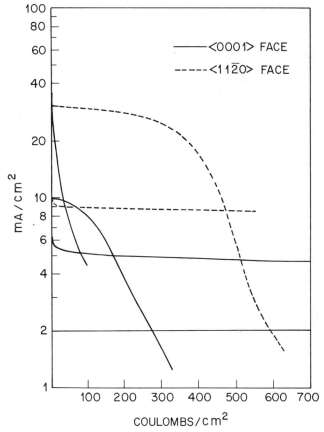

The Electrochemical Society, Inc.

Figure 5. Formation of a CdS surface layer on CdSe results in a decrease in the short circuit current in the n-CdSe|Na₂S–S°–NaOH|C cell. The decrease is much faster at higher light intensities where the rate of hole transport to (oxidation of) the redox couple no longer matches the rate at which holes arrive at the interface. The curves show short circuit currents as a function of charge passage for light intensities defined by the zero time current level (11).

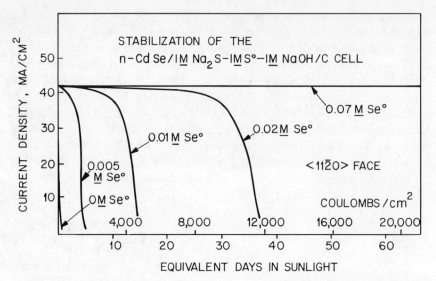

Figure 6. Effect of Se addition on the short-circuit output stability of the n-CdSe|Na₂S–S°–NaOH|C cell. In the presence of 0.07M Se, dissolved as SeS²⁻, the blocking CdS layer is no longer formed and the optput is stable.

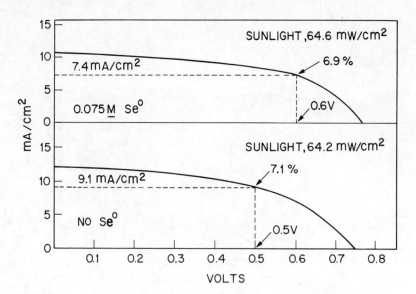

The Electrochemical Society, Inc.

Figure 7. Current–voltage curves, under 64 mW/cm² solar irradiance, for the same n-CdSe crystal run in 1M Na₂S–1M S°–1M NaOH with (top) and without (bottom) 0.075M Se present. Maximum efficiency parameters are indicated (11).

(Figure 6), the equivalent of 2 months of operation in sunlight (*11*). The current–voltage characteristics of the 7% efficient Se° stabilized and fresh, unstabilized cells are shown in Figure 7 (*10*).

The problem of ion exchange can be avoided by choosing systems in which the anions of the semiconductor lattice and the solution are identical or systems in which the compound resulting from the reaction of the lattice cation and the solution anion does not grow epitaxially on the semiconductor lattice. An example of a cell with a common ion, in which the semiconductor is stable to both photocorrosion and ion exchange is the n-CuInS$_2$ | 1MNa$_2$S–1MS°–1MKOH | C cell (*14*). No deterioration in cell performance is observed following passage of 2×10^4 C/cm^2. The band gap of n-CuInS$_2$ is 1.53 eV, close to the optimum for solar energy conversion, and the efficiency of the cell at 70°C is 6%. Because of poor kinetics, the efficiency of this cell decreases at lower temperatures (Figure 8).

An example of a cell in which the product of the lattice cation and the solution anion does not grow on the semiconductor, is the n-GaAs | K$_2$Se–KOH | C cell. Here the selenium-containing layer remains less than a few monolayers thick after passage of 2×10^4 C/cm^2.

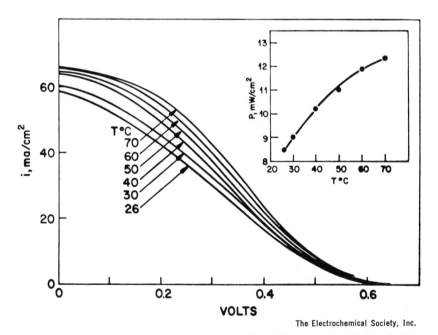

The Electrochemical Society, Inc.

Figure 8. Current–voltage curves for the cell n-CuInS$_2$|2M Na$_2$S–3M S°–2M NaOH|C *at different cell temperatures. The insert shows the maximum power plotted vs. temperature* (*15*).

Surface States and Stability to Adsorbed Impurities

Semiconductor-based devices are invariably sensitive to impurities and are critically sensitive to changes in the chemistry at their junctions. Thus, at first glance it appears doubtful that the power output of an exposed-junction semiconductor–liquid junction solar cell, made with common 85% borosilicate glass and a strong base and other chemicals (reagent grade), could be stable for a reasonable period of time. Indeed one does note rapid deterioration in performance of most cells in the presence of certain impurities. For example, traces of selenium in solution damage the output of the n-CuInS$_2$ | Na$_2$S–S°–NaOH | C cell and the performance of the n-GaAs | K$_2$Se–K$_2$Se$_2$–KOH | C cell is adversely affected by cations such as Bi^{3+} or Pd^{2+}.

We find, however, that incorporation of some impurities in the surface has a beneficial effect (*1, 2, 8*). Incorporation of ruthenium in the surface of n-GaAs photoanodes produces a higher open circuit potential and a higher fill factor, and considerably reduces hysteresis. Maximum effect is gained by dipping the n-GaAs electrode that had previously been in the selenide electrolyte into a Ru(III) solution. Addition of Ru(III) at the 10^{-5}–^{6}M level to the cell electrolyte produces a slow improvement for an untreated electrode by Ru adsorption, though not to the level or permanence of the pretreatment method.

We attribute the improvement to an alteration of the surface states at the GaAs–solution interface by strongly chemisorbed ruthenium. The presence of surface states at adsorbate-free and chemically modified GaAs surfaces is well recognized (*see* Ref. 16). Recently, surface states at GaAs in nonaqueous electrolytes have been invoked to explain voltammetric behavior with redox couples (*17, 18*). The surface of etched n-GaAs in our cells has at least three types of chemical entities with which surface states may be associated: hydroxides, selenides, and nonstoichiometric compositions of gallium and arsenic. If one of the resulting surface states occupies a position (E_{ss}) below the edge of the conduction band, as shown in Figure 9, a loss in photovoltage or fill factor may result if electrons tunnel from the conduction band of the illuminated photoanode ($E_{cb}{}^*$) to the surface state, provided the thickness of the barrier (Δ) allows such tunneling. In the absence of such a surface state, the position of the conduction band in the bulk of the semiconductor moves from E_{cb} to $E_{cb}{}^*$ and a photovoltage V_{ph} results. (A voltammetric curve, for a well-behaved photoanode is shown schematically in Curve a of Figure 10). If tunneling can take place and is very rapid, a perfect shunt is created and the photovoltage will not exceed $E_{ss} - E_{cb}$ (Figure 9), as electrons raised above E_{ss} will spill into the surface state. Such a situation is referred to in Schottky junction cells as "pinning" by surface

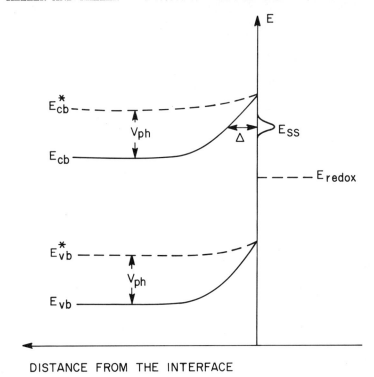

Figure 9. Shunting of a liquid junction solar cell by a surface state near the conduction band. cb, vb, ss, and redox refer to the conduction band, valence band, surface state, and solution redox potential, respectively. The positions of the energy levels in an intensely illuminated cell are shown in dashed lines and are marked by an asterisk. V_{ph} is the highest photovoltage that can be reached in the absence of a surface state (8).

states and leads to the voltammetric behavior shown, schematically, in Figure 10, Curve b. In our experiments, we found such behavior with n-GaAs photoanodes dipped in Pd(II) solutions, a treatment resulting in metallic Pd on the surface.

If the rate at which electrons transfer to the surface state is not sufficiently rapid, a photovoltage V_{ph} of $E_{cb}{}^* - E_{cb}$ may still be approached, but current will be lost and the fill factor will be reduced. Hysteresis will also be observed in cyclic scans, schematically represented by Figure 10, Curve c, and in practice, with freshly etched but otherwise untreated n-GaAs photoanodes (Figure 3, Curves 1 and 3).

A chemisorbed ion may change the position of a surface state either by electrostatic interaction with the surface species or by forming a bond in which electrons are shared.

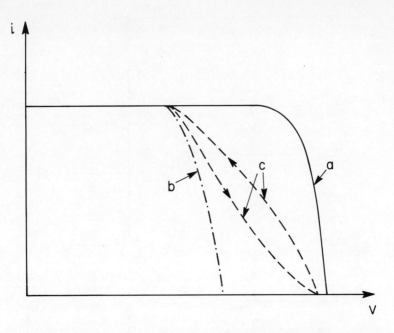

Figure 10. Current–voltage curves for idealized cells with (a) no surface states near the conduction band, (b) perfectly shunting surface states, and (c) imperfectly shunting surface states (8).

If the electrostatic interaction is weak, or if the splitting that results from the formation of the new bond is slight (Figure 11) the photovoltage and the fill factor may deteriorate further. The reason for this is the introduction of a new surface state closer to e_{cb}, the position of the conduction band in the dark in the bulk of the semiconductor, which is the potential of the counterelectrode, neglecting iR effects. The tunneling thickness (Δ) widens for the new lower state and thus the photovoltage is reduced and hysteresis is increased. Since the interaction is weak, the adsorbed species may be slowly desorbed. These phenomena are indeed observed with ions such as Bi(III). Bi(III) lowers the photovoltage, increases hysteresis (Figure 12), and is slowly desorbed.

If the electrostatic interaction is strong, or if a strong chemical bond is formed by sharing of electrons, strong chemisorption is implied. If the interaction is electrostatic, the surface state will move substantially closer to the valence band. If electrons are shared, a splitting results that may be adequate to raise the new upper state to a position above the edge of the conduction band and to lower the lower state to a position where Δ becomes excessive for tunneling to take place (Figure 13). If such a situation is achieved, major improvements in cell performance are expected as a consequence: the photovoltage and fill factor increase; hysteresis decreases; and the sensitivity to weakly interacting impurities, which cause performance deterioration, also decreases. All of the above are indeed observed when Ru(III) is chemisorbed on GaAs: the photovoltage and fill factor increase; hysteresis decreases (Figure 3 Curves 2 and 4, and Figure 12); the improvement persists; and exposure to Bi(III) no longer causes the deterioration of performance as seen in Figure 12.

In summary, ruthenium incorporation in n-GaAs improves cell performance by stabilizing a new surface composition that produces a shift of surface state energies so as to defeat the power loss mechanisms. Essentially, the GaAs surface is made to closely approach the ideal of no interfering surface states in the band gap by deliberate chemical modifi-

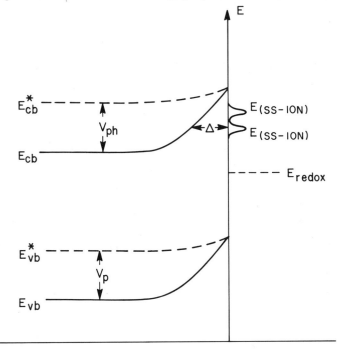

DISTANCE FROM THE INTERFACE

The Electrochemical Society, Inc.

Figure 11. Weak interaction of a surface entity with an ion splits the original state into two new states with energies $E'_{(ss, ion)}$ and $E_{(ss, ion)}$. If tunneling from the conduction band to E_{ss} is possible, the fill factor is reduced or the photovoltage is limited to $E_{(ss, ion)} - E_{cb}$, a value lower than $E_{ss} - E_{cb}$ by about 1/2 of the splitting (8).

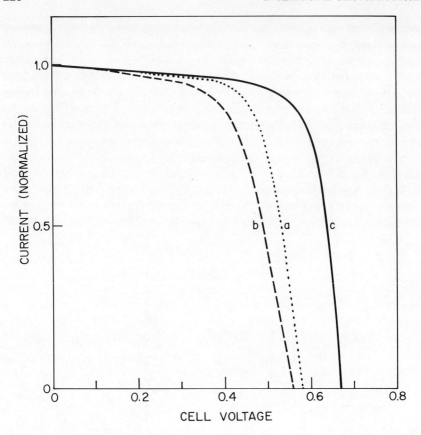

Figure 12. Normalized current–voltage curves for the n-GaAs|0.8M K_2Se–0.1M K_2Se_2–1M KOH|C cell after (a) etching, (b) etching, dip in the redox couple solution, followed by dip in 0.01M Bi(III)–0.1M HNO_3, and (c) etching, dip in the redox couple solution, followed by dip in 0.01M Ru(III)–0.1M HNO_3 (2).

cation. Figure 14 shows the actual current–voltage curve for the n-GaAs | $0.8M$ K_2Se–$0.1M$ K_2Se_2–$1M$ KOH | C cell with the ruthenium-treated photoanode (1, 2). The cell converted sunlight to electrical power with a 12% efficiency and maintained this performance for several months. Ultimate cell failure was caused by leakage of the plastic encapsulant of the photoanode.

The 12% external efficiency was realized while the overall quantum (current) efficiency was only 65–70%. The loss in quantum efficiency is mainly traceable to absorption by the electrolyte and residual reflection at the air–system and semiconductor–liquid interfaces. If all incident photons with energies exceeding the band gap were actually absorbed by the semiconductor, the solar to electrical conversion efficiency would have exceeded 17%. While lower than the efficiency of several solid–solid junction solar cells the 12% external conversion efficiency realized is the highest reported for photoelectrochemical or photochemical systems, including photobiological systems.

Although the possible combinations of semiconductors and redox couples seem very large, the various requirements for output and stability severely reduce the list of candidates. Therefore, close scrutiny of the latter to obtain the maximum stability and efficiency along the lines of the experience reported above will be important in the future. We anticipate that a very close understanding of the interfacial chemistry will be required for achievement of practical viability.

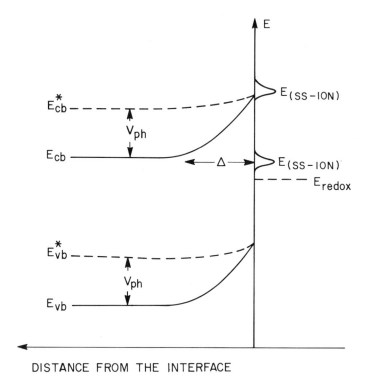

The Electrochemical Society, Inc.

Figure 13. Strong interaction of a surface entity with an ion splits the original state to $E'_{(ss, ion)}$ and $E_{(ss, ion)}$. Electrons cannot tunnel to $E'_{(ss, ion)}$ because the barrier is too thick. $E'_{(ss, ion)}$ is above the conduction band edge and cannot capture electrons. The split surface state no longer reduces the photovoltage or the fill factor (8).

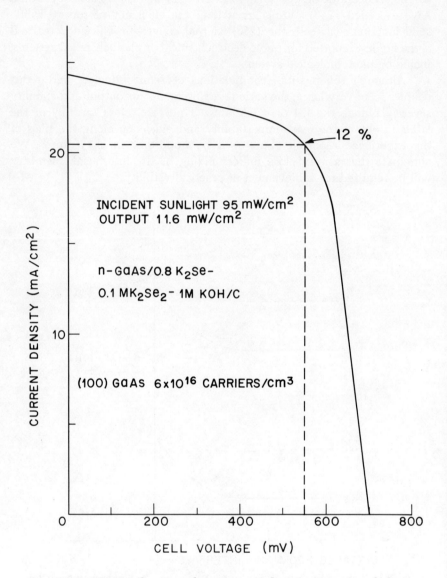

Figure 14. Current density–voltage curve under 95 mW/cm² sunlight for the cell n-GaAs|0.8M K_2Se–0.1M K_2Se_2–1M KOH|C with matte-etched, Ru(III) treated n-GaAs(100) face and 6×10^{16} cm⁻³ carrier concentration.

Acknowledgment

The authors enjoyed their collaboration on the subject of this article with their colleagues, Klaus J. Bachmann, Kuang-chou Chang (now at Polaroid), Shalini Menezes, Bruce A. Parkinson, Murray Robbins, Bertram Schwartz, Gary P. Schwartz, and Richard G. Vadimsky.

Literature Cited

1. Parkinson, B.; Heller, A.; Miller, B. *Conf. Rec. IEEE Photovoltaic Spec. Conf.* **1978**, *13*, 1253–1254.
2. Parkinson, B. ;Heller, A.; Miller, B. *Appl. Phys. Lett.* **1978**, *33*, 521.
3. Hovel, H. J. "Semiconductors and Semimetals Volume II: Solar Cells"; Academic: New York, 1977; pp. 226–227.
4. Gerischer, H. *J. Electroanal. Chem.* **1975**, *58*, 236.
5. Reiss, H. *J. Electrochem. Soc.* **1978**, *125*, 937.
6. Chang, K. C.; Heller, A.; Schwartz, B.; Menezes, S.; Miller, B. *Science* **1977**, *196*, 1097.
7. Harvey, W. W. *J. Electrochem. Soc.* **1967**, *114*, 472.
8. Parkinson, B. A.; Heller, A.; Miller, B. *J. Electrochem. Soc.* **1979**, *126*, 954.
9. Ellis, A. B.; Kaiser, S. W.; Wrighton, M. S. *J. Am. Chem. Soc.* **1976**, *98*, 6855.
10. Hodes, G.; Manassen, J.; Cahen, D. *Nature* **1976**, *261*, 403.
11. Heller, A.; Schwartz, G. P.; Vadimsky, R. G.; Menezes, S.; Miller, B. *J. Electrochem. Soc.* **1978**, *125*, 1156.
12. Cahen, D.; Hodes, G.; Manassen, J. *J. Electrochem. Soc.* **1978**, *125*, 1623.
13. Gerischer, H.; Gobrecht, J. *Ber. Bunsenges. Phys. Chem.* **1978**, *82*, 520.
14. Noufi, R. N.; Kohl, P. A.; Rogers, J. W., Jr.; White, John M.; Bard, A. J. "Extended Abtracts," Spring Meeting of the Electrochemical Society, Seattle, Washington, May 1978; Abstract No. 417.
15. Robbins, M.; Bachmann, K. J.; Lambrecht, V. G.; Thiel, F. A.; Thomson, J., Jr.; Vadimsky, R. G.; Menezes, S.; Heller, A.; Miller, B. *J. Electrochem. Soc.* **1978**, *125*, 831.
16. Morrison, S. R. "The Chemical Physics of Surfaces"; Plenum: New York, 1977; Chapter 8.
17. Frank, S. N.; Bard, A. J. *J. Am. Chem. Soc.* **1975**, *97*, 7429.
18. Bard, A. J.; Kohl, P. A. "Semiconductor Liquid Junction Solar Cells"; Heller, A., Ed.; The Electrochemical Society, Inc.: Princeton, NJ, 1977; pp. 222–230.

RECEIVED October 25, 1978.

13

Photochemical Processes at the Solid–Gas Interface

The Adsorption and Reactions of Gaseous CO_2 and H_2O on Pt–$SrTiO_3$ Single-Crystal Sandwiches

J. C. HEMMINGER[1], R. CARR, W. J. LO[2], and G. A. SOMORJAI

Materials and Molecular Research Division, Lawrence Berkeley Laboratory, Department of Chemistry, University of California, Berkeley, CA 94720

> *We have observed the photoassisted production of CH_4 from CO_2 and H_2O vapor in contact with a sandwich consisting of single-crystal $SrTiO_3$ and Pt foil. In an attempt to elucidate the fundamental chemical processes involved in this reaction, we have studied the chemisorption of H_2O, O_2, CO, and CO_2 on $SrTiO_3$ (111) crystals. The mechanism of CH_4 production is discussed in light of the chemisorption results.*

The purpose of our studies is to carry out thermodynamically uphill chemical reactions at solid surfaces with light as the energy source. There are two ways by which these reactions may occur. They may be carried out by excitation of the solid by light. This creates mobile electrons or electron–hole pairs that may be available to the adsorbates at the surface. By transfer of the excited electrons to or from the adsorbed molecules the surface chemical reactions become energetically feasible. These reactions also may occur via direct excitation of the adsorbed molecules by light. When the stable molecules are placed in an excited electronic or vibrational state they may become energetic enough to make their surface reactions thermodynamically feasible.

[1] Present address: Department of Chemistry, U. C. Irvine, CA 92717.
[2] Present address: University of West Virginia, Morgantown, WV 26506.

One of the simplest reactions of the first type described above is the dissociation of H_2O:

$$H_2O \rightleftharpoons H_2 + \frac{1}{2} O_2 \quad \Delta G° = 2.46 \text{ eV}; \Delta\epsilon° = 1.23.$$

In this reaction the reactant molecule is both oxidized and reduced to produce H_2 and O_2.

We have chosen to investigate those surface reactions where light is used to excite the solid near the surface. By using band gap radiation, light can be converted efficiently to excited electron–hole pairs at semiconductor surfaces that may be trapped by the adsorbed molecules before their recombination could occur. The photographic process is a good example of such a surface reaction where the photoexcited mobile charge carriers cause decomposition of the solid in the near-surface region. After charge capture leading to oxidation and reduction that produce silver and halogen atoms, a diffusion-controlled aggregation of silver atoms and the removal of the halogen molecules leads to irreversible photodecomposition. In our case, however, we attempt to transfer charges to and carry out reactions of adsorbed molecules instead of atoms in the solid surface.

Another reaction that is more complex than the H_2O dissociation is the one between gaseous H_2O and CO_2.

$$CO_2 + 2H_2O = CH_4 + 2O_2 \quad \Delta G° = 8.30 \, \frac{\text{eV}}{\text{mol}}; \Delta\epsilon° = 1.04 \, \frac{\text{eV}}{\text{electron}}$$

$$CO_2 + 2H_2O = CH_3OH + \frac{3}{2} O_2 \quad \Delta G° = 7.15 \, \frac{\text{eV}}{\text{mol}}; \Delta\epsilon° = 1.19 \, \frac{\text{eV}}{\text{electron}}$$

$$CO_2 + H_2O = H_2CO + O_2 \quad \Delta G° = 5.32 \, \frac{\text{eV}}{\text{mol}}; \Delta\epsilon° = 1.33 \, \frac{\text{eV}}{\text{electron}}$$

$$CO_2 + H_2O = HCOOH + \frac{1}{2} O_2 \quad \Delta G° = 2.98 \, \frac{\text{eV}}{\text{mol}}; \Delta\epsilon° = 1.49 \, \frac{\text{eV}}{\text{electron}}$$

There are several different types of organic molecules that may be produced by this photosynthetic process, though O_2 is always liberated. Although the standard free energies ($\Delta G°$) and the free energies per electron transfer ($\Delta\epsilon°$) are all positive they are of different magnitudes. Unlike photosynthesis that leads to the production of complex high-molecular-weight carbohydrates, the above reactions lead to the formation of small organic molecules of high vapor pressure. Since all of the reactants and products are gaseous, these reactions may be readily investigated at the solid–vapor interface.

Much of the past research that is concerned with thermodynamically uphill photochemical reactions has concentrated on photosynthesis—the oxidation and reduction cycles and the elementary charge transfer steps.

Only recently has photoelectrochemistry provided a new direction for the investigations of positive free energy processes (1–5). Nozik recently has reviewed extensively the field of photoassisted processes in electrochemical cells (3).

Surprisingly, very little effort has been made to study photoassisted processes that produce small molecules at the solid–vapor interface. There are major advantages in carrying out photoreactions using gaseous reactants and products as compared with studies at the solid–liquid interface. The surface composition and its changes can be analyzed readily by the modern diagnostic techniques of surface science that require high vacuum. Thus the reaction mechanisms can be studied more easily. The chemical dissolution of the active surface and the absorption of the incident light by the electrolyte are absent. Moreover, the diffusion of reactants and products to and from the surface is more rapid, perhaps an important consideration affecting the rate of the photochemical reaction.

Unlike photoelectrochemical reactions at the solid–liquid interface, however, one cannot apply a variable external potential to facilitate chemical reactions at the solid–gas interface. Recently, there have been successful attempts to dissociate H_2O in an electrochemical cell by using only light instead of an external potential (3). O_2 evolves at the oxide ($SrTiO_3$, TiO_2) anode and H_2 at the cathode (Pt or p-type GaP) from a basic aqueous electrolyte using band gap radiation ($\sim$ 3 eV) to illuminate the anode surface. In fact, H_2 and O_2 evolve at the aqueous electrolyte–solid interface even when the anode and cathode are short circuited, i.e., touching each other.

The success of the photoelectrochemical dissociation of H_2O without using an external potential indicated the intriguing possibility of carrying out photochemical surface reactions with gaseous reactants. The production of H_2 and O_2 with both separated and touching oxide (anode)–metal (cathode) surfaces shows that there may be several mechanisms for the same photochemical reaction. In explaining the operation of the photoelectrochemical cell, the importance of a Schottky barrier that forms at the oxide–electrolyte interface is invoked. Since photodissociation of H_2O occurs at the surface when Pt and $SrTiO_3$ crystallites are mixed and pressed into a pellet, i.e., in a short-circuit configuration, charge transfer at the metal–semiconductor interfaces also may cause sufficient charge separation and the formation of a space charge barrier at the surface to induce H_2O photodissociation. Thus, the electrolyte may not be necessary as an active electrical circuit component when the oxide semiconductor and the metal are touching.

Nevertheless, using gaseous H_2O as a reactant instead of an aqueous electrolyte solution is a major departure from the experimental conditions used in photoelectrochemistry. While large concentrations of OH^- ions are present in the basic solution, the H_2O molecules adsorbing on the oxide surface from the gas phase must be dissociated; this added reaction step is required for the success of the photochemical process. In addition, certain reaction intermediates may be stabilized in solution that would not be stable in the vapor phase.

We have investigated the adsorption characteristics of CO_2, H_2O, CO, and O_2 on $SrTiO_3$ and TiO_2 single-crystal surfaces by a combination of techniques that include electron loss spectroscopy (ELS), UV photoelectron spectroscopy (UPS), Auger electron spectroscopy (AES), and low-energy electron diffraction (LEED). Simultaneously, we have investigated the surface chemical reactions of gaseous CO_2 and H_2O at the surfaces of an oxide–metal sandwich ($SrTiO_3$–Pt). This was carried on in 30-torr ambient pressure of the reactants at 300 K while illuminating the oxide by band gap (> 3 eV) radiation from a filtered mercury lamp. We have used a small-surface-area (~ 1 cm^2) single-crystal oxide sample that was cleaned and well characterized by AES before and after the experiments. The use of well characterized single-crystal surfaces instead of higher-surface-area powders was deemed necessary because of the possible side reactions caused by contaminants, mainly carbon. The formation of CH_4 by the nearly thermoneutral reaction $2C + 2H_2O \rightarrow CH_4 + CO_2$, $\Delta G° = +2.87$ kcal/mol has been observed (16). While large-surface-area powders or polycrystalline deposits are always carbon contaminated, the small-area, high-purity, single-crystal sample is readily cleaned by a combination of ion bombardments and chemical treatments before the experiments, thus eliminating the possibility of uncontrolled side reactions. However, as a result of the small surface area of our single-crystal oxide samples it is difficult to obtain products in detectable concentrations. The photon-assisted reaction is carried out at near atmospheric pressure in a specially constructed isolation cell that is located in the center of an ultrahigh vacuum (UHV) chamber (see Figure 1). When the cell is open the sample could be cleaned in situ by ion bombardment or by chemical treatments at low pressures and the surface composition could be analyzed by AES. The cell is closed around the sample, then it is pressurized by introducing the reactants (CO_2 and H_2O). The chemical reaction is carried out at any desired surface temperature while the sample is illuminated through a sapphire window.

Using this experimental configuration we have found evidence for the production of CH_4 from gaseous CO_2 and H_2O at the surfaces of the $SrTiO_3$–Pt sandwich. To verify the elementary steps of the complex photoreaction the chemisorption studies will be described first. Then the photochemical reactions of CO_2 and H_2O will be discussed along with the blank experiments that were carried out to identify the photochemically active system.

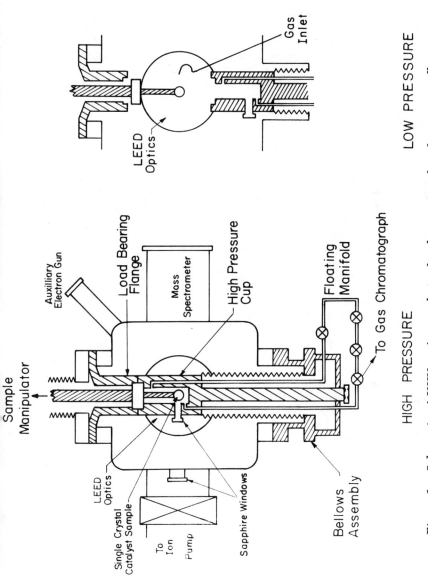

Figure 1. Schematic of UHV surface analysis chamber equipped with reaction cell

Experimental

Equipment. All the chemisorption experiments were performed in a stainless steel UHV chamber evacuated by ion- and water-cooled titanium sublimation pumps. The base pressure in the system was in the low 10^{-10}-torr range.

A double-pass cylindrical mirror analyzer (CMA) with a coaxial electron gun was the primary electron source and electron energy analyzer by which electrons emitted into a conical segment with a half angle of 42.3° from the axis of the analyzer were collected. In all AES, the CMA was operated at a constant resolution of 1.5 eV. In this way we could resolve detailed structures of the peaks in the dN/dE Auger spectra and give reliable estimates of strontium-to-oxygen and oxygen-to-titanium ratios under different conditions of surface preparation.

A primary electron beam with energies between 60–180 eV was used in the electron energy loss experiments. The ELS were obtained either by directly measuring the electron energy distribution $N(E)$ as a function of energy loss or its second derivative $-d^2N/dE^2$. Both methods yielded the same loss peaks except that the features in the ELS $-d^2N/dE^2$ spectra were enhanced. The CMA was operated in the retarded mode with pass energy at 50 eV. However, since the primary electron beam was not energy analyzed, the ultimate resolution was limited by the thermal energy spread of the incident electrons, which was about 0.5 eV. The energy positions of the loss peaks were independent of incident electron energies in the range of 60–180 eV. All energy loss spectra reported in this chapter were obtained with incident energies of approximately 100 eV. The features of these spectra were very surface sensitive.

In the UPS studies, the cold cathode discharge lamp was operated to generate the HeI spectral line at 21.2 eV. A two-stage differential pumping manifold was used to minimize the helium leak flow from the discharge lamp into the UHV chamber which maintained a pressure of $1–2 \times 10^{-9}$ torr during all UPS experiments. The mass spectrometer indicated that the pressure rise was caused by the increase of helium partial pressure in the chamber. The specimen was positioned with its surface normal coincident with the axis of CMA. The angle of incidence of the photons on the specimen was 75° from the normal. The analyzer was operated with a constant resolution of 0.035 eV. Typically, a spectrum could be obtained within 5 min.

Sample Preparation and Experimental Procedure. For the chemisorption studies the specimen used was a 99.99% undoped $SrTiO_3$ single crystal with perovskite structure. Samples of (111) orientation, as determined by the Laue back reflection technique, were cut from this crystal and mechanically polished using 0.05-μm Al_2O_3 powders. The specimen then was rinsed in distilled H_2O and mounted on a high-density alumina holder, which had a tungsten heater wire located at the back of the sample to allow radiative heating of the crystal. A LEED optics made by Physical Electronics was used to study the surface structure after annealing. Two samples of (111) orientation had been prepared and both gave essentially the same results.

For Ar^+ bombardment of $SrTiO_3$ surfaces, the vacuum chamber was back-filled with Ar to a pressure of 6×10^{-5} torr. With an accelerating

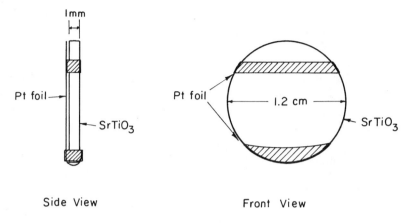

Figure 2. Schematic of the $SrTiO_3(111)$–Pt foil sample

voltage of 2 keV, the ion bombardment typically could deliver an Ar^+ beam of 20 μA to the sample surface.

For the photochemical reaction studies, the $SrTiO_3$ crystal samples (1-cm diameter and 1-mm thick disks) were mounted on a disk of 0.001-in. thick Pt foil and were held in place by two straps of Pt foil (1-mm wide) that were spot welded to the Pt foil backing (see Figure 2). Although the oxide–metal contact was secure, it was only a relatively poor mechanical electrical contact. The oxide–metal sandwich so prepared was mounted in the UHV chamber. The sample was cleaned by ion bombardment of both sides followed by heating in O_2. This heating has been used routinely for removing carbon and sulfur from Pt samples in this laboratory, and was the only cleaning technique applicable to the side of the Pt foil in contact with the $SrTiO_3$. Following the cleaning procedure, the sample was annealed by heating in vacuum. Auger spectra of both sides of the sample were then taken to determine the surface composition. At this point the isolation cell was closed and pressurized with the reactant gases. The CO_2 was Matheson Research Grade and was used without further purification. It contained a CH_4 impurity that was small compared with the CH_4 produced. The pure H_2O was obtained from Scientific Products and was used after several cycles of outgassing by freezing with liquid N_2 while pumping on the sample. No impurities were detectable in the H_2O by gas chromatography (GC) or mass spectroscopy (MS).

Band gap radiation was provided by a 500-W high-pressure mercury lamp in a water-cooled housing. Light from the lamp was collimated and focused on the sample with two quartz lenses. An IR filter consisting of a quartz cell filled with $NiSO_4$ solution was placed between the mercury lamp and the reaction cell to absorb the heat and transmit the near UV. With this arrangement we were able to obtain a photon flux of approximately 10^{17} photons/sec on the 1-cm^2 crystal surface.

To obtain quantitative determination of the reaction product concentration, the gas chromatograph was calibrated using pure CH_4.

Results

Selected Properties of the Clean SrTiO₃ Crystal Surface. Stoichiometric $SrTiO_3$ is an insulator, transparent to visible light (band gap = 3 eV). It is cubic (perovskite structure) at 300 K, but becomes tetragonal at about 110 K. When the 99.99% undoped $SrTiO_3$ obtained from the National Lead Company is reduced by heating in a flow of H_2 at 1000 K for 2 hr it turns black and exhibits a dark conductivity of approximately 1.0 $(ohm/cm)^{-1}$. The AES and UPS spectra were obtained after ion bombardment at 300 K to remove the impurities (mostly carbon) followed by annealing at 900 K. The AES spectra obtained at 300 K give peak-to-peak ratios of $Sr(65\ eV)/O(510\ eV)$ and $O(510\ eV)/Ti(389\ eV)$ of 1.5 and 2.0, respectively, for both the stoichiometric and reduced samples. However, it appears that the surface composition of the stoichiometric oxide is altered by electron bombardment and it becomes reduced. Thus the similarity of the AES spectra of the stoichiometric and reduced crystals may be caused by electron bombardment reduction of the stoichiometric crystal during the time necessary to obtain the spectra. This does not affect the use of the AES as an analytical tool for the determination of the surface cleanliness. The surface composition of $SrTiO_3$ is temperature dependent as shown in a recent study (6). The (111) crystal face exhibits a (1×1) LEED pattern when ordered. Ordering, however, requires heating to 900 K after ion bombardment. The chemisorption studies were carried out mostly on disordered surfaces owing to our inability to anneal the oxide surface at a high enough temperature because of poor thermal contacts. The Auger peak-to-peak ratios were not affected by the different degree of ordering of the (111) surface.

The UPS spectra were different for the stoichiometric and reduced (111) surfaces of $SrTiO_3$ as shown in Figure 3. Both the stoichiometric and reduced crystals have large concentrations of Ti^{3+} ions. This is different from the results found for TiO_2, where the stoichiometric sample has no observable Ti^{3+} concentration. The transition in the ELS (*see* Figure 4) caused by Ti^{3+} in the reduced $SrTiO_3$ sample is significantly broadened, indicating the possibility of band formation. This broadening is also apparent in the UPS spectra, which is why the Ti^{3+} transition at -0.6 V is not as obvious in the spectrum obtained from the reduced sample (*see* Figure 3). There is an additional transition in the UPS (*see* Figure 3) for the reduced sample near 11 eV indicating differences in the valence band structure.

Chemisorption of H_2O, O_2, CO, and CO_2 on the $SrTi_3O$ (111) Crystal Face in Dark and in Light. When gaseous H_2O is introduced into the vacuum chamber at pressures of 10^{-6} torr, adsorption on the oxide surface occurs. Typical exposures were about 10^4 Langmuir (L). The UPS $N(E)$

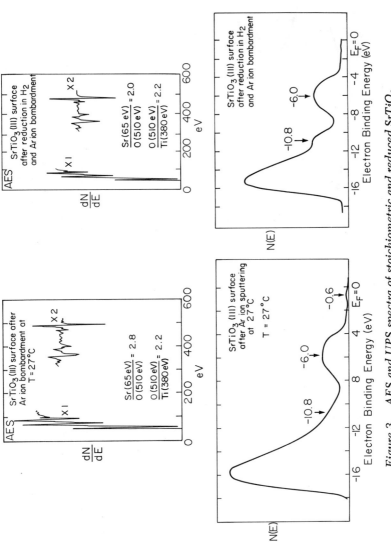

Figure 3. AES and UPS spectra of stoichiometric and reduced $SrTiO_3$

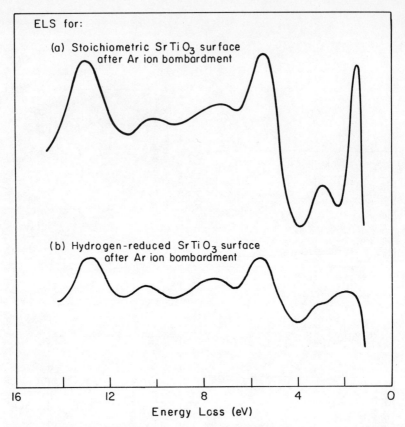

Figure 4. ELS spectra in the region of the Ti^{3+} transition for stoichiometric and reduced $SrTiO_3$

spectra in the region of the Ti^{3+} transition are shown in Figure 5. The signal caused by the presence of Ti^{3+} ions in the surface decreases upon H_2O adsorption, indicating that much of the Ti^{3+} is oxidized to Ti^{4+} by the adsorbed H_2O molecule. The work function increases 0.4 eV as a result of H_2O adsorption. This may be caused by an increase in the band bending near the surface. This is in striking contrast to H_2O adsorption on TiO_2, which causes a decrease of 0.8 eV in the work function (7).

When H_2O adsorption is followed by illumination of the surface with band gap radiation, the Ti^{3+} signal is regenerated only partially, as shown in Figure 5. The work function is not affected.

The UPS difference spectra for H_2O adsorption on $SrTiO_3$ (111) (1 × 1) and a TiO_2 surface that was reduced by Ar^+ bombardment are shown in Figure 6. The two spectra are quite similar, but they are quite different from that of undissociated H_2O (7). Recent calculations by

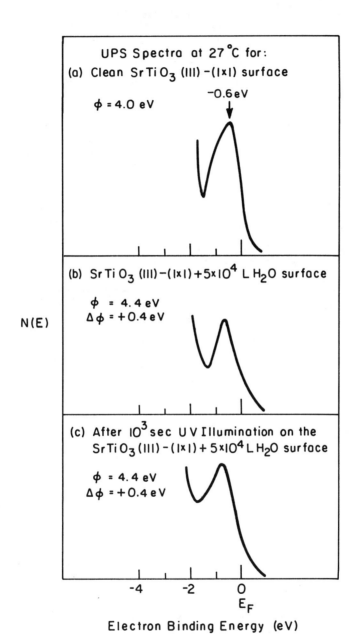

Figure 5. UPS N(E) spectra in the region of the Ti^{3+} transition for H_2O adsorbed on $SrTiO_3$, before and after illumination

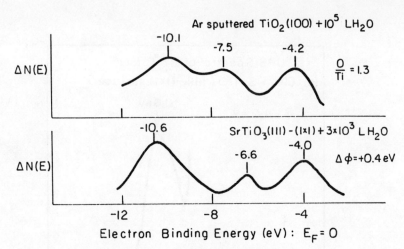

Figure 6. UPS difference spectra for H_2O adsorbed on TiO_2 reduced by Ar^+ bombardment and on stoichiometric $SrTiO_3$

Tsukada et al. indicate that OH^- may be formed on the reduced TiO_2 surface (8). From our results it appears that the H_2O adsorbs dissociatively on surfaces of TiO_2 and $SrTiO_3$ that have significant Ti^{3+} concentrations (6, 7). Thus, H_2O adsorbs dissociatively on reduced TiO_2 and both stoichiometric and reduced $SrTiO_3$. In the process of dissociative adsorption of H_2O, the Ti^{3+} sites are oxidized to Ti^{4+}.

O_2 chemisorption on the $SrTiO_3$ (111) crystal face after exposure to approximately 5×10^4 L increases the work function by $\Delta\phi = 0.9$ eV. The Ti^{3+} signal is removed completely by chemisorbed O_2. This is shown in the UPS $N(E)$ spectra that is displayed in Figure 7. Illumination of the surface with band gap radiation, after O_2 adsorption, regenerates a significant amount of Ti^{3+} as shown in Figure 7. However, upon illumination of the O_2-covered surface the work function is reduced by only 0.2 eV, which is still 0.7 eV greater than the work function of the clean stoichiometric sample. The fact that the work function does not return to the value of the clean stoichiometric sample upon illumination indicates that not all of the adsorbed O_2 is removed by photodesorption.

Both CO and CO_2 give rise to the same UPS difference spectra when adsorbed on the $SrTiO_3$ surface. The spectrum is shown in Figure 8. The measurements were taken after exposure of the surface to 5×10^4 L of these gases. The work function increases 0.4 eV upon adsorption of CO_2. The Ti^{3+} signal is decreased significantly. Upon illumination there is no noticeable change in either the work function or the Ti^{3+} signal.

While heating the sample after CO adsorption only CO_2 is observed to desorb, indicating that CO may be efficiently converted to CO_2 on the oxide surface.

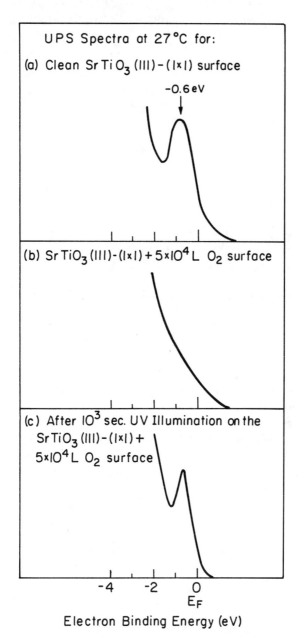

Figure 7. UPS N(E) spectra in the region of the Ti^{3+} transition for O_2 adsorbed on $SrTiO_3$, before and after illumination

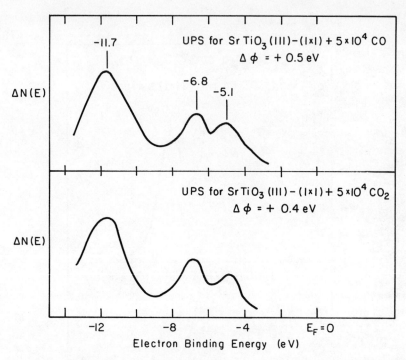

Figure 8. UPS difference spectra for CO_2 and CO adsorbed on stoichiometric $SrTiO_3$.

There is considerable experimental evidence indicating that CO is readily oxidized to CO_2 over several transition metal oxide surfaces in addition to $SrTiO_3$ (9). An oxygen ion from the crystal lattice is likely to be used for this purpose. The reduction of CO_2 to CO that commonly is observed on many transition metal surfaces does not appear to occur on the oxide surface. Another reaction that may occur that converts CO to CO_2 is its disproportionation where $2CO \rightarrow C + CO_2$. We cannot rule out this reaction as a possible path to convert CO to CO_2, although we do not see the accumulation of much carbon on the oxide surface. Under reaction conditions carbon never accumulates on the reduced $SrTiO_3$ that presents an O_2-deficient surface to the incoming reactants. Work on photodesorption from oxide surfaces indicates that CO_2 (rather than CO) is the primary species observed (16, 17). This combined evide e leads us to conclude that conversion of CO_2 to CO is not invol in the mechanism of CH_4 formation.

In summary, H_2O dissociates at least partially upon adsorption as indicated by the electron spectra and oxidizes Ti^{3+} to Ti^{4+} in the process. Ti^{3+} is only partly regenerated by band gap radiation. O_2 adsorbs in several states on the surface. It desorbs from the state that is associated with the oxidation of Ti^{3+} to Ti^{4+} and the Ti^{3+} sites are regenerated. Other types of chemisorbed O_2 remain on the surface during illumination, as indicated by the work function change. CO and CO_2 give rise to the same changes in the electron spectra, indicating that they form the same surface species on the oxide surface. It appears that CO is converted to CO_2 by a slow surface reaction.

Chemisorption of H_2O, O_2, CO, and CO_2 on Pt Surfaces. H_2O is known to chemisorb only poorly on Pt surfaces at low pressures. It remains molecular on the surface and is bound only weakly (*19, 20, 21*). Evacuation of the reaction chamber after exposure of the Pt to high pressures (several torr) of H_2O resulted in total removal of the H_2O at room temperature, clearly indicating weak binding. There was no evidence for H_2O dissociation on the Pt surface even at these higher pressures.

O_2 chemisorbs on Pt single-crystal and polycrystalline foil surfaces at low and high pressures. There is evidence for the presence of chemisorbed O_2 as well as for the formation of oxide under appropriate experimental conditions (*10, 11*). Under conditions of the photochemical reaction studied here, O_2 should be mostly in the chemisorbed state.

Chemisorbed O_2 also interacts strongly with carbon or adsorbed CO on the Pt surface. The kinetics of oxidation of C or CO on Pt have been studied extensively (*12, 13*).

CO chemisorbs on Pt with a sticking probability of unity even at very low pressures ($\sim 10^{-9}$ torr). It stays molecular but strongly bound on the metal and it exhibits many binding states with distinguishable heats of adsorption that vary from 32 to 14 kcal/mol (*14*). Only surface kink sites will dissociate CO to C and O on Pt as indicated by PES studies (*15*). Since the concentration of kink sites is a small fraction of the total number of surface sites, most of the CO adsorbed on the Pt surface should remain molecular.

CO_2 adsorbs poorly with a low sticking probability on Pt at low pressures ($\sim 10^{-6}$ torr) (*22*). At pressures of several torr, which are used in our photochemical studies, CO_2 adsorbs as indicated by subsequent thermal desorption when only a Pt foil sample is used. Only CO_2 is observed in our thermal desorption experiments. No carbon buildup on the Pt is observed by AES after exposure to 15-torr CO_2 in the dark. Thus it appears that CO_2 remains largely molecular on the Pt surface in the absence of H_2 and light and is weakly bound. We find no evidence for the dissociation of CO_2 to CO and O.

The Photochemical Reaction of Gaseous H_2O and CO_2 to Produce CH_4 over $SrTiO_3$–Pt Sandwiches. The reduced-oxide $SrTiO_3$–Pt sandwich, after cleaning by ion bombardment and heat treatments, is analyzed by AES and then enclosed in the isolation cell. The cell is filled with 15-torr H_2O and 15-torr CO_2. The gas composition is analyzed by GC and then the experiment commences. The oxide side of the sandwich is illuminated by light of band gap or larger energy using a 500-W Hg lamp. A $NiSO_4$ filter is used to screen out the IR radiation to avoid heating the sample. The thermocouple that is attached to the sandwich registers less than a 10° temperature rise throughout the experiments at 300 K.

CH_4 gas is produced for the first 10 min of the illumination and is detected readily by the chromatograph. The number of CH_4 molecules formed is plotted as a function of time in Figure 9. The initial rate of formation is 2×10^{14} molecules/min, which corresponds to a quantum yield of one molecule of CH_4 per 10^4 photons. The total amount of CH_4 formed is approximately 10^{-9} mol, which corresponds to about one monolayer. The production of CH_4 slows down with time and then stops after 10 min. This reaction inhibition is caused by a tenacious "poison," since pumping out the reactants and reintroducing fresh H_2O and CO_2 does not regenerate the chemical activity of the surfaces. The photochemical activity is regenerated, however, by renewed ion bombardment and annealing (i.e., complete cleaning of the metal and oxide surfaces). AES indicates the buildup of a monolayer of carbon on the Pt.

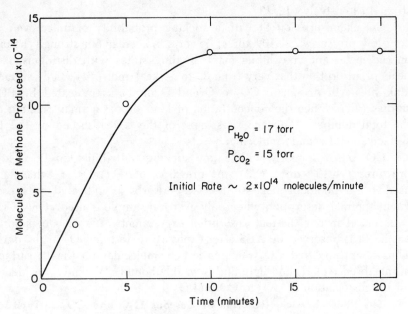

Figure 9. CH_4 production as a function of time of illumination of the $SrTiO_3$–Pt sandwhich in the presence of CO_2 and H_2O

There is a thermal generation of CH_4 when a reduced $SrTiO_3$ sample is used in the oxide–metal sandwich. Upon heating to 600 K in the dark, a monolayer of CH_4 can be generated just as in light at 300 K. In both cases CH_4 production stops after the formation of about a monolayer. It appears that the poisoning reaction is the same in the light-driven and thermally driven reactions, and we detect a monolayer deposit of carbon on the Pt in both cases. Substitution of CO for CO_2 in the reaction mixture does not increase the CH_4 yield. This observation, combined with the evidence favoring CO_2 over oxides in the CO/CO_2 equilibrium, seems to rule out CO as an important intermediate in the reaction paths of CH_4 formation.

The following experiments were carried out to prove that CH_4 production was indeed the property of the metal–oxide contact. The following experiments all produced no detectable amount of CH_4 above background:

(1) the $SrTiO_3$–Pt sample with only CO_2 (no H_2O) in the presence of light and by heating the sample to 450 K;

(2) the $SrTiO_3$ crystal without the Pt foil using CO_2 and H_2O mixtures and illumination;

(3) the Pt foil alone using CO_2 and H_2O mixtures, both with illumination, and by heating the Pt foil to 450 K; and

(4) the $SrTiO_3$–Pt sample in a CO_2 and H_2O mixture using illumination with light of energy less than the band gap of $SrTiO_3$ (light with energy less than the band gap was obtained by placing a Corning glass filter between the $NiSO_4$ solution filter and the reaction cell).

The last experiment indicates that band gap radiation is necessary for the photoassisted production of CH_4 from CO_2 and H_2O. This is consistent with the observed necessity for band gap radiation to dissociate H_2O in the electrochemical cell experiments. These experiments establish that the photon-assisted reaction is a property of the reduced oxide–metal contact system.

Discussion

There are several important observations that help us to understand the photon-assisted process leading to the formation of CH_4 from CO_2 and H_2O. Both the oxide $SrTiO_3$ and the metal are needed to observe the production of CH_4. The photochemical reaction poisons in 10 min at 300 K in the reactant mixture. Surface analysis indicates carbon deposition at the Pt surface while the reduced-oxide surface remains free of carbon. A thermal reaction exists that also produces CH_4. H_2O adsorbs dissociatively on the oxide surface while it remains molecular on the Pt surface. CO is oxidized to CO_2 on the oxide surface. O_2 chemisorbs on both oxide and metal surfaces. Part of the chemisorbed O_2 that is used for the oxidation of Ti^{3+} ions to Ti^{4+} photodesorbs and the Ti^{3+} sites are regenerated. Band gap radiation is necessary to carry out the photochemical reaction. The energy needed to carry out the production of CH_4 from CO_2 and H_2O is much larger than the band gap of $SrTiO_3$, indicating that the process must occur in a stepwise manner.

The presence of the metal, Pt in our case, appears to be necessary for both catalytic and electrical reasons. Separation of the photon-generated electrons and holes at the oxide surface is likely to be enhanced by the metal–oxide contact. Pt appears to participate in the CO_2 reduction process, as indicated by the carbon deposition on the metal surface, most likely through its hydrogenation of the molecule, which probably occurs through one or more partially reduced intermediates.

While the chemisorption studies indicate that H_2O dissociates on the oxide surface, at this point there is little information on the mechanism of CO_2 reduction. Since carbon accumulates at the metal surface it appears that at least part of the carbon cycle involved in the reduction of CO_2 to CH_4 takes place at the metal surface. One possible reduction scheme would be the conversion of CO_2 to CO and then into dissociated oxygen and carbon. The rehydrogenation of carbon to CH_4 that is obtained by the dissociation of CO on Ni, Fe, and Rh surfaces appears to be an important mechanism for methanation from CO and H_2 on these transition metal surfaces. Pt, however, dissociates CO only at kink sites that are present in low surface concentrations, and this process is not likely to account for the formation of the carbon monolayer. Moreover, when CO was substituted for CO_2 the photochemical reaction was not enhanced and there is evidence for the oxidation of CO to CO_2 at the $SrTiO_3$ surface. Thus we are tempted to rule out this reduction scheme.

A more likely reaction path leading to CH_4 formation may be the hydrogenation of CO_2 to formic acid ($HCOOH$) or formaldehyde (H_2CO) with further reduction to CH_4. Future studies will be directed toward verifying the reaction intermediates—the carbon cycle leading to CH_4 production. In this scheme, the carbon layer on the Pt would be formed by fragmentation of the as yet unidentified hydrocarbon intermediates. The competition between fragmentation and further hydrogenation then would be an important factor in the rate of buildup of the carbon layer.

We propose a tentative reaction sequence leading to the formation of CH_4 from CO_2 and H_2O on the oxide–metal sandwich. H_2O adsorbs dissociatively on the $SrTiO_3$ surface. The photoelectrons and holes produce H atoms (or H^+ ions) and O_2 molecules.

$$Ti^{3+} + H_2O = Ti^{4\cdot} + OH^- + H(H_2)$$

$$\mathrm{Ti^{4+} + 2OH^- + }h\nu = \mathrm{Ti^{3+}} + \frac{1}{2}\mathrm{O_2 + H_2O}$$

While O_2 photodesorbs, at least in part (there is evidence from our UPS studies that not all of the chemisorbed O_2 is removed by photodesorption), the H atoms or H^+ ions migrate onto the Pt surface where the hydrogenation of CO_2 and subsequently that of the reaction intermediates occur. Based on the presently available experimental evidence we cannot rule out the possibility that the reduction of CO_2 may be occurring at the oxide surface. The deposition of carbon on the Pt surface is indicative of the importance of the Pt in the CO_2 chemistry.

CH_4 may be produced in the dark over reduced $SrTiO_3$ metal sandwiches. Since the poisoning mechanism seems to be the same as during the photon-assisted process, further investigations may permit the steady-state production of CH_4 in this circumstance as well. Under certain conditions it can be advantageous to carry out this reaction in the thermal mode instead of under illumination. An added problem, of course, is the need to stabilize the nonstoichiometric composition in the near-surface region. Our UPS studies revealed the likely presence of a Ti^{3+} impurity band that appears to be the cause of the thermally generated electrons and holes that provide the driving force for this thermodynamically uphill reaction.

Perhaps the main difference in carrying out the photochemical surface reaction with adsorbed H_2O vapor as compared with a basic aqueous solution is the necessity for dissociative adsorption. Hydroxyl ions that are already present in large concentration in the electrolyte have to be produced first when gaseous H_2O adsorbs. It is important to compare the rate of H_2O dissociation at the solid–gas and solid–liquid interfaces using the same experimental geometry to evaluate this effect. At present we are tempted to view the photochemical process leading to the formation of CH_4 as consisting of two parts: (1) H_2O dissociation to O_2 and H_2 and (2) the reduction of CO_2 with the H_2. Although the first part yields perhaps the same net reaction as the photoelectrochemical process, its mechanism may be entirely different. Both oxidation and reduction may occur on the same oxide surface similar to that in the photographic process. The second process is similar to a methanation reaction: $CO_2 + 4H_2 \rightleftharpoons CH_4 + 2H_2O$, $\Delta G° = -1.4$ eV and is thermodynamically allowed. Its mechanism may be very complex, consisting of several steps.

There are many future experiments necessary to verify the mechanism of this photosynthetic reaction over the oxide–metal contacts. The role of the metal, electronic or catalytic, should be verified. The effect should be tested using other oxides and other metals as the photochemistry may be changed markedly in this way.

It is likely that other molecules, in addition to CH_4, may be produced also using appropriate oxide–metal contacts, light of suitable wavelength, and gas mixtures of CO_2, H_2O, and N_2. Indeed, light-assisted reactions over oxide–metal contacts may provide a new route for the production of many different small molecules.

Since the formation of CH_4 from CO_2 and H_2O is obviously a multistep process, semiconductors with smaller band gaps should not be ruled out as photoassisted agents in these types of reactions.

Acknowledgment

This work was supported by the Division of Basic Energy Sciences of the United States Department of Energy.

Literature Cited

1. Wrighton, M. S.; Wolczanski, P. T.; Ellis, A. B. *J. Solid State Chem.* **1977**, 22, 17.
2. Wrighton, M. S.; Ginley, D. S.; Wolczanski, P. T.; Ellis, A. B.; Morse, D. L.; Linz, A. *Proc. Natl. Acad. Sci. U.S.A.* **1975**, 72, 1518.
3. Nozik, A. J. *Annu. Rev. Phys. Chem.* **1978**, and references therein.
4. Wrighton, M. S.; Willis, A. B.; Wolczanski, P. T.; Morse, D. L.; Abrahamson, H. B.; Ginley, D. S. *J. Am. Chem. Soc.* **1976**, 98, 2774.
5. Jujishima, A.; Honda, K. *Nature* **1972**, 238, 37.
6. Lo, W. J.; Somorjai, G. A. *Phys. Rev. B.* **1978**, 17, 4942.
7. Lo, W. J.; Chung, Y. W.; Somorjai, G. A. *Surf. Sci.* **1978**, 71, 199.
8. Tsukada, M.; Satoko, C.; Kawai, T., personal communication.
9. Steinbach, F.; Harborth, R. *Discuss. Faraday Soc.* **1974**, 58, 143.
10. Matsushima, T.; Almy, D. B.; White, J. M. *Surf. Sci.* **1977**, 67, 89.
11. McCabe, R. W.; Schmidt, L. D. *Surf. Sci.* **1977**, 65, 189.
12. Hopster, H.; Ibach, H.; Comsa, G. *J. Catal.* **1977**, 46, 37.
13. Baker, R. T. K.; France, J. A.; Rouse, L.; Waite, R. J. *J. Catal.* **1976**, 41, 22.
14. Ertl, G.; Neumann, M.; Streit, K. M. *Surf. Sci.* **1977**, 64, 393.
15. Iwasawa, Y.; Mason, R.; Textor, B.; Somorjai, G. A. *Chem. Phys. Lett.* **1976**, 44, 468.
16. Lichtman, D. *Crit. Rev. Solid State Sci.* **1974**, 395.
17. Shapira, Y.; Cos, S. M.; Lichtman, D. *Surf. Sci.* **1976**, 54, 43.
18. Rabo, J., personal communication.
19. Firment, L. E.; Somorjai, G. A. *J. Chem. Phys.* **1975**, 63, 1037.
20. Firment, L. E.; Somorjai, G. A. *Surf. Sci.* **1976**, 64, 1934.
21. Firment, L. E. Ph.D. Thesis, University of California, Berkeley, CA 1977.
22. Witerbottom, W. L. *Surf. Sci.* **1972**, 36, 205.

RECEIVED October 2, 1978.

14

Titanium Dioxide and Platinum/Platinum Oxide Chemically Modified Electrodes with Tailormade Surface States

H. O. FINKLEA, H. ABRUNA, and ROYCE W. MURRAY

Department of Chemistry, University of North Carolina, Chapel Hill, NC 27514

Nitroaromatic and ruthenium bipyridine complexes are covalently linked to PtO and TiO_2 electrodes using organosilane reagents. Cathodic electrochemical reactions of these species occur at potentials more negative than the TiO_2 flat band potential and corresponding reductions occur at similar potentials on both Pt and TiO_2 electrodes. There are, however, marked chemical differences in the chemical stability of the reduction products on the two electrodes.

Chloro- and alkoxy-organosilanes react with and bind to metal oxide electrode surfaces, including SnO_2 (1–8), TiO_2 (3, 5, 6), RuO_2 (9), Pt/PtO (10, 11), Au/AuO (12), and Si/SiO (13). The oxide thickness on the SnO_2, TiO_2, and RuO_2 electrodes is relatively large; these display electronic properties characteristic of the bulk oxide. The oxide layer on the Pt/PtO, Au/AuO, and Si/SiO electrodes is, on the other hand, very thin, possibly monomolecular, and the electronic properties of these electrodes seem to be dominated by the bulk material (e.g., Pt, Au, and Si). Electrode characteristics can thus be chosen in electrode modification work ranging from metallic (Pt, Au), highly conducting oxide (RuO_2, highly doped SnO_2), large-band-gap semiconductor oxide (SnO_2, TiO_2), and small-band-gap semiconductor (Si). It is useful to consider exploiting properties that can be achieved by proper combinations of electrode material and chemical substance attached to the oxide surface using the silanization reaction. This paper describes attachment chemistry for some electroactive moieties on Pt/PtO and n-type TiO_2 and some electrochemical results obtained in the potential range where TiO_2 is metal-like, that is, negative of its flat band potential, V_{FB}.

Silanization can be written as a reaction with surface hydroxyls

$$M\text{---}OH + X\text{---}SiX_2R \rightarrow M\text{---}OSiR$$

where the number of surface –MOSi– bonds has been speculated (14) to be two. The –MOSi– bond seems to be hydrolytically stable although its real stability limits on the various oxide surfaces are as yet only vaguely defined. The "R" group can bear amine, pyridyl, cyano, acid chloride, etc., functionalities; surface structures and shorthand designations of the former two are:

$$M\text{---}OSi(CH_2)_3NH(CH_2)_2NH_2 \qquad MO\text{---}OSi(CH_2)_2\text{---}\langle\bigcirc\rangle N$$

(MO/*en* surface)   (MO/*py* surface)

Subsequent surface synthetic steps can include amidization of the amine and metal coordination of the pyridine.

Investigation of the silanization chemistry is materially aided by x-ray photoelectron spectroscopy (XPES). Figure 1 illustrates typical before and after spectra for TiO_2. The polycrystalline material was vacuum oven-dried and reacted with 10% v/v 3-(2-aminoethylamino)propyl-trimethoxysilane (*en* silane) in anhydrous benzene at room temperature followed by thorough rinsing with fresh benzene. Under these mild reaction conditions, approximately a monolayer of silane is thought to be attached (15) without silane degradation (14). Consistent with this expectation, a reduction of about 50% occurs in the Ti $2p$ peak (area) intensity on the TiO_2/*en* surface. In addition to silicon and nitrogen peaks, a higher binding energy O $1s$ peak appears on TiO_2/*en;* this has been assigned to silicon-bonded oxygen (14). Bonding of more than a monolayer of silane as by formation of siloxane polymer is detected in an approximate way by a more severe diminution of the Ti $2p$ peaks and the presence of large Si $2p$ peaks.

Highly precise information about the silane coverage is difficult to obtain because of combined uncertainties of absolute band intensity reproducibility, photoelectron escape depths, and contaminant background peaks. More reliable information is obtained from relative peak areas on a given sample than by comparing intensities on different samples. For example, the ratio of nitrogen to silicon peak areas on *en* silanized oxide can, after corrections for cross-sections, be used to ascertain whether the silane is stoichiometrically intact or if degradation leading to nitrogen-

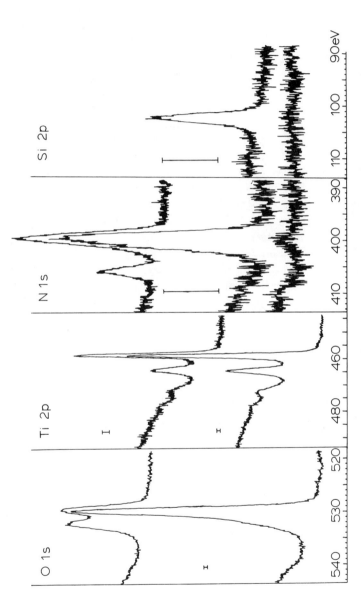

Figure 1. XPES spectra of polycrystalline TiO_2 electrode before (lower traces) and after (upper traces) reaction with en silane. Top trace of N 1s is after further reaction with 3,5-dinitrobenzoic acid. Spectra recorded on DuPont 650B spectrometer with Mg anode. Bars represent unit sensitivity

poor or silicon-rich surfaces has occurred (*14*). For *en*-silanized TiO_2, an N:Si atom ratio of 1.51 ± 0.37 (15 samples) was measured as compared with the 2.0 stoichiometric ideal. The determined N:Si atom ratio on Figure 1 is 1.62. Similar results were obtained for *py*-silanized TiO_2, where N:Si was 0.84 ± 0.19 (9 samples) compared with a stoichiometric ideal of 1.0.

Relative XPES peak intensities can also be used to study coupling reactions such as amidization when the added moiety adds a distinctive new peak to the XPES spectrum. The percentage of active amine in alkylaminesilane layers upon amidization with 3,5-dinitrobenzoic acid or its acid chloride is thus determined. The top trace in the N 1s panel of Figure 1 illustrates the appearance of nitro N 1s at 406.2 eV binding energy, well separated from the (amine plus amide) NR_2 peak at about 400 eV. The ratio of these peak areas yields the amide yield, which in Figure 1 is about 24%. For *en* silane, the percentage of active amine seldom exceeds 50%, suggesting that the terminal amine is much more reactive toward amidization than the secondary amine (*14*).

When TiO_2 is irradiated with photons of energy greater than the band gap (3.0 eV = 420 nm), an excess population of holes is generated at the surface. At potentials more positive than V_{FB}, the holes migrate to the surface and react with available reducing agents. Because of the large band gap, the holes have an oxidizing potential on the order of + 2.0 V vs. SCE and are capable of oxidizing water, supporting electrolyte, and other solutes. Consequently, a study of the stability of the silane surface layer toward hole generation was made. In both water (*16*) and acetonitrile, XPES spectra reveal no significant changes in the Ti 2p, O 1s, Si 2p, or N 1s peaks after extensive hole generation (> 6×10^{17} holes/cm^2) at a potential 2 V more positive than V_{FB}. In particular, there is, as shown in Figure 2, essentially no loss of Si 2p peak area or shape change in the O 1s doublet. Consequently, the Ti–O–Si– linkage appears to resist hole oxidation, an important fact when sensitizing molecules are attached to semiconductors (*5, 6, 8, 11*).

This laboratory's research on various silanized metal oxide electrodes is evolving a library of surface synthetic and electrochemical information within which it is of interest to ask (1) whether synthetic chemistry developed on one metal oxide can be readily transposed to another, and (2) to what extent are the electrochemical and chemical properties of attached species affected differently, if at all, by the different metal oxide substrates. These questions were explored with comparison experiments on silanized Pt/PtO and TiO_2 surfaces.

The silanized Pt/PtO surface is well behaved for studies of electron transfer reactions of surface bound species. Controlled anodization generates the thin oxide layer to which the silane is bound. An electro-

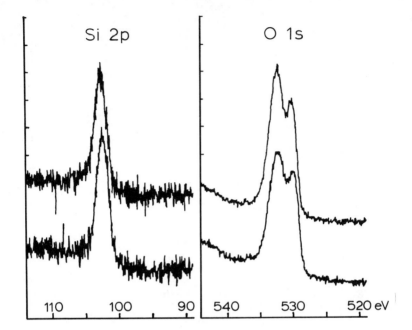

Figure 2. XPES spectra of polycrystalline TiO_2 reacted with en silane. (bottom trace) Freshly silanized; (top trace) after $> 6 \times 10^{17}$ holes/cm^2 of photocurrent generated in 0.1M TEAP/CH_3CN.

chemical window of about 4 V is available in acetonitrile solvent, since silanization suppresses the oxide reduction at negative potentials. Some of the most stable surface couples to date have been synthesized on Pt. An example of surface couple cyclic voltammetry is given in Figure 3. The Pt/PtO surface has been reacted with en silane and then amidized with a known composition mixture of two iso-nicotinic acid complexes of ruthenium, cis-(bpy)$_2$RuCl(iso-nic)$^{1+}$ and cis-(bpy)$_2$Ru(iso-nic)$_2^{2+}$, (bpy = 2,2'-bipyridine), in the presence of the coupling reagent dicyclohexylcarbodiimide (DCC). The surface waves exhibit the characteristic symmetrical shape and small peak potential separation, with formal potentials $E°'$ close to the solution values. The solution electrochemistry corresponds to a $Ru^{3+,2+}$ couple, and it is assumed that the electrochemistry of the immobilized complexes is the same. Coverages are readily obtained from peak areas. After correcting for the 1.3 surface roughness factor obtained from integrated hydrogen adsorption waves (11), Γ for the two waves of Figure 3 is 0.53 (more cathodic wave) and 0.71×10^{-10} mol/cm^2, which is about 40% richer in the bis-(iso-nic) complex than the reactant solution mixture. This is a small reactivity difference considering that the bis-(iso-nic) complex has two potential coupling sites, and a different charge.

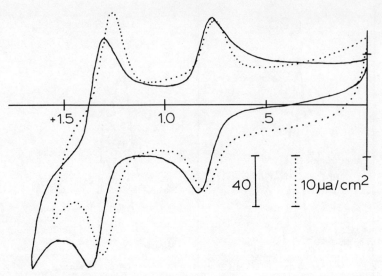

Figure 3. Cyclic voltammograms of a mixture of $(bpy)_2Ru(\text{iso-nic})(Cl)^{1+}$ and $(bpy)_2Ru(\text{iso-nic})_2^{2+}$ in solution (———) and bound to Pt/PtO/en (· · ·); 0.1M TEAP/CH_3CN, 0.2 V/sec, V vs. SSCE.

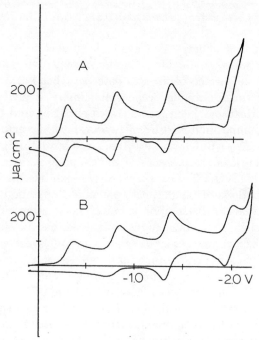

Figure 4. Cyclic voltammograms of $Cr(bpy)_3^{3+}$ at Pt (Curve A) and polycrystalline TiO_2 (Curve B) electrode. 0.1M TEAP/CH_3CN, 0.1 V/sec, V vs. SSCE.

TiO$_2$ electrodes differ considerably from Pt/PtO electrodes in electrochemical behavior because of the fundamental differences between a semiconductor and a metal. For an n-type semiconductor, electron transfer across the electrode–electrolyte interface is blocked when the electrode is biased positive of the flat-band potential (V_{FB}) (18, 19). Only when the applied potential moves negative of V_{FB} does one obtain quasireversible electron transfer. Since V_{FB} for TiO$_2$ is typically near the hydrogen reduction potential in aqueous electrolyte (20), electrochemical reactions are best observed in nonaqueous electrolytes at negative potentials. Figure 4 illustrates this with cyclic voltammograms of the Cr(bpy)$_3^{3+}$ complex dissolved in acetonitrile. With a Pt electrode, three reversible waves are present with formal potentials of -0.26, -0.78, and -1.37 V vs. SSCE (NaCl-saturated SCE). A fourth reduction step at about -1.95 V appears to cause decomposition of the complex. On TiO$_2$, the two most positive reoxidation peaks are poorly developed. Only the third electron transfer step proceeds facilely in both directions, with peak potentials closely matching those on Pt. Consequently quasireversible electron transfer can only be obtained for couples with redox potentials negative of around -1.0 V, the approximate V_{FB} of TiO$_2$ in acetonitrile (19).

Immobilized Nitroaromatics

Coupling of 3,5-dinitrobenzoic acid with a Pt/PtO/*en* surface by exposure to an acetonitrile solution of the acid plus excess DCC gives (10) reversible surface waves with $E^{\circ\prime}$ of -0.77 and -1.14 V vs. SCE. The solution (amide) analog is reduced to anion radical and dianion at -0.72 and -1.12 V vs. SCE, respectively. Surface coverage of the coupled nitroaromatic amounts to ca. 2×10^{-10} mol/cm^2. The surface waves decay to background in 20–50 potential cycles and persist longer if only the radical anion is generated. When the same coupling chemistry is carried out on TiO$_2$, only broad and ill-defined reduction waves appear on the initial potential scan (Figure 5A). These waves are gone on the second cathodic sweep. Reduction to the more stable anion radical only is thwarted by the low surface electron density on the TiO$_2$ at potentials more positive than -1.0 V, while generation of the dianion leads to immediate loss of the nitro group electrochemistry as stated. When examined by XPES, the cycled electrode yields spectra demonstrating retention of the silane but complete loss of the nitro N 1s, as was shown to be the case on the Pt/PtO surface (10).

One possible mechanism for the rapid loss of the nitroaromatic may be amide bond cleavage. To avoid any destabilizing effect of the nitroaromatic anion on the amide bond, the three nitrophenylacetic acid isomers were attached to TiO$_2$, in effect inserting an isolating methylene linkage. In Figure 5, a cyclic voltammogram of surface-bound *m*-nitrophenylacetic acid is illustrated. Both cathodic and anodic peaks are now

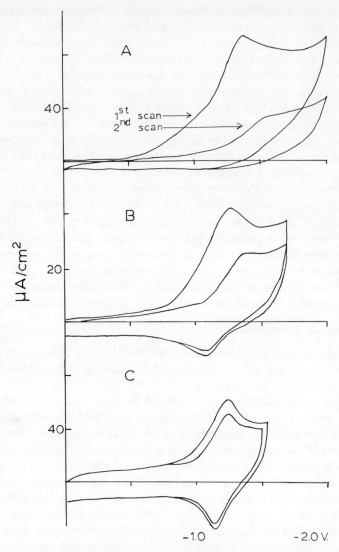

Figure 5. Electrochemistry of surface-attached nitroaromatics on polycrystalline TiO_2/en electrodes. (Curve A) 3,5-Dinitrobenzamide; (Curve B) 3-nitrophenylacetamide; (Curve C) 3-nitrophenylacetamide on spin-coated TiO_2/en electrode. 0.1M TEAP/CH_3CN, 0.1 V/sec, V vs. SSCE.

obtained, although these decay to invisibility in five cycles. The $E^{o\prime}$ (−1.17 V) obtained on the first sweep matches the solution formal potential of *m*-nitrotoluene (−1.18 V). The ortho- and para-isomers were also examined and produced surface reduction waves, but only on the first scan, with no reoxidation waves apparent. In Curve 5B, the integrated peak area on the first scan corresponds to a coverage of 1.7 $\times$ 10^{-9} mol/cm^2. This rather high value was calculated assuming unity roughness factor, and may only correspond to a monolayer on the polycrystalline TiO$_2$.

A marked improvement in electrochemical stability was achieved by forming silane multilayers. Although siloxane polymer formation in our laboratory has mainly been induced by water introduction into the silanization reaction medium, spin coating may be a more controllable technique for building multilayers. This technique has only recently been introduced to chemically modified electrodes (21). An example of its application to silanes is given in Figure 5C. One drop of a 5% solution of *en* silane in xylene was placed on the TiO$_2$ surface and the electrode spun at approximately 25,000 rpm for 15 sec. It was exposed to air and cured at 50°C in vacuum for one hour, then reacted with *m*-nitrophenylacetic acid with DCC as coupling agent. The cyclic voltammograms exhibit the same halfwave potential ($-$ 1.19 V vs. SSCE) as the "monolayer" version, but show smaller peak potential separation ($\Delta E_p = 80$ mV on the first scan as opposed to 185 mV in Figure 5B) and greater electrochemical stability. After 20 cycles, the current peaks have decayed to about 20% of their original height.

XPES spectra indicate that the outlined spin coating procedure deposits a silane layer somewhat thicker on the average than the anhydrous solution reaction procedure. In particular, the Ti 2p peak areas are reduced to about 20% of their original value. The original NO$_2$:NR$_2$ peak area ratio is 0.1. Thus the silane layer is 2–3 times thicker, but the coupling yield is similar (20% of terminal amine groups couple). It is interesting that the quantity of electroactive nitroaromatic from the spin-coated surface (Curve 5C) measured at 0.1 V/sec, $\Gamma = 1.2 \times 10^{-9}$ mol/cm^2, remains about the same as that from the monolayer version (Curve 5B), $\Gamma = 1.7 \times 10^{-9}$ mol/cm^2.

While the XPES-detected amide coupling of nitroaromatics seems to proceed similarly on Pt/PtO/*en* and TiO$_2$ surfaces, and the electrochemical formal potentials of immobilized couples are similar, the stability of the reduced nitroaromatic is substantially lower on the TiO$_2$ surface than on the Pt/PtO surface. The reasons for this are not known but solution experiments on RuO$_x$ (22) show that nitroaromatic reduction is very sensitive to water, and nitroaromatics immobilized on SnO$_2$ (23) by similar chemistry are equally unstable. When the nitroaromatic is immobilized at sites physically more remote from the surface, as we infer is the case on the spin coated TiO$_2$/*en* electrode, its stability rises. The kinetic prerequisites for charge transport through a siloxane polymer matrix and through other electrodes coated with the electroactive polymers (12, 13, 14, 24–27) are of obvious future interest.

Immobilized Ruthenium Complexes

Bis-(bipyridyl)L$_2$Ru complexes offer a wide range of reversible couples encompassing both oxidation and reduction. The reductions meet the cathodic V_{FB} limit requirements of TiO$_2$. Attempts to synthesize surface-bound Ru complexes have demonstrated differences in successful procedures for TiO$_2$ and Pt/PtO surfaces. In Table I, two synthetic philosophies are exemplified by Reactions 1–4 and 5–8. The first method entails amide coupling of the amine moiety on an attached *en* silane with a suitable carboxylic acid on the Ru complex. While successful on Pt/PtO/*en*, this chemistry produced only trace coverages on TiO$_2$. The more successful procedure on TiO$_2$ involved displacing a labile ligand in a

Table I. Syntheses of

	Reaction[c]	
(1)	MO/*en* + (bpy)$_2$Ru(2,2'-bpy-4,4'-(COOH)$_2$)$^{2+}$	$\xrightarrow[\text{DCC}]{\text{CH}_3\text{CN}}$
(2)	MO/*en* + (bpy)$_2$Ru(2,2'-bpy-4,4'-(COCl)$_2$)$^{2+}$	$\xrightarrow[\text{or BuCN}]{\text{CH}_3\text{CN}}$
(3)	MO/*en* + (bpy)$_2$Ru(*iso*-nic)$_2$$^{2+}$	$\xrightarrow[\text{DCC}]{\text{CH}_3\text{CN or BuCN}}$
(4)	MO/*en* + (2,2'-bpy-4,4'-(COOH)$_2$)$_2$Ru(Cl)$_2$	$\xrightarrow[\text{DCC}]{\text{CH}_3\text{CN}}$
(5)	MO/*en* + (bpy)$_2$Ru(DME)$^{2+}$	$\xrightarrow{\text{DME}}$
(6)	MO/*en* + (bpy)$_2$Ru(acetone)$_2$$^{2+}$	$\xrightarrow{\text{acetone}}$
(7)	MO/*py* + (bpy)$_2$Ru(Cl$_2$)Ru(bpy)$_2$$^{2+}$	$\xrightarrow[\text{or acetone}]{\text{propylene carbonate}}$
(8)	MO/*py* + (bpy)$_2$Ru(DME)$^{2+}$	$\xrightarrow[\text{(Cl}^-\text{)}]{\text{DME}}$

[a] Observable Ru 3$d_{5/2}$ peak.
[b] Electrochemical surface waves present at potentials matching a solution analog.
[c] Pt/PtO or TiO$_2$ silanized with 3-(3-aminoethylamino)propyltrimethoxysilane (*en* silane) is Mo/*en* and with trichlorosilyl-4-ethylpyridine (*py* silane) is MO/*py*.

suitable complex with a pyridine ligand already attached to the silanized electrode surface. In Reactions 5, 6, and 8, the Ru complex was prepared by reacting $(bpy)_2Ru(Cl)_2$ with a stoichiometric quantity of $AgPF_6$ in the appropriate degassed solvent. Detection of Ru by XPES is complicated by overlap of its $3d$ bands with the ubiquitous C $1s$ peak and the Ru $3p_{3/2}$ with the Ti $2p$ peaks, but as Figure 6 demonstrates, the reaction with Pt/PtO/py as well as TiO_2/py yields a clearly observable Ru $3d_{5/2}$ peak at 280.6 eV, consistent with the binding energy of powder samples of $(bpy)_2Ru(Cl)(py)^{1+}$ (280.7 eV). On Pt/PtO, electrochemical detection by the Ru oxidation wave is possible and aids in confirming the surface coordination as illustrated in Figure 3. Some of the Pt/PtO/py bound complexes exhibit extraordinary stability, persisting for over 15,000

Surface-Attached Ru Complexes

TiO_2—Ru Present?		Pt/PtO—Ru Present?	
ESCA[a]	EC[b]	ESCA[a]	EC[b]
No	No	Yes	Yes
No	No	—	—
No?[d]	Yes?[d]	Yes	Yes
—	—	—	Yes
—	—	—	Yes
Yes?[d]	No?[d]	—	—
Yes	Yes	—	Yes?[d]
Yes	Yes	Yes	Yes

[d] Queries represent equivocal XPES or electrochemical results. bpy = 2,2'-bipyridine, DME = 1,2-dimethoxyethane, DCC = dicyclohexylcarbodiimide.

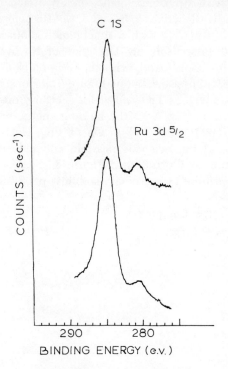

Figure 6. C 1s and $3d_{5/2}$ XPES spectra of Pt/PtO/py (upper trace) and TiO_2/py (lower trace) electrodes. Both surfaces were reacted with $(bpy)_2$-$Ru(DME)^{2+}$ as in Reaction 8.

complete conversions at $0.1\ sec^{-1}$ between the Ru^{3+} and Ru^{2+} oxidation states (28). On the other hand, reduction of the complexes leads to moderately rapid decay. Blank experiments performed with nonsilanized TiO_2 and Pt/PtO electrodes have generally shown little or no electrochemical waves due to adsorption of the Ru complexes.

Reaction 7 in Table I involves symmetrical cleavage of the bridging chlorides in the Ru dimer followed by insertion of the desired ligand:

Interestingly, Reactions 7 and 8 seem to lead to the same surface complex. On Pt/PtO/py, Reaction 8 results in a surface wave with oxidation $E^{\circ\prime}$ at $+0.76$ V vs. SSCE; its potential and general behavior are appropriate for the chlorocoordinated Pt/PtO/py–Ru(bpy)$_2$(Cl)$^{1+}$ species. The solution complex (bpy)$_2$Ru(Cl)(py)$^{1+}$ has $E^{\circ\prime} = +0.80$ V vs. SSCE in acetonitrile. Potentials in Figure 3 are also very similar to this. The solution species, (bpy)$_2$Ru(Cl)(py)$^{1+}$ exhibits (Figure 7A) two reduc-

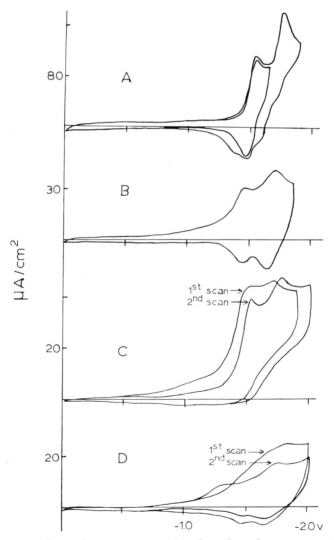

Figure 7. Electrochemistry of surface-bound ruthenium complexes. (Curve A) Solution of (bpy)$_2$Ru(py)(Cl)$^{1+}$ at Pt; (Curve B) product of Reaction 8 with Pt/PtO/py; (Curve C) product of Reaction 7 with TiO$_2$/py single crystal; (Curve D) product of Reaction 8 with polycrystalline TiO$_2$/py. 0.1M TEAP/CH$_3$CN, 0.1 V/sec, V vs. SSCE.

Table II. Summary of $E^{o\prime}$ Values

Species

(bpy)$_2$Ru(py)$_2^{2+}$ (soln) [b]
(bpy)$_2$Ru(Cl)(py)$^{1+}$ (soln) [b]
(bpy)$_2$Ru(Cl)(py–CH$_2$CH$_2$Si–O–⧸Pt)$^{1+}$ (Curve 7B)
(bpy)$_2$Ru(Cl)(py–CH$_2$CH$_2$Si–O–⧸Ti)$^{1+}$ (Curve 7C)

(bpy)$_2$Ru(Cl)(py–CH$_2$CH$_2$Si–O–⧸Ti)$^{1+}$ (Curve 7D)

[a] Potentials in V vs. SSCE (NaCl-saturated SCE), measured in 0.1M TEAP/CH$_3$CN. E_p and $E^{o\prime}$ are respectively the cathodic sweep reduction peak potentials and formal potentials. Formal potential $E^{o\prime}$ is average of cathodic and anodic peak potentials.

tion waves at -1.54 and -1.77 V, the second typically being irreversible. The reduction $E^{o\prime}$ values on Pt/PtO/en following Reaction 8 are consistent with the solution data, although slightly shifted anodically (Figure 7B and Table II). Sources of the chloride in Reaction 8 could be residual chloride from the chlorosilane or excess (bpy)$_2$Ru(Cl)$_2$ left over from preparation of the DME complex. We have not as yet observed a *bis*-surface pyridine-coordinated product from Reaction 8. The chloride can however be subsequently displaced by other ligands from the immobilized complex which is synthetically useful.

For TiO$_2$, the product of Reaction 7 also has two reduction waves (Figure 7C), totally irreversible, with peaks sufficiently close to the (bpy)$_2$Ru(Cl)(py)$^{1+}$ values as to rule out the alternative possibility of a coupled *bis*pyridine complex (Table II). Subsequent cyclic scans reveal a rapid decay and cathodic shift of the waves. Stirring the solution accelerates the decay. As with nitroaromatics, the reduced, immobilized Ru species exhibits a lowered stability on TiO$_2$ as compared with Pt/PtO. Interference by adsorbed complex cannot be entirely ruled out as a trace of adsorbed material is detectable by XPES on TiO$_2$ surfaces exposed to the Ru complex but bearing no silane. Oxidation of the surface complex on TiO$_2$ is not electrochemically observable as its halfwave potential ($+0.76$ V) occurs well positive of V_{FB}.

In Figure 7D, it is evident that on TiO$_2$ lower coverage was obtained in Reaction 8, and the sequential reductions are not distinct. However, traces of reoxidation waves are now visible. The estimated formal potentials indicate that again the dominant surface species is the (bpy)$_2$Ru(Cl)-(py)$^{1+}$ complex. As before, very rapid decay of the surface waves is observed.

In answer to the two questions posed earlier, synthesis of a specific surface species via silane linkages is often readily transposable from one metal oxide surface to another. Although redox formal potentials of and Reduction Peak Potentials[a]

$E_p[1]$	$E_p[2]$	$E°'[1]$	$E°'[2]$
−1.38	−1.59	−1.35	−1.56
−1.54	−1.77	−1.51	(−1.73)[c]
−1.46	−1.77	−1.45	−1.66
−1.48[d]	−1.70[d]		−1.7
−1.52[e]	−1.76[e]		
−1.6	−1.8	−1.5	

[b] Measured at Pt electrode.
[c] Irreversible reduction.
[d] First cyclic scan.
[e] Second cyclic scan.

immobilized species remain constant upon surface transposition, major differences in chemical and electrochemical stability become apparent between the TiO_2 and Pt/PtO surfaces. The extent to which this is a function of intrinsic surface properties of the electrode as opposed to differences in the silane structure at the two interfaces remains unclear. Certainly surface species on TiO_2 exhibit an undesirable level of reductive instability. It does not necessarily follow however that oxidations of surface species on TiO_2 will be similarly unstable. Provided oxidative stability of attached molecules on TiO_2 is achieved, the surface synthetic picture appears good for synthesis of molecular surface states to act as photosensitizers, to catalyze electron transfer between solution species and the conduction of valence bands, or to inhibit surface decomposition in liquid junction solar cells (14).

Acknowledgment

This research was supported in part by a grant from the National Science Foundation.

Literature Cited

1. Moses, P. R.; Wier, L.; Murray, R. W. Anal. Chem. 1975, 47, 1882.
2. Armstrong, N. R.; Lin, A. W. C.; Fujihira, M.; Kuwana, T. Anal. Chem. 1976, 48, 741.
3. Moses, P. R.; Murray, R. W. J. Am. Chem Soc.. 1976, 98, 7435.
4. Fujihira, M.; Matsue, T.; Osa, T. Chem. Lett. 1976, 875.
5. Osa, T.; Fujihira, M. Nature 1976, 264, 349.
6. Fujihira, M.; Osishi, N.; Osa, T. Nature 1977, 268, 226.
7. Srinivasan, V. S.; Lamb, W. J. Anal. Chem. 1977, 49, 1639.
8. Fujihara, M.; Osa, T.; Hursh, D.; Kuwana, T. J. Electroanal. Chem. 1978, 88, 285.
9. Moses, P. R.; Murray, R. W. J. Electroanal. Chem. 1977, 77, 393.
10. Lenhard, J. R.; Murray, R. W. J. Electroanal. Chem. 1977, 78, 195.
11. Lenhard, J. R.; Murray, R. W. J. Am. Chem. Soc. 1978, 100, 7870.

12. Wrighton, M. S.; Austin, R. G.; Bocarsly, A. B.; Bolts, J. M.; Haas, O.; Legg, K. D.; Nadjo, L.; Palazzotto, M. C. *J. Electroanal. Chem.* **1978**, *87*, 429.
13. Wrighton, M. S.; Palazzotto, M. C.; Bocarsly, A. B.; Bolts, J. M.; Fischer, A. B.; Nadjor, L. *J.Am. Chem. Soc.* **1978**, *100*, 7264.
14. Wrighton, M. S.; Austin, R. G.; Bocarsly, A. B.; Bolts, J. M.; Haas, O.; Legg, K. D.; Nadjo, L.; Palazzotto, M. C. *J. Am. Chem. Soc.* **1978**, *100*, 1602.
15. Moses, P. R.; Wier, L. M.; Lennox, J. C.; Finklea, H. O.; Lenhard, J. R.; Murray, R. W. *Anal. Chem.* **1978**, *50*, 576.
16. Untereker, D. F.; Lennox, J. C.; Wier, L. M.; Moses, P. R.; Murray, R. W. *J. Electroanal. Chem.* **1977**, *81*, 1309.
17. Finklea, H. O.; Murray, R. W. *J. Phys. Chem.* **1979**, *83*, 353.
18. Gerischer, H. *Adv. Electrochem. Electrochem. Eng.* **1961**, *1*, 139.
19. Frank, S. N.; Bard, A. J. *J. Am. Chem. Soc.* **1975**, *97*, 7427.
20. Dutoit, E. C.; Cardon, F.; Gomes, W. P. *Ber. Bunsenges. Phys. Chem.* **1976**, *80*, 475.
21. Tachikawa, H.; Faulkner, L. R. *J. Am. Chem. Soc.* **1978**, *100*, 4379.
22. Rolison, D. R.; Kuo, K.; Umana, M.; Brundage, D.; Murray, R. W. *J. Electrochem. Soc.* **1979**, *126*, 407.
23. Wier, L. Ph.D. Thesis, University of North Carolina, 1977.
24. Davis, D. G. *J. Electroanal. Chem.* **1977**, *78*, 383.
25. Miller, L. L.; Van De Mark, M. R. *J. Am. Chem. Soc.* **1978**, *100*, 639.
26. Merz, A.; Bard, A. J. *J. Am. Chem. Soc.* **1978**, *100*, 3222.
27. Flanagan, J. B.; Margel, S.; Bard, A. J.; Anson, F. C. *J. Am. Chem. Soc.* **1978**, *100*, 4248.
28. Abruna, J.; Meyer, T. J.; Murray, R. W. *Inorg. Chem.*, in press.

RECEIVED October 2, 1978.

15

Chemically Derivatized Semiconductor Photoelectrodes

A Technique for the Stabilization of n-Type Semiconductors

MARK S. WRIGHTON[1], ANDREW B. BOCARSLY, JEFFREY M. BOLTS, MARK G. BRADLEY, ALAN B. FISCHER, NATHAN S. LEWIS, MICHAEL C. PALAZZOTTO, and ERICK G. WALTON

Department of Chemistry, Massachusetts Institute of Technology, Cambridge, MA 02139

Pretreated Au, Pt, n-type Si, and n-type Ge can be derivatized with trichlorosilylferrocene, (1,1'-ferrocenediyl)dichlorosilane, and 1,1'-bis(triethoxysilyl)ferrocene to yield electroactive, surface-attached, oligomeric ferrocene material. Derivatized, n-type semiconductors exhibit photoeffects expected for such an electrode material; irradiated derivatized n-type Si can be used to effect the oxidation of solution reductants by mediated electron transfer, unique proof for which comes from the semiconductor electrode that responds to two stimuli, light and potential. The sustained, mediated oxidation of $Fe(CN)_6^{4-}$ in aqueous solution in an uphill sense by irradiation of derivatized n-type Si is possible whereas a naked n-type Si undergoes decomposition to SiO_x at a rate too fast to allow sustained energy conversion. This establishes the principle of manipulating interfacial charge-transfer kinetics for practical applications.

Semiconductor-based photoelectrochemical cells have proved to give the highest efficiency optical to chemical (1, 2, 3) and electrical (4) energy conversion of any wet chemical system. The highest solar energy conversion efficiency claimed (4) thus far is 12% for an n-type GaAs-

[1] Author to whom inquiries are to be addressed.

based cell employing an aqueous electrolyte solution of $Se_n{}^{2-}$. Sustained conversion of light to electricity in a liquid junction device employing a non-oxide n-type semiconductor has depended on the discovery (5–12) of reductants that are capable of capturing photogenerated holes at a rate that precludes photoanodic decomposition of the semiconductor. Photoanodic decomposition is energetically possible (13, 14) for any n-type semiconductor immersed in a liquid electrolyte exposed to band gap, or greater, energy light. For all non-oxides studied thus far, H_2O is not oxidized interfacially at a rate that competes with the omnipresent photoanodic decomposition.

The technique of n-type semiconductor stabilization by adding reductants to the solution does not allow the electrode to be used to directly drive any other oxidation reaction other than the oxidation of the additive. By way of contrast, the kinetically inert n-type semiconducting oxides such as TiO_2 can be used to effect a number of oxidations including oxidation of H_2O (15–20) and halides (21, 22). But the oxides suffer from either having such a large band gap that only short wavelength light is effective or having a set of energy levels improperly matched to the redox reactions of interest (23, 24). Inert oxide-based cells do not require rigorous protection from O_2 as do cells employing highly reduced stabilizing reagents such as S^{2-}, Se^{2-}, and Te^{2-} (4, 5–11). Nonaqueous electrolyte solutions may offer some advantages with respect to both energetics and kinetics at the interface (12, 13,25–28), but lower solution conductivity compared with aqueous electrolyte systems is a potential difficulty. Also, in nonaqueous electrolyte systems protection from both O_2 and H_2O may be required for long term constant operation.

In this chapter we outline our results concerning a new technique aimed at ultimately yielding stable semiconductor–liquid interfaces for optical energy transduction. Basically, our approach is to covalently attach a reducing reagent (A) to the surface of the semiconductor such that the photogenerated hole rapidly yields A^+, which in turn oxidizes some solution reductant (B) forming B^+, thus regenerating the surface reductant. The crucial difference between a "naked" and a "derivatized" electrode is that the net oxidation, $B \to B^+$, is effected by a hole localized in semiconductor electronic levels in the former and by a discrete molecular oxidant in the latter. While the photogenerated hole in either case is a sufficiently powerful oxidant that the $B \to B^+$ reaction is thermodynamically possible, the kinetics for net oxidation of B to B^+ will be different in the two cases. An important advantage of the derivatized electrodes is that the structure of A can be very well known, since A may be a small molecule. Manipulating the nature of A, and hence the nature of the surface exposed to the solution, may result in major changes in the kinetics of the net interfacial charge-transfer reactions.

Electroanalytical (*25–28*) and surface analysis (*29, 30, 31*) indicate that electronic levels at the surface facilitate electron transfer to solution species. Such surface states may control both energetics and kinetics. Derivatizing electrode surfaces with electroactive molecules can be viewed as a designed introduction of surface states that can be well characterized from the point of view of structure, kinetics, and energetics.

Derivatized surfaces are interesting in additional ways related to the stabilization and efficient utilization of semiconductor–liquid interfaces. First, derivatization of the surface with molecules endows the surface with molecular specific properties. As a prototype, note that derivatization of a conventional electrode surface with a chiral molecule results in an ability to produce optically active electrochemical products (*32*). Second, the achievement (*4*) of a 12%-efficient, n-type GaAs-based liquid junction cell depends on a surface pretreatment that changes the interface electronic states so that a higher output voltage can be obtained. Derivatization may yield similar effects, in addition to allowing designed manipulation of charge-transfer kinetics. Finally, by physically separating the semiconductor surface from the liquid and/or taking advantage of hydrophobic–hydrophilic barriers to charge transfer and solvent accessibility, derivatization of surfaces with polymers (electroactive or not) may allow the design of interfaces having very different properties. For example, micelles have demonstrated (*33*) efficient photoinduced electron transfer with slow back reaction, but have not been practically useful since the photoseparated charges could not be collected. The derivatized electrode surfaces may allow the exploitation of such effects.

In our studies to date we have been mainly concerned with derivatizing small band gap materials with the aim of manipulating charge-transfer kinetics to prevent photoanodic decomposition of the semiconductor. Other workers have undertaken the derivatization of inert, but wide band gap, oxides such as SnO_2 and TiO_2 with visible-light-absorbing dye molecules (*34–37*). The particular emphasis in these systems has been to sensitize the wide band gap oxides to visible light for the H_2O splitting reaction, but the oxidized form of dye molecules produced by optical excitation on the surface is generally incapable of evolving O_2 from H_2O. Further, in one pass of the light, thin layers of dye molecules are incapable of absorbing a large fraction of the incident irradiation. It is also not clear whether the efficiency of generation of separated electron–hole pairs will be high, owing to the fact that the electron–hole pairs are generated in the dye molecule and transfer of an electron to the conduction band of the semiconductor may not be completely efficient. In our experiments thus far the aim has been to derivatize electrodes with a redox couple A^+/A that is transparent to visible light so that light absorption results in electron–hole pair generation in a region of high field near the surface of the semiconductor. The systems which have received detailed study at this point are n-type Ge (38) and Si (39, 40) having band gaps of 0.7 and 1.1 eV (41), respectively, derivatized

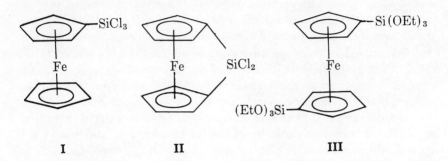

I II III

with the hydrolytically unstable ferrocenes trichlorosilylferrocene (I), (1,1'-ferrocenediyl)dichlorosilane (II), and 1,1'-*bis*(triethoxysilyl)ferrocene (III). For purposes of comparison in terms of electrochemical behavior, Au (42) and Pt (43, 44) electrodes have been derivatized with I, II, and III and characterized by electroanalytical techniques.

Working Hypotheses

We set out to illustrate the principles of photoelectroactivity of surface-attached redox couples where the surface is an n-type semiconductor. Our working hypotheses center about the model for the energetics of the n-type semiconductor–liquid interface (44). Figure 1 shows a comparison of the interfacial energetics for a naked n-type semiconductor and for the same semiconductor derivatized with a redox couple, A^+/A, where E_{BG} is the semiconductor band gap, E_f is the fermi level, or electrochemical potential of the semiconductor, E_{redox} (A^+/A) and E_{redox} (B^+/B) are the electrochemcial potentials of attached and solution couples, respectively, and E_{VB} and E_{CB} are the valence-band and conduction-band positions, respectively, on an electrochemical potential scale. We assume here an ideal situation where there are no surface states between E_{VB} and E_{CB} and the position of E_{VB} and E_{CB} is independent of whether or not the A^+/A couple is attached. At charge-transfer equilibrium in the dark, E_f, E_{redox} (A^+/A), and E_{redox} (B^+/B) must all be the same, but upon illumination at open-circuit with photons with potentials greater than E_{BG} the value of E_f approaches the so-called flat-band potential, E_{FB}, and $|E_{redox}$ (B^+/B) $- E_{FB}|$ represents the maximum photovoltage. The value of E_{redox} (A^+/A) can be no more positive than E_{VB}, and represents the maximum oxidizing power that can be achieved under illumination. The position of E_{redox} (A^+/A) is discussed in more detail below. As is often found, we assume that the n-type semiconductor electrode is

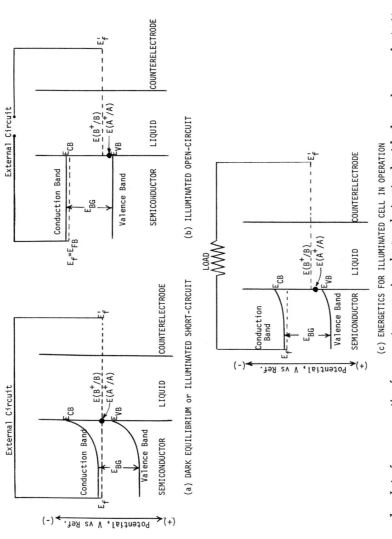

Figure 1. Interface energetics for an n-type semiconductor derivatized with the redox couple A^+/A and immersed in a solution of B^+/B

blocking to oxidation reactions in the dark, even for the surface-attached couple. Owing to the barrier to electron transfer, reductions at the n-type semiconductor are sluggish for E_f more positive than E_{FB}, even though the value of E_f may be more negative than E_{redox}. This rectifying junction is similar to a Schottky barrier where the liquid plays the role of a metal having $E_f = E_{redox}(B^+/B)$. In this ideal model, the maximum outputs for the naked and derivatized electrodes are the same. At short circuit under illumination, the interfacial energetics are the same as in the dark at charge transfer equilibrium, for fast equilibration of A^+/A with B^+/B.

When actually operating, the objective is to maximize the value of quantum yield for electron flow (Φ_e) times the output voltage $|E_f - E_{redox}(B^+/B)|$, where E_f is more negative than $E_{redox}(B^+/B)$ on the electrochemical scale. Since there must be some amount of band bending to separate electron–hole pairs and to prevent back electron transfer, E_f must be somewhat more positive than E_{FB}. Net $B \rightarrow B^+$ conversion is in competition with direct electron–hole recombination in the semiconductor and back electron transfer to reduce B^+ back to B. Semiconductor–liquid interface diagrams for a cell in operation are included in Figure 1. For the derivatized electrode, the value of $E_{redox}(A^+/A)$ during operation must be situated between $E_{redox}(B^+/B)$ and E_{VB}, but the position will depend on the rate of equilibration with the B^+/B couple relative to the rate of oxidation by photogenerated holes. It is desirable that the reaction $A^+ + B \rightarrow B^+$ proceeds at a rate faster than that for the hole oxidation of B for the naked semiconductor case. But the ultimate advantages and utility do not necessarily depend on this property, since molecular specificity, for example, need not involve fast rates.

Rationale for Choice of Systems Studied

Several factors governed our choice of initial systems for study. First, studies in this laboratory (12) established that ferrocene is capable of capturing photogenerated holes at n-type Si at a rate that would preclude photoanodic surface reaction to produce insulating SiO_x layers in a nonaqueous electrolyte solution. Further, the surface of Si bears functional groups that allow covalent attachment of substances such as **I, II,**

$$\text{surface—OH} + R_1R_2R_3Si\text{—Cl} \xrightarrow{\Delta} \text{surface—O—}SiR_1R_2R_3 + HCl \quad (1)$$

$$\text{surface—OH} + R_1R_2R_3Si\text{—}OR_4 \xrightarrow{\Delta} \text{surface—O—}SiR_1R_2R_3 + R_4OH \quad (2)$$

or **III** by the general reaction indicated in Equations 1 or 2. Such surface derivatization chemistry has ample precedence in a number of areas (45, 46, 47) including derivatization of reversible electrodes (48, 49, 50).

Thus, we have combined the general chemistry represented in Equations 1 and 2 with the finding that ferrocene neutralizes holes at Si at a fast rate (12). We also began with the knowledge that simple derivatives of ferrocene and ferrocene itself do not have substantially different redox properties (kinetics and energetics) (51). Examining derivatized Ge (38) was an outgrowth of the work on Si (49, 40), and the work on Au (42) and Pt (42, 43) surfaces was undertaken to establish some points of reference with respect to energetics and kinetics of electron transfer of attached forms of ferrocene from reaction of I, II, or III with the functionalized surface.

The systems that are discussed here are prototypic; a number of different systems can now be envisioned and work is underway to elaborate the concepts described herein. Our aim has been to illustrate some new techniques for potential utilization, conversion, and storage of optical energy.

Results and Discussion

Derivatized Au and Pt Electrodes (42, 43). Reaction of pretreated (anodized) Au or Pt electrode surfaces with isooctane solutions of I, II, or III at room temperature results in the attachment of electroactive material. This is determined by cyclic voltammetry of the derivatized electrodes in nonaqueous electrolyte solutions containing no deliberately added electroactive materials. Some representative electroanalytical data (42, 43) are included in Table I, and the cyclic voltammetric scans in Figure 2 are typical. The essential findings are as follows: the electroactive material is likely oligomeric, essentially reversibly electroactive, and persistently attached; in most respects, the properties of the derivatized surfaces are as expected (52) for a surface-attached, reversible,

Table I. Anodic Peak Positions for Derivatized Electrodes

Electrode Material	Derivatizing Reagent	E_{PA} (V vs. SCE)
Pt[a]	I	0.53 (avg. of 7 electrodes)
Pt[a]	II	0.51 (avg. of 10 electrodes)
Pt[a]	III	0.60 (avg. of 7 electrodes)
Au[a]	II	0.47 (avg. of 20 electrodes)
n-type Ge[b]	I	~0.3
n-type Ge[b]	II	~0.3
n-type Si[b]	I	−0.1
n-type Si[b]	II	−0.05

[a] Metal electrodes are for $CH_3CN/0.1M$ [n-Bu$_4$N]ClO$_4$ electrolyte solution. Cyclic voltammetry reveals reversible behavior.
[b] For semiconductor electrodes the most negative photoanodic peaks are given for EtOH/0.1M [n-Bu$_4$N]ClO$_4$ electrolyte solution.

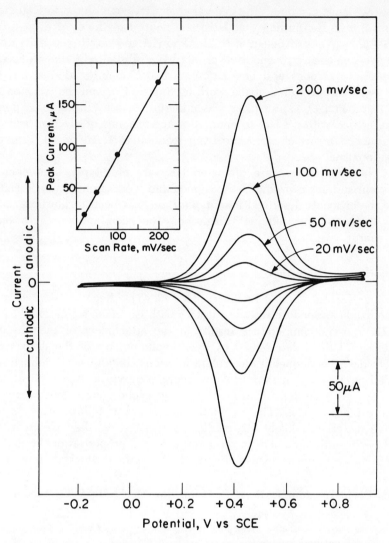

*Figure 2. Cyclic voltammograms for Au derivatized with **II** as a function of scan rate in 0.1M [n-Bu_4N]ClO_4 in CH_3CN at 298 K. Coverage of electroactive material is 6.2 × 10^{-9} mol/cm^2. The inset shows a plot of peak anodic current against scan rate.*

one-electron redox couple. Cyclic voltammetry reveals that electroactive surfaces remain essentially unchanged for eight weeks of shelf storage and can be cycled between oxidized and reduced form thousands of times without deterioration. For typical electrodes the $E°$ for the attached material is within 100 mV of the value (51) for $E°$ (ferricenium/ferrocene) in solution, as has been generally found for other derivatized electrodes. Generally the peak current is directly proportional to scan rate, at least up to 200 mV/sec. Though the separation of the anodic and cathodic current peaks generally increases at fast scan rates, the peak-to-peak separation can be nearly zero and well below 60 mV for scan rates as high as 500 mV/sec.

There is at least one property of the derivatized Pt and Au electrodes that is not in accord with the notion of a reversible, one-electron redox couple bound to the surface of a reversible electrode—the full width of the cyclic voltammetric waves at half height is typically in the range of 200–300 mV rather than the 90 mV theoretically expected (52). Values near the theoretical value of 90 mV have been obtained in other systems; in particular, polyvinylferrocene on Pt gives cyclic voltammetric waves that are considerably sharper (53) than those we have found for derivatization of Au and Pt with I, II, and II.

Pt and Au electrodes derivatized with II have been subjected to analysis by electron spectroscopy (54); these results and those from the electroanalytical characterization are in accord with a surface that has oligomers of the electroactive ferrocene units linked by –Si–O–Si– bonds. The electron spectroscopy of derivatized surfaces and elemental analyses of substances formed by exposing II to air and moisture indicate that the ferrocene units do not remain completely intact upon reaction. The data indicate loss of iron from the material, a fact not inconsistent with known susceptibility of ferrocenes to decomposition in nonaqueous solutions containing nucleophiles (55). Nonetheless, the cyclic voltammetry demands the presence of electroactive material persistently attached to the surface. The loss of iron and the various possible oligomeric structures likely cause the rather broad cyclic voltammetric waves on Au and Pt. We have adopted the interpretation that there is a variety of ferrocene units on the surface, essentially independent of surface coverage, each with its own $E°$.

Derivatized n-Type Si (39, 40). Reaction of single crystal Si and Ge surfaces with I, II, and III results in the persistent attachment of electroactive material. Table I includes some electroanalytical data for these derivatized surfaces. Figures 3, 4, and 5 show a representative electroanalytical characterization of a derivatized n-type Si electrode; Figure 6 shows the comparable data for a naked n-type Si electrode in the same electrolyte solution. The cyclic voltammograms of the naked electrode in the electrolyte solution illustrate the problem with n-type photoanodes; the photoanodic current corresponds to the growth of an insulating layer of oxide material (SiO_x) that cannot be reduced over the potential range scanned. It is this reaction that must be suppressed in the use of n-type Si as a photoanode. The solvent in Figure 6 is CH_3CN and the source of the oxide oxygen is trace H_2O in solution. Also, naked

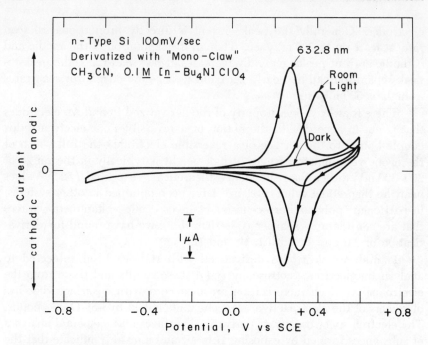

Figure 3. Cyclic voltammograms for derivatized n-type Si showing qualitative effects of light intensity. "Mono-Claw" is reagent I.

Si bears an oxide layer of some thickness that apparently does not preclude electron transfer. Under the conditions shown, less than 10^{-2} C/cm^2 effectively passivates the surface to the flow of photoanodic current. Consequently, protection of n-type Si from surface photoanodic reaction depends on extremely competitive hole capture processes. For example, assuming that ambient solar intensity would yield about 40 mA/cm^2 of current density, a hole capture process that was only 99.99% efficient could still lead to SiO$_x$ formation at a rate that would render the cell useless in approximately 1 hour.

The cyclic voltammetry data included in Figures 3, 4, and 5 can be repeated a number of times without significant variation in the essential properties, evidencing protection from gross oxide growth found for the naked electrode under the same conditions. We will amplify this point below.

Derivatized n-type Si exhibits little or no anodic current in the dark, but illumination with light of greater than E_{BG} results in the flow of anodic current. The cathodic return peak is observed whether the light is on or not, but the cathodic current peak position is light dependent, since the net current flow when the light is on is the sum of dark cathodic plus photoanodic current. At a sufficiently high light intensity, the peak anodic current is directly proportional to scan rate, as expected for a surface-attached redox couple, whereas the peak current varies with the square root of the scan rate for a naked Si electrode illuminated in a nonaqueous electrolyte solution containing ferrocene.

The peak potentials can be substantially more negative than $E°$ (ferricenium/ferrocene) for illuminated n-type Si, in contrast to the situation for Pt or Au. The peak potentials correspond closely to those for naked n-type Si in a nonaqueous electrolyte solution of ferrocene. The extent to which the surface-attached ferrocene material can be oxidized at a potential more negative than on a reversible electrode is a measure of the output voltage for a cell using such a photoelectrode. If the photoanodic current peak is symmetrical, the peak potential represents the potential at which the surface-attached material is 50% oxidized and is thus the electrode potential, E_f, at which E_{redox} for the surface-attached species is equal to the formal potential, $E°$. We have observed photoanodic peak potentials for derivatized n-type Si as negative as approximately -0.1 V vs. SCE, but the value is typically around $+0.1$ V vs. SCE. Anodic peak potentials for derivatized Pt and Au are in the range $+0.4$–0.6 V vs. SCE. Thus, the output voltage for derivatized n-type Si photoelectrodes is in the range 300–700 mV, assuming that the thermodynamics for the various surface-attached substances are independent of the surface. This assumption should be valid within 100 mV (*56*).

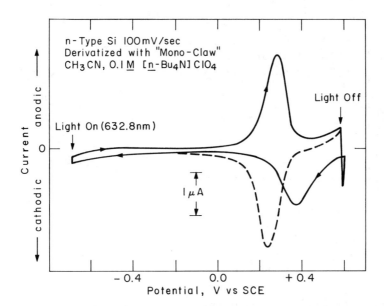

Figure 4. As in Figure 3 except light is turned off at the anodic limit (+ 0.6 V) to show that the cathodic peak does not require illumination; (– – –) result obtained when the light is left on for the entire scan.

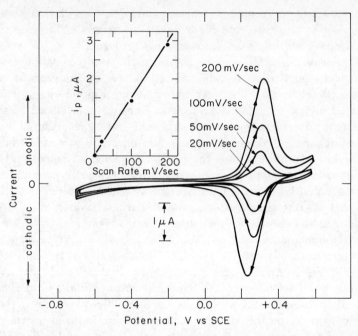

Figure 5. Scan rate dependence of cyclic voltammograms for same electrode and electrolyte solution as in Figures 3 and 4.

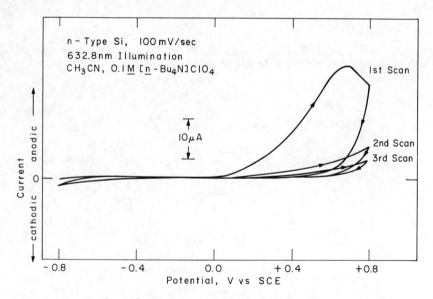

Figure 6. Cyclic voltammograms for naked n-type Si in same electrolyte and under same illumination conditions as for derivatized electrode in Figure 5. Note the lack of a cathodic wave and the declining photoanodic current on the successive scans reflecting SiO_x growth.

The energetics established above correspond well with those found for a naked n-type Si exposed to a nonaqueous electrolyte solution of ferricenium/ferrocene. At least for this semiconductor photoelectrode, then, the covalent attachment of the electroactive material does not significantly alter the interface energetics, as is generally true for reversible electrode systems (56). The potential onset for photoanodic current at the highest light intensity is a reasonable approximation of E_{FB}; we often find that the cathodic current peak in the dark is substantially more positive than E_{FB}. This observation in other systems (25, 26, 27) has been taken to indicate that there are interface or surface states of the semiconductor that can be filled with electrons at a potential more positive than E_{FB}. We adopt this interpretation and note that the crucial interface states seem to be present for both naked and derivatized Si. This finding is consistent with the conclusion that such interface states are likely associated with the SiO_x layer that is invariably present on Si and on which the derivatizing layer is built. Ideally, such interface states could be manipulated to preclude reduction of ferricenium at potentials more positive than E_{FB}; photoanodic and cathodic peaks would be situated more or less symmetrically about E_{FB}. The importance of an oxide layer between n-type GaAs and Au is evident in a Schottky barrier solar cell, where significant changes in solar efficiency were found with variation in the oxide layer (57).

The shape of the cyclic voltammetric waves for derivatized Si is quite variable, depending on the exact procedure for derivatization, scan rate, and light intensity. Peak widths at half height as narrow as 110 mV have been observed, but more typical values are in a range similar to that for derivatized Pt and Au. The peak potentials vary with scan rate, depending on coverage and light intensity; higher coverages and lower light intensity yield photoanodic current peaks that are more anodic at the faster scan rates. The dark current–potential properties are also quite variable, depending on the preparation procedure; generally there is no significant dark anodic current for potentials more negative than about +0.6 V. Some derivatized electrodes have been prepared that exhibit no dark oxidation current at potentials as positive as +10.0 V vs. SCE. But typically, surface-attached ferrocenes can be oxidized in the dark at potentials of around +1.0 V vs. SCE, still substantially more anodic than for Pt or Au.

Clearly, derivatization of n-type Si can protect the surface from significant photoanodic SiO_x formation. This conclusion is based on the observation that the cyclic voltammetric waves for the illuminated derivatized surface are essentially unchanged for many scans. If the SiO_x layer were growing significantly the photoanodic peak would be more anodic with the thicker oxide layer, while the cathodic peak would be more negative. These changes are observed for prolonged cycling and are particularly gross at very anodic potentials under illumination. These observations suggest that the oxide growth rate is, not unexpectedly, a function of both light intensity and potential. The important finding is that substantially more than 10^{-2} C/cm^2 of photoanodic current can pass through the derivatized electrode interface without passivating the electrode. Less than 10^{-2} C/cm^2 passivates the naked electrode.

Protection from SiO$_x$ growth during cell operation will be discussed below, but the rationale for the durability of the derivatized surface will be mentioned here. Photogenerated holes in the valence band of Si are transferred rapidly to the surface-attached ferrocene centers. The rate of transfer to the ferrocenes is apparently fast enough to preclude prompt oxide formation. When the light intensity is low enough and the coverage of ferrocene great enough this rapid hole transfer to form ferricenium centers can be regarded as essentially irreversible, since E_{VB} is much more positive than E_{redox} for the attached species when there is low fractional conversion of ferrocene to ferricenium. But for situations where E_{redox} for the surface-attached species is positive enough, there may be a significant steady-state hole concentration on the Si. For this situation we conclude that oxide growth is kinetically inhibited by the inaccessibility of the holes to attack by H$_2$O. Thus, the oligomeric ferrocene layer physically protects the underlying Si/SiO$_x$ surface, as well as providing a system that can rapidly accept photogenerated holes.

In the limit of very large, polymeric coverages or at sufficiently low light intensity, the derivatized n-type Si electrode behaves like a naked electrode exposed to an electrolyte solution that contains a reductant capable of irreversibly and rapidly capturing every photogenerated hole. For low coverages of surface-attached material we typically observe "break in" changes in the cyclic voltammetry that correspond to oxide growth and passivation in areas of the surface where there is no, or low, coverage. The first few scans show large photoanodic currents, but there is no similar amount of cathodic current; the amount of such photoanodic current declines with each successive scan until the current–voltage properties become essentially constant and the integrated photoanodic and cathodic currents are equal. The large photoanodic current likely corresponds to SiO$_x$ growth on nonderivatized areas of the electrode surface. The fraction of the surface that is nonderivatized is, of course, variable but can be insignificant as determined by the relative current density at naked and derivatized electrodes. The oxide that grows on the nonderivatized electrode areas does not significantly alter the properties of the attached electroactive material; the amount of material and the peak potentials are relatively constant during the break-in of a good electrode as determined by the position and area under the cathodic peak.

Persistent attachment of electroactive material and constancy of energetics for the derivatized n-type Si have been illustrated in several ways. First, derivatized electrodes have good shelf life; they have been stored for weeks and have still exhibited photoelectroactivity associated with an attached ferrocene derivative. Second, illuminated, derivatized electrodes can be cycled at 100 mV/sec between potential limits corresponding to cyclical oxidation and reduction of the attached material. The attached electroactive material is lost slowly (hundreds of cycles), evidenced by declining integrated peak areas, without significant shift in the peak positions. Generally, though, the effect of prolonged use is a shift of the photoanodic peak to more positive potentials and of the cathodic peak to more negative potentials. The peak shifting is probably a consequence of the growth of some undermining SiO_x layer, while the simple loss of electroactive material is likely due to decomposition of the ferricenium form. The durability of the surface is also illustrated by chopping the excitation beam while holding the derivatized electrode at a potential where photoanodic current and dark cathodic current occur. Figure 7 illustrates the results for a chopping frequency of about 1 Hz. While much information can be gleaned from such data, the main point here is that response of the derivatized electrode is essentially the same for a large number of cycles where there is essentially complete oxidation of all surface-attached electroactive material when the light is turned on.

All discussion thus far has concerned results for derivatized Si characterized in nonaqueous solutions. The derivatized electrodes are durable in aqueous electrolyte solutions as well, since persistent cyclic voltammetric waves can be observed in aqueous electrolyte solutions. Alkaline

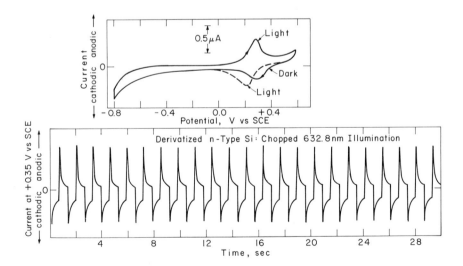

Figure 7. Cyclic voltammograms for n-type Si derivatized with II (top) and current at + 0.35 V vs. SCE against time while chopping the illumination source at ~1 Hz in CH_3CN solution of 0.1M [n-Bu_4N]ClO_4.

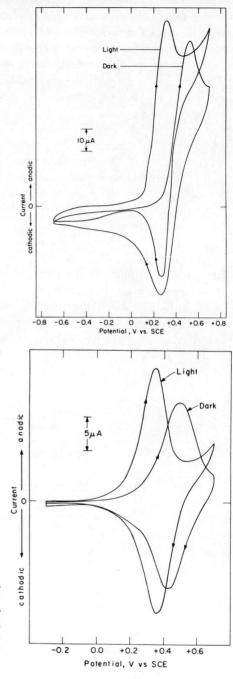

*Figure 8. Cyclic voltammograms for naked n-type Ge in the presence of 2×10^{-3}M ferrocene (top) and for n-type Ge derivatized with **I** in absence of any electroactive solution species. Electrolyte solution is $CH_3CN/0.1M$ [n-Bu_4N]ClO_4.*

solutions are avoided owing to the known sensitivity of ferricenium to basic media (55). The persistent photoelectroactivity of the derivatized electrode in aqueous solution is particularly noteworthy since H_2O is very likely the source of the oxygen in SiO_x formation.

Derivatized n-Type Ge (37). n-Type Ge can be derivatized in a manner similar to that for n-type Si. But unlike Si, the derivatized n-type Ge surfaces that we have studied exhibit sufficient dark currents that good cyclic voltammetric waves for attached ferrocene can be observed in the dark. Illumination results in a more negative anodic current onset, and the peak of the photoanodic wave can be shifted by up to around 200 mV, representing the maximum output photovoltage attainable for derivatized n-type Ge. This value accords well with the value of E_{FB} found from the maximum open-circuit photovoltage for a naked n-type Ge electrode exposed to a solution of ferricenium/ferrocene. Further, the naked n-type Ge exhibits dark anodic current in the presence of ferrocene with a cyclic voltammetric peak nearly the same as that for the derivatized electrode (Figure 8). Thus, as for Si, derivatization does not appear to appreciably alter interface energetics or kinetics for ferrocene oxidation and the durability of derivatized n-type Ge is similar to that for n-type Si.

The observation of significant dark anodic current at n-type Ge likely reflects the presence of a high density of surface states at the Ge/GeO_x interface that are not influenced by the attachment of ferrocene derivatives to the surface oxide. While ferrocene in solution protects n-type Ge from deleterious oxide formation, the low output voltage precludes an efficient optical energy conversion device; we have accordingly concentrated more effort on the use of derivatized n-type Si to illustrate the potential utility of molecularly modified photoelectrodes in energy conversion experiments.

Proof of Mediated Electron Transfer Using Derivatized n-Type Si. One of the unique features of a derivatized semiconductor photoelectrode, compared with a derivatized metal electrode, is that the ratio of oxidized to reduced material depends on both light and potential. By definition, the ratio of oxidized to reduced material on the surface of a reversible electrode depends only on potential. The two-stimuli response of the semiconductor depends on its rectifying property and allows us to obtain direct evidence for mediated oxidation of solution reductants. By mediated we mean that the oxidation of the solution species proceeds by electron transfer to a hole localized on the surface-attached molecule. In the case of the semiconductor, the hole is generated by photoexcitation and transferred from the valence band to the surface-attached species.

Mediated oxidation is in contrast to a direct electron transfer to a hole localized on the electrode surface; when the electrode is derivatized either mechanism for net oxidation of the solution species may yield a rate that is different than for a naked electrode. A priori, a true mediated electron transfer mechanism would seemingly yield a surface having greater molecular specificity, particularly if the mediated oxidation occurs by what would be analogous to an inner-sphere electron transfer mechanism involving prior complexation of the solution reductant and the surface redox reagent.

Figure 9 shows electroanalytical proof for mediated oxidation of solution ferrocene by derivatized n-type Si. The key points are as follows. First, illumination of the derivatized n-type Si in the electrolyte solution in the absence of ferrocene yields the usual cyclic voltammogram that is independent of whether or not the electrolyte solution is stirred, since the electroactive material is attached to the electrode surface. Second, the addition of ferrocene to the solution results in enhanced photoanodic current such that diffusion limited current is obtained in quiet solution, while a hole (light intensity) limited current is reached when the solution is stirred. And third, when the light is switched off at the anodic limit and the electrolyte is stirred, there is little detectable surface-attached oxidized material since there is little observable cathodic current. That is, when the light is switched off at the anodic limit, hole generation ceases and the reductant in solution reacts with the surface-attached oxidant at a rate that is fast compared with the return scan time in the dark. At sufficiently fast scan rates and low concentration of solution reductant a cathodic return peak for the surface-attached oxidant is observed. In principle, such data will allow measurement of heterogeneous electron transfer rates between the solution reductant and attached oxidant. Measuring the consumption of attached oxidant in this manner on a reversible electrode is impossible, since when an attached oxidant reacts with a solution reductant the ratio of surface-attached oxidized to reduced material is instantaneously re-established to a value that depends only on the electrode potential.

Proof of mediated oxidation of solution reductants other than ferrocene has been obtained. In particular, derivatized n-type Si can be used to effect the photoelectrochemical oxidation of $Fe(CN)_6^{4-}$ in an aqueous electrolyte solution (Figure 10). Using the derivatized electrode in the aqueous electrolyte solution is interesting for several reasons: efficient direct oxidation of anything in H_2O using naked n-type Si is unlikely owing to the SiO_x problem; ferrocene could not be used as a solution mediator, since it is insoluble in H_2O; and an aqueous electrolyte solution has higher conductivity than nonaqueous systems.

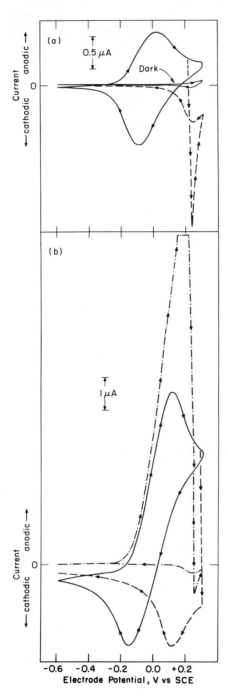

Figure 9. (a) Cyclic voltammograms for n-type Si derivatized with **II** in stirred EtOH/0.1M [n-Bu$_4$N]ClO$_4$ solutions; (– – –) light switched off at + 0.25 V vs. SCE. (b) Cyclic voltammetry for electrode in (a) in same electrolyte solution but containing 5×10^{-4}M ferrocene: (———) quiet solution with illumination for entire scan; (– – –) quiet solution light switched off at + 0.25 V; (– · – ·) stirred solution light switched off at + 0.25 V.

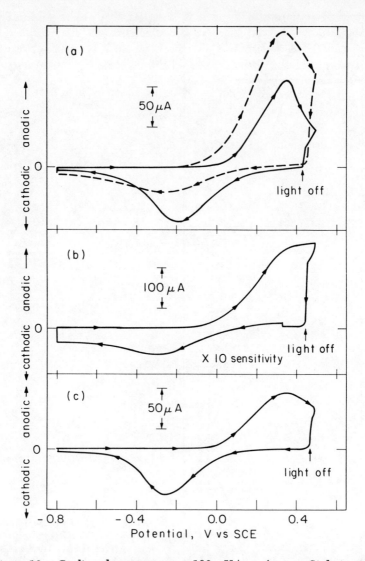

Figure 10. Cyclic voltammograms at 100 mV/sec of n-type Si derivatized with **II** in aqueous electrolytes. The electrode is illuminated with a tungsten halogen source during the anodic scans; the cathodic, return scans are in the dark. (a) Cyclic voltammograms in quiet solutions: (———) 0.1M $NaClO_4/H_2O$ showing surface waves at $+ 0.3$ V (photooxidation) and $- 0.1$ V (reduction); (- - -) with 9×10^{-4}M $K_4Fe(CN)_6$ added. (b) The effect of stirring on cyclic voltammogram in 9×10^{-4}M $K_4Fe(CN)_6$ electrolyte. Note the reverse scan current scale is expanded by 10 times the forward scan scale. (c) Repeat of (a); (———) after completing (b).

Use of Derivatized n-Type Si in a Full Cell Configuration: Equilibrium Current–Potential Curves.

In sufficiently dry EtOH electrolyte solutions containing ferricenium/ferrocene we have been able to sustain the conversion of light to electricity in a cell like that depicted in Scheme I using a naked n-type Si photoelectrode. A comparison of the equilibrium current–potential properties of a naked and derivatized electrode are given in Figure 11 for such a cell. The potential is given relative to the SCE but the photocurrent at E_{redox} can be taken as the usual short-circuit value; current at any potential more negative than E_{redox} reflects

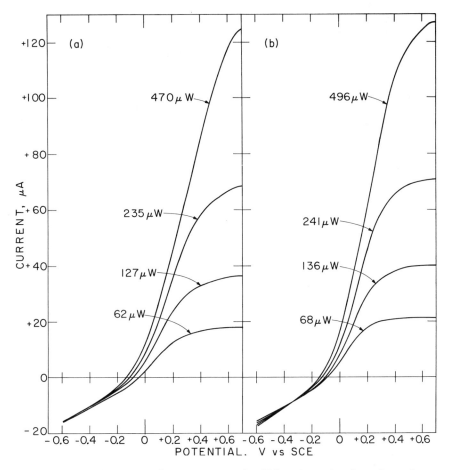

Figure 11. Current–voltage curves at 2 mV/sec for n-Si photoelectrode in stirred EtOH solution of 5×10^{-2}M ferrocene, 2.9×10^{-3}M ferricenium as the PF_6^- salt and 0.1M [n-Bu_4N]ClO_4. Uniform irradiation with 632.8-nm light at the indicated power. Solution $E_{redox} = +0.32$ V vs. SCE. (a) Naked electrode freshly etched with HF; (b) same electrode after derivatization with II.

conversion of light to electricity. From data like that in Figure 11 we conclude that the derivatized electrode is certainly no worse, and perhaps a little better, than the naked electrode in terms of efficiency. Each electrode suffers from low output voltages at low light intensity and from poor fill factors and quantum efficiencies at the higher intensities. Importantly, the derivatized electrode surfaces remain intact (coverage and peak positions) during the experimentation represented in Figure 11, as determined by cyclic voltammetry before and after recording the data in Figure 11. Assuming that all of the photocurrent occurs by mediation, we claim that each attached redox center can conservatively pass more than 10^4 electrons without deterioration of properties.

Figure 12 shows the photocurrent against time for a naked and a derivatized n-type Si photoelectrode in an aqueous solution of $Fe(CN)_6^{4-}$. The improvement in the constancy of the output parameters in the derivatized case is obvious. Equilibrium current–potential curves are shown in Figure 13; such curves are not obtainable for the naked electrode, since the formation of SiO_x is too rapid. The optical-to-electrical energy conversion efficiency is still low in the aqueous electrolyte solution, but it should be emphasized that the derivatized electrode allows sustained energy conversion whereas the naked electrode undergoes decomposition

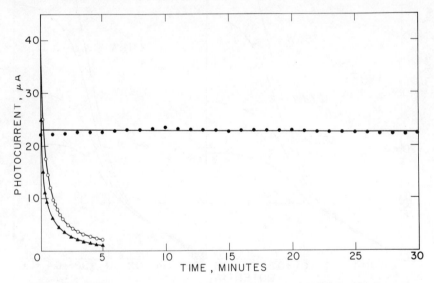

Figure 12. Plots of photocurrent against time for a single n-Si electrode illuminated with 632.8-nm light at ~ 6 mW. Photoelectrode held at + 0.2 V vs. SCE in stirred solutions. Supporting electrolyte is 0.1M $NaClO_4$ in doubly distilled, deionized H_2O. (▲) Run 1, HF-etched naked electrode in supporting electrolyte only; (○) Run 2, naked electrode re-etched with HF, in supporting electrolyte plus 4×10^{-3}M $Fe(CN)_6^{4-}$; (●) Run 3, electrode derivatized with II in same solution as Run 2.

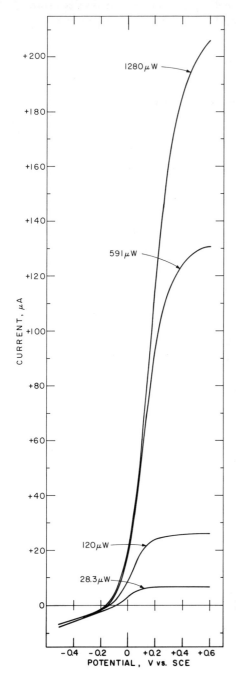

Figure 13. Current–voltage curves for n-Si photoelectrode derivatized with **II** in stirred 0.1M $Fe(CN)_6^{4-}$, 0.01M $Fe(CN)_6^{3-}$, in doubly distilled, deionized H_2O. n-Si illuminated at 632.8 nm at the indicated power. Solution $E_{redox} = +0.13$ V vs. SCE; scan rate 5 mV/sec.

at a rate that precludes sustained cell operation. These results establish the principle of manipulating interface properties for practical applications.

Summary

n-Type Si and Ge semiconductors and Au and Pt metal electrodes can be derivatized with hydrolytically unstable ferrocenes such that oligomeric amounts of essentially reversibly electroactive material are persistently attached. In all cases it appears that attachment leads to little change in the energetics for electron transfer. Derivatized n-type Si has been used to provide direct evidence for mediated electron transfer, the first such evidence for any derivatized electrode. Most importantly, derivatized n-type Si can be used to illustrate the manipulation of interfacial charge transfer kinetics in such a way that the derivatized electrode is useful in energy conversion applications under conditions where the naked electrode is not.

Acknowledgments

We thank the United States Department of Energy, Office of Basic Energy Sciences for support of this research. MSW acknowledges support as a Dreyfus Teacher–Scholar, 1975–1980, NSL as a John and Fannie Hertz Foundation Fellow, 1977–present, and MGB as an M.I.T. Cabot Solar Energy Fellow, 1978–present.

Literature Cited

1. Wrighton, M. S.; Ellis, A. B.; Wolczanski, P. T.; Morse, D. L.; Abrahamson, H. B.; Ginley, D. S. *J. Am. Chem. Soc.* **1976**, *98*, 2774.
2. Watanabe, T.; Fujishima, A.; Honda, K. *Bull. Chem. Soc. Jpn.* **1976**, *49*, 355.
3. Mavroides, J. G.; Kafalas, J. A.; Kolesar, D. F. *Appl. Phys. Lett.* **1976**, *28*, 241.
4. Heller, A.; Parkinson, B. A.; Miller, B. *Conf. Rec. IEEE Photovoltaic Spec. Conf.* **1978**, *13*, 1253.
5. Ellis, A. B.; Kaiser, S. W.; Wrighton, M. S. *J. Am. Chem. Soc.* **1976**, *98*, 1635, 6418, 6855.
6. Ellis, A. B.; Kaiser, S. W.; Bolts, J. M.; Wrighton, M. S. *J. Am. Chem. Soc.* **1977**, *99*, 2839, 2848.
7. Hodes, G.; Manassen, J.; Cahen, D. *Nature (London)* **1976**, *261*, 403.
8. Ellis, A. B.; Bolts, J. M.; Wrighton, M. S. *J. Electrochem. Soc.* **1977**, *124*, 1603.
9. Miller, B.; Heller, A. *Nature (London)* **1976**, *262*, 680.
10. Chang, K. C.; Heller, A.; Schwartz, B.; Menezes, S.; Miller, B. *Science* **1977**, *196*, 1097.
11. Heller, A.; Chang, K. C.; Miller, B. *J. Electrochem. Soc.* **1977**, *124*, 697.
12. Legg, K. D.; Ellis, A. B.; Bolts, J. M.; Wrighton, M. S. *Proc. Natl. Acad. Sci. U.S.A.* **1977**, *74*, 4116.

13. Bard, A. J.; Wrighton, M. S. *J. Electrochem. Soc.* **1977**, *124*, 1706.
14. Gerischer, H. *J. Electroanal. Chem.* **1977**, *82*, 133.
15. Fujishima, A.; Honda, K. *Nature (London)* **1972**, *238*, 37.
16. Wrighton, M. S.; Ginley, D. S.; Wolczanski, P. T.; Ellis, A. B.; Morse, D. L.; Linz, A. *Proc. Natl. Acad. Sci. U.S.A.* **1975**, *72*, 1518.
17. Nozik, A. J. *Nature (London)* **1975**, *257*, 383.
18. Hardee, K. L.; Bard, A. J. *J. Electrochem. Soc.* **1975**, *122*, 739.
19. Keeney, J.; Weinstein, D. H.; Haas, G. M. *Nature (London)* **1975**, *257*, 383.
20. Fujishima, A. J.; Kohayakawa, K.; Honda, K. *J. Electrochem. Soc.* **1975**, *122*, 1487.
21. Frank, S. N.; Bard, A. J. *J. Am. Chem. Soc.* **1977**, *99*, 4667.
22. Fujishima, A.; Honda, K. *J. Chem. Soc. Jpn.* **1971**, *74*, 355.
23. Bolts, J. M.; Wrighton, M. S. *J. Phys. Chem.* **1976**, *80*, 2641.
24. Wrighton, M. S.; Wolczanski, P. T.; Ellis, A. B. *J. Solid State Chem.* **1977**, *22*, 17.
25. Kohl, P. A.; Bard, A. J. *J. Am. Chem. Soc.* **1977**, *99*, 7531.
26. Frank, S. N.; Bard, A. J. *J. Am. Chem. Soc.* **1975**, *97*, 7427.
27. Laser, D.; Bard, A. J. *J. Phys. Chem.* **1976**, *80*, 459.
28. Nakatani, K.; Matsudaira, S.; Tsubomura, H. *J. Electrochem. Soc.* **1978**, *125*, 406.
29. Chung, Y. W.; Lo, W. J.; Somorjai, G. A. *Surf. Sci.* **1977**, *64*, 588.
30. Ibid., **1978**, *71*, 199.
31. Henrich, V. E.; Dresselhaus, G.; Zeiger, H. *J. Phys. Rev. Lett.* **1976**, *36*, 1335.
32. Watkins, B. F.; Behling, J. R.; Kariv, E.; Miller, L. L. *J. Am. Chem. Soc.* **1975**, *97*, 3549.
33. Razena, B.; Wong, M.; Thomas, J. K. *J. Am. Chem. Soc.* **1978**, *100*, 1679.
34. Fujihira, M.; Ohishi, N.; Osa, T. *Nature (London)* **1977**, *268*, 226.
35. Clark, W. D. K.; Sutin, N. *J. Am. Chem. Soc.* **1977**, *99*, 4676.
36. Gerischer, H.; Willig, F. *Top. Curr. Chem.* **1976**, *61*, 33.
37. Gleria, M.; Memming, R. *Z. Phys. Chem. (Frankfurt am Main)* **1975**, *98*, 303.
38. Bolts, J. M.; Wrighton, M. S. *J. Am. Chem. Soc.* **1978**, *100*, 5257.
39. Wrighton, M. S.; Austin, R. G.; Bocarsly, A. B.; Bolts, J. M.; Haas, O.; Legg, K. D.; Nadjo, L.; Palazzotto, M. C. *J. Am. Chem. Soc.* **1978**, *100*, 1602.
40. Bolts, J. M.; Bocarsly, A.; Palazzotto, M. C.; Walton, E. G.; Lewis, N. S.; Wrighton, M. S. *J. Amer. Chem. Soc.*, **1979**, *101*, 1378.
41. Pankove, J. I. "Optical Processes in Semiconductors"; Dover Publications: 1971.
42. Wrighton, M. S.; Palazzotto, M. C.; Bocarsly, A. B.; Bolts, J. M.; Fischer, A. B.; Nadjo, L. *J. Am. Chem. Soc.* **1978**, *100*, 7264.
43. Wrighton, M. S.; Austin, R. G.; Bocarsly, A. B.; Bolts, J. M.; Haas, O.; Legg, K. D.; Nadjo, L.; Palazzotto, M. C. *J. Electroanal. Chem.* **1978**, *87*, 429.
44. Gerischer, H. In "Physical Chemistry: An Advanced Treatise"; Eyring, H., Henderson, D., Jost, W., Eds.; Academic: New York, 1970; Vol. 9A, Chapter 5.
45. Grushka, E., Ed. "Bonded Stationary Phases in Chromatography"; Ann Arbor Science Pub.: Ann Arbor, MI, 1974.
46. Weetall, H. H. *Science* **1969**, *166*, 615.
47. Allum, K. G.; Hancock, R. D.; Howell, I. V.; McKenzie, S.; Pitkethly, R. C.; Robinson, P. J. *J. Organomet. Chem.* **1975**, *87*, 203.
48. Lenhard, J. R.; Murray, R. W. *J. Electroanal. Chem.* **1977**, *78*, 195.
49. Moses, P. R.; Murray, R. W. *J. Electroanal. Chem.* **1977**, *77*, 393.
50. Moses, P. R.; Weir, L.; Murray, R. W. *Anal. Chem.* **1975**, *47*, 1882.
51. Janz, G. J.; Tomkins, R. P. T. "Nonaqueous Electrolytes Handbook"; Academic: New York, 1973; Vol. II.

52. Laviron, E. *J. Electroanal. Chem.* **1972**, *39*, 1.
53. Merz, A.; Bard, A. J. *J. Am. Chem. Soc.* **1978**, *100*, 3222.
54. Fischer, A. B.; Wrighton, M. S.; Umana, M.; Murray, R. W. *J. Amer. Chem. Soc.* **1979**, *101*, 3442.
55. Becker, E.; Tsutsui, M., Eds. "Organometallic Reactions"; Wiley: New York, 1972; Vol. IV.
56. Lenhard, J. R.; Rocklin, R.; Abrun, H.; Willman, K.; Kuo, K.; Nowak, R.; Murray, R. W. *J. Am. Chem. Soc.* **1978**, *100*, 5213.
57. Stirn, R. J.; Yeh, Y. C. M. *Appl. Phys. Lett.* **1975**, *27*, 95.

RECEIVED October 2, 1978.

INDEX

INDEX

INDEX

A

Absorption
  band of sensitizers, singlet 3
  coefficient 159
  cross section 117
  in depletion region 202
  of polystyrene immobilized
    sensitizer 7–9
  Q band region of α-NiPc 149f
  of silica-functionalized sensitizer 7–9
  spectra 84f, 86f, 93f
  of phthalocyanines 148–152
  of stearaldehyde–MBTH 75f
  of β-ZnPc 151f
  spectroscopy 115, 122
  spectrum of sandwich cell .. 128f, 129f
Absorptivity of CdS 200
Acceptor impurity 124
Acetonitrile solution 54
Acid quenching 81
Actinometry, Reineckate chemical . 143
Action spectra of sandwich cell ... 129f
Activation of polymer-anchored
  catalysts, photo- 13–25
Active semiconducting electrodes,
  photo- 155
Addition, reductive photo- 63
Adduct formation, time dependence
  of radical 181f
Adsorption
  characteristics of CO_2, H_2O, CO,
    and O_2 on $SrTiO_3$ 237
  characteristics of CO_2, H_2O, CO,
    and O_2 on TiO_3 236
  of CO_2 and H_2O on Pt–
    $SrTiO_3$ 233–252
  heat 38
  of CO on Pt 247
  H_2O 242
  on $SrTiO_3$ reduced by Ar^+
    bombardment, UPS dif-
    ference spectra 242–244
  on TiO_2 surface reduced by
    Ar^+ bombardment, UPS
    difference spectra 242–244
Adsorptive affinity of polystyrene 10, 11f
Aerosol OT (AOT) 175, 179
Air–water interface 47–48, 53,
                            63, 70, 87
Airy equation 159–160
Albery–Archer criteria 139–143
Alcohols, photoionization in 101–102
Alkanes, photoionization in 100–101
Alkene chemistry 19
Alkene reactions, iron-carbonyl-
  catalyzed 13–25

π-Allyl hydride intermediate for
  isomerization 23–24
Aluminum
  Cl_3 dissolved in (1,2-dimethoxy-
    ethane), titantium tetraiso-
    propoxide 165
  –H_2Pc–AuAl–ZnPc–Au cell,
    current–voltage character-
    istics 126
Amide bond cleavage 259
Amidization, XPES study of 256
Aminoperylene 102, 104
  cation polarization energy for .. 108
Aminopyrene (AP) 101–109
  cation polarization energy for ... 108
  laser photoionization of 102–104
  after laser photolysis 103f
  pulse radiolysis of 102
  singlet state 104
  triplet state 102–103
Anion alteration 91
Anionic micelles, quantum yields
  for photoionization in 106t
Anodic
  current, dark 281, 285
  current, photo- 282–283, 286
  decomposition, photo-
    of cadmium sulfide 186
    prevention of 271
    of n-type semiconductor 270
Anodization, controlled 256
Antenna chromophores 116–117
Antenna effect in thin-film systems 117
Anthracene 98, 123
  ionization potential of 109
Antibonding states 41
AOT (Aerosol OT) 175, 179
AP (see Aminopyrene)
Aquo complexes of metal ions 58
Arachidic acid 50–53
Arenes, triplet state of 107
Argon ion bombardment ... 29, 238–239
Argon ion laser 191
  UPS difference spectra for H_2O
    adsorption on TiO_2 surface
    reduced by 242–244
  UPS spectra for H_2O adsorption
    on $SrTiO_3$ reduced by .. 242–244
Array, close-packed 91
Assays, film composition 115
Assembly–solution interface, solute
  penetration at 55–59
Atmosphere, effect of exposure of
  cells to 142f
Atomic desorption 43f
Atomic emission spectroscopy 70

Auger
 decay of a core hole 41
 electron spectroscopy
 (AES) 29–30, 236, 238, 247
 peak-to-peak ratios of $SrTiO_3$ 240
 spectra of stoichiometric and
  reduced $SrTiO_3$ 241f
 transitions 42
 intraatomic 41
Azo dye coupling reaction 55

B

Back electron transfer 60
Band
 bending 159, 206
 effect on emission from doped
  CdS 198
 of p-GaP, valence 164
 parallel 197
 potential 164
 energy 160, 163
 to prevent back electron
  transfer 274
 to separate electron–hole pairs 274
 gap of CdS 194
 gap energy 41
 gap of H_2O 158
Barrier to electron transfer 274
Benzene excimer 54
Benzoic acid, p-benzoyl 2
Benzoquinone, photolysis in ... 106–107
Binding energy of $(bpy)_2Ru(Cl)$-
 $(py)^{1+}$ 263
Binding energy, excitation 189–191
Biological membrane models 74
Bipyridine ligands 92
4,4'-Bipyridine 132
Blocking contact 124
Bond cleavage, metal–metal 14
Borate buffer 73–94
Born equation 107–108

C

Cadmium arachidate 55
Cadmium selenide (CdSe) cell,
 short-circuit output stability of 222f
Cadmium selenide (CdSe), current–voltage curves for 222f
Cadmium sulfide (CdS) 175
 absorbtivity of 200
 band gap of 194
 difference between in- and out-
  of-circuit emission for ... 200, 202
 doped
 current–voltage–emission
  curves for 202f, 203f, 205f
 effect of band bending on
  emission from 198
 effect of electrolytes on
  emission from 197–198

Cadmium sulfide (CdS) (continued)
 doped (continued)
 efficiency of 206–209
 electrode, Te– 189–211
 electrodes, stability of 191–194
 emission spectra of 196f, 197f,
  198f, 208f
 energy conversion characteristics for 207t
 optical density of polycrystalline 195f
 wavelength dependence of
  emission and photocur-
  rent from 199t
 effect of temperature on emission from 207
 electrodes, optical properties of . 194
 photoanodic decomposition to .. 186
 spin trapping on aqueous dispersion of 177
 valence band of 219
Calibration curve for stearaldehyde
 concentration 78
Carbon
 dioxide (CO_2)
 chemisorption on $SrTiO_3$ 244
 and CO adsorbed on stoichiometric $SiTiO_3$, UPS difference spectra for 246f
 CO disproportionation to 246
 H_2O, CO, and O_2 on $SrTiO_3$,
  adsorption characteristics
  of 236
 H_2O, methane production
  from 236
 and H_2O on Pt–$SrTiO_3$,
  adsorption and reactions
  of 233–252
 on Pt surfaces, chemisorption
  of 247
 on $SrTiO_3(111)$ crystal face,
  chemisorption of H_2O,
  O_2, CO and 240–247
 over $SrTiO_3$–Pt, CH_4 production from H_2O and ... 247–249
 over transition metal surfaces,
  CO oxidation to 246
 monoxide (CO)
 adsorbed on stoichiometric
  $SiTiO_3$, UPS difference
  spectra for CO_2 and 246f
 desorption 28–36
 rates 30
 from stainless steel 33–34
 from W 33
 disproportionation to CO_2 ... 246
 extrusion of 19
 from nickel 28–36
 oxidation to CO_2 over transition metal surfaces 246
 Pc films 135
 Pd-catalyzed oxidation of 36–37
 on Pt, chemisorption of 247
 on Pt, heats of adsorption of . 247

INDEX
299

Carbon (*continued*)
  monoxide (*continued*)
    on SrTiO$_3$, adsorption
      characteristics of CO$_2$,
      H$_2$O, O$_2$, and 236
    on SrTiO$_3$(111) crystal face,
      chemisorption of H$_2$O,
      O$_2$, CO$_2$, and 240–247
    and sulfur removal from Pt
      samples by heating in O$_2$. 239
Carbonyl transitions, monitoring of   51
Carrier(s)
  collisions, phonon- 163–164
  density of p-GaP 164
  effective mass of minority 160
  pairs, photoredox 123–124
  -phonon collisions 160
  separation 124
  thermal velocity 162
Catalysis
  heterogeneous 28–43
  homogeneous 13
  photo- 156
  organometallic 13–25
Catalyst(s)
  light-generated 14
  photoactivation of polymer-
    anchored 13–25
  precursors 21–23
  resin 17
  turnover rate 22
Catalytic chemistry, probe 15
Catalytic cells, energy level diagrams for photoelectrolytic
  and photo- 157f
Cathode discharge lamp 238
Cathodes, photoenhanced reduction
  of N$_2$ on p-GaP 165–168
Cathodes, p-GaP photo- 164–165
Cathodic current 282, 286
  dark 283
  peak 281
Cathodic return for surface-
  attached oxidant 286
Cation(s)
  photoproduction of 106
  polarization energy(ies) 101t, 107–109
  solvent interaction 108
  yield, intensity dependence of .. 104f
Cationic micelles 98, 109
Cationic surfactant 87
CdS (*see* Cadmium sulfide)
Cell(s)
  atmospheric exposure, effect of . 142f
  configuration, derivatized n-type
    Si in 289–292
  CuInS$_2$, current-voltage curves . 223f
  current-voltage curves for
    idealized 226f
  diffusion-controlled photo-
    galvanic 139–141
  dissociation of H$_2$O in an electro-
    chemical 235
  electrochemical photovoltaic .155–156

Cell(s) (*continued*)
  fabrication 143
  ideal 142
  inert oxide-based 270
  performance, improvements
    in 226–228
  photoelectrochemical 155–156,
                          165, 185–211
  solar 215–230
  photoelectrosynthetic 155–156
  reaction kinetics, controlled .. 216–218
  regenerative semiconductor 140
  Schottky Junction 124–125
  short-circuit output stability of
    CdSe 222f
  solar, liquid junction 155–156
  shunting of by surface state .. 225f
  solar, photoelectrochemical ..215–230
  solar, Schottky barrier 281
  solar energy conversion efficiency
    for n-type GaAs ...269–270, 271
  Teflon 143
Cetyltrimethylammonium bromide
  (CTAB) 98, 175, 179
Characterization
  chromatographic and spectral .. 77
  of electrodes 191–209
  luminescent, of excited states ... 189
Charge
  delocalization of 153
  separation at interfaces 124
  space-, components 133
  space-, region 125, 179
  transfer 161
    equilibrium 272–274
    at metal–semiconductor inter-
      faces 235
    reactions, manipulation of
      interfacial 270–271
    reverse 163
  semiconductor–electrolyte
    interfaces 155–169
Chelate, diphosphine 22
Chelating reagent 19
Chemical electron transfer agents,
  photo- 141
Chemical properties, spectroscopy
  and photo- 152
Chemisorption
  of H$_2$O, O$_2$, CO, and CO$_2$ on Pt
    surfaces 247
  of H$_2$O, O$_2$, CO and CO$_2$ on
    SrTiO$_3$ crystal face 240–241
  on Pt polycrystalline foil surfaces,
    O$_2$ 247
  on SrTiO$_3$. CO$_2$ and O$_2$ 244
Chemistry, photo-, radical intermediates 173–183
Chemistry, photo-, solid–gas interface 233–252
Chloride salts, FDMS of perchlorate and 90f
Chromatographic analysis 92

Chromium complex, voltammograms of 258f
Chromium, deposition of 132
Chromophore(s)
  antenna 116–117
  aryl ketone 55
  dimerization of 87
  hydrophilic 48
  orientation 86
  polar 48
  styrene 55
CO (see Carbon monoxide)
CO_2 (see Carbon dioxide)
Compression
  –expansion cycling 76
  –expansion hysteresis 88
  rate 87
Concentration quenching 92
Condensed phase effect on photoionization 97–110
Conduction band 178, 224
  dark 226
  of p-GaP 164
Conduction, thermal 40
Conductivity, dark 240
Conductivity dependence on lattice modification 152
Conversion efficiency 216–218
Coordination sphere, Fe 19–24
Copper-InS_2 cell, current–voltage curves for 223f
Core hole, Auger decay of a 41
Corey–Pauling–Koltun (CPK) molecular models 85–86
Corrosion, photo-, of GaAs 216
Corrosion, photo-, reaction 182
Counterelectrode potential 226
Counterelectrode, Pt foil 201
Counterion effects 87–94
Counterion exchange 79
Coupling agent 257, 261
Cross
  -linking agent 18–19
  section(s)
    absorption 117
    desorption 40
    photodesorption 30, 33
    -sectional areas, orthogonal 86
Crosslinkage, interchain 74
Crystal
  modification, phthalocyanine .147–148
  ordering 240
  photoreactions 54
α-Crystal modification 150–152
β-Crystal modification 148, 151
Crystalline linear paraffins, packing of 51
CTAB (Cetyltrimethylammonium bromide) 98, 175, 179
C terms 152
Current
  dark anodic 278
  density 127
    reverse saturation 127–128
    voltage curve for GaAs 230f

Current (continued)
  efficiency (see also Quantum efficiency)
    flow, electrode weight loss after external 192
  photo-
    density, short-circuit 130
    deterioration 221
    effect on emission intensity .203–204
    effect of excitation intensity on 202
    enhancement of 146
    from n-type Si electrode, time dependency from 290f
    -potential curves, equilibrium .289–292
    -potential properties for derivatized Si, dark 281
  short circuit 221f
  –voltage
    characteristics of sandwich cells 126f, 127f
    curves 143
      for CdSe 222f
      for $CuInS_2$ cell 223f
      for derivatized n-Si photoelectrode 291f
      for etched GaAs 228f
      for idealized cells 226f
      for n-SnO_2 electrodes ... 144–148
    –emission curves for doped Cd 202f, 203f, 205f
    –emission properties of doped CdS 200–206
Cyanine dye in multilayer assemblies, fluorescence quenching of 64
Cylindrical mirror analyzer (CMA) 238

D

Dark
  conductivity 240
  electrochemistry 132
  electrode 140
  semiconductor electrode 187f
Davydov model 150
Deactivation, excited state
  mechanisms for 188–189
  rate constants 200
  routes, competition between .200–209
  of semiconductor electrodes 188f
Defect channels in multilayer assemblies 58–59
Degenerate electrolysis 206
Degenerate excited state 148–149
Degree of congestion in microenvironment 56
Degree of order in multilayers ... 51
Delocalization of charge 153
Density filters 30
Depletion region
  absorption in 202
  at electrode–solution interface 133–134
  at n-type semiconductor–electrolyte interface 187f

INDEX

Desorption
  efficiencies 39–40
  energy 42
  models, microscopic 40
  photo- 38–39
    cross sections 30, 33
    from oxide surfaces 246
    via quantum effect 28–36
    threshold 42
1,1-Dialkyl-4,4' bypyridinium^{2+} ... 60
Diazonium salt 55
Dicyclohexylcarbodiimide 257
Dielectric constant 107
Differential capacitance of H_2Pc
  electrodes 134f
Diffusion
  -controlled photogalvanic cell . 139–141
  lengths, exciton 201
  limited current 286
  times, p-GaP electron 164
  times, n-TiO_2 hole 162
N,N'Diheptyl-5,5' dipyridine (heptylviologen) 145
Dimerization 53
  of chromophores 87
  photo- 53–54
  of thiazine dyes 87
Dimethylaminobenzophenone
  chromophore 2
  as a sensitizer 3–5
N,N-Dimethylaniline (DMA) 61–63, 123
5,5-Dimethyl-1-pyroline-1-oxide
  (DMPO) 174
  hydroxyl radical adduct of ... 175–176
  super oxide radical adduct of . 176–177
Diphenylvinylphosphine 16
Dipolar relaxation 163
Dipole terms, forbidden 150
Dipole transition 152
  moments 150
Discharge lamp, cathode 238
Dismutation of HO_2 182
Dispersion(s)
  application of spin trapping
    to 177–179
  of CdS, spin trapping on
    aqueous 177
  of phthalocyanine, spin trapping
    on aqueous 177
  of pigments, ultrasonic 175
  prevention of flocculation in
    pigment 179
  spin trapping in oxygenated
    nonaqueous 178
Dissolution, photoanodic 191
Ditelluride electrolyte 192
Dithionite 74
DMPO (see 5,5-Dimethyl-1-
  pyroline-1-oxide)
Docosyl trimethylammonium
  bromide 88
Donor–acceptor pairs 60
Dye films, optical densities of
  monolayer 87

E

Effective mass of minority carriers . 160
Efficiency
  of doped CdS 206–209
  optical-to-electrical energy
    conversion 290–292
  for n-type GaAs cell, solar energy
    conversion 269–271
Ejection, photo-, of electrons 108
Ejection, photo-, process 98
Electric field 40
Electric effect at H_2Pc electrode,
  photo- 135f
Electrochemical
  cell, dissociation of H_2O in an .. 235
  cells, photo- ... 155–156, 165, 185–211
    solar 215–230
  cells, photovoltaic 155–156
  selectivity 85
  stability by silane multilayers,
    improving 261
  storage 114
Electrochemistry, dark 132
  photo-
    of metal phthalocyanine
      films 139–154
    at phthalocyanine electrodes 131–135
    physics of 186
Electrode(s)
  characterization of 191–209
  composite 132
  corrosion, prevention of 69
  current–voltage curves
    for Mg 145f
    for NiPc 145f
    for n-SnO_2 144–148
    for α-ZnPc 145f
    for β-ZnPc 147f
  dark 140
  semiconductor 187f
  derivatization, effect of 270
  derivatized
    anodic peak positions for 275t
    equilibrium current-potential
      properties of naked and . 289
    interfacial energetics for ... 272–274
    in nonaqueous electrolyte,
      cyclic 275–277
  output voltage for n-type
    Si 279
  Pt 275–277
  semiconductor 269–292
  shelf life of 283
  etching 192, 217–219
  excited state deactivation of
    semiconductor 188f
  fluoride sensitive 99
  GaAs 216
  p-GaP, enhanced reduction on
    photoelectrolysis 167f
  H_2Pc, differential capacitance
    of 134f
  H_2Pc, electric effect at photo- .. 135

Electrode(s) (continued)
  Mg, current–voltage curves for . 145f
  NiPc, current–voltage curves for 146f
  paramagnetic resonance (EPR)
    spectra characterization ... 106
  photo-
    active semiconducting 155
    electric effect at H_2Pc 135f
    electrochemistry at phthalo-
      cyanine 131–135
    electrolysis 156, 168
    electrolytic and photocatalytic
      cells, energy level dia-
      grams for 157f
    electrosynthetic cells 155–156
    elimination processes, Type I
      and II 49–52
    enhanced reduction on p-GaP 167f
    enhanced reduction on p-type
      semiconductors, energetics
      of 167–168
    enhancement of current 146
    generated hole in n-type semi-
      conductor 161f
    luminescent properties of semi-
      conductor 185–211
  potential on luminescence, effect
    of 204
  potential on photocurrent and
    luminescence efficiency,
    effect of 201
  preparation 143
  properties of thin film 115
  Pt, derivatized 275–277
  Pt–PtO modified 253–267
  quantum efficiencies at high
    intensities 290
  reactions, effects of light
    intensity on 278f
  reactions, effect of oxide layer
    on 281–282
  semiconducting 139–154
  p-type semiconductor 132
  shelf life of derivatized 283
  n-type Si
    cyclic voltrammetry of 277–280
    output voltage for 279
    time dependency of photo-
      current from 290f
  n-SnO_2, current–voltage curves
    for 144–148
  –solution interface, depletion
    region at 133–134
  stability to adsorbed
    impurities 224–230
  surface modification 215
  surface states 224–230
  syntheses of 131–132
  tailormade surfaces, modified .253–267
  TiO_2
    electrochemical behavior of .. 259
    modified 253–267
    XPES, spectra before and after
      silanization 255f

Electrode(s) (continued)
  weight loss after derivatized
    electrodes 275–277
  ZnPc, current–voltage curves
    for 145, 147
Electrolysis, degenerate 206
Electrolyte
  cyclic voltammetry of
    derivatized electrodes in
    nonaqueous 275–277
  ditelluride 192
  on emission from doped CdS,
    effect of 197–198
  Fermi level, electrolyte redox
    potential and 158
  interfaces, charge transfer
    semiconductor- 155–169
  interface, depletion region at
    n-type semiconductor- 187f
  nonaqueous 165
  polysulfide 193
  redox potential and electrolyte
    Fermi level 158
  redox reactions 188
  supporting 132–133
Electrolytic fixation of N_2 166
Electron
  affinity of H_2O 158
  affinity of solutes 109
  bombardment, surface composi-
    tion alteration from 240
  density 28
  diffraction (LEED), low-energy 236
  diffusion times, p-GaP 164
  energy(ies) 38, 124
    distribution 238
  excitation of an insulator 39
  excitation of metal 39
  flow in external circuit, quantum
    efficiency for 206
  flow, quantum yield for 274
  gun, coaxial 238
  –hole pair(s) 141, 178, 200, 233
    band bending to separate ... 274
    recombination 188, 200
    inhibition of 186
    separation of 271
  injection, hot 164–165
  loss spectroscopy (ELS) .236, 238, 240
    spectra for stoichiometric and
      reduced $SrTiO_3$ 242f
  micrograph of etched GaAs 218f
  -phonon collisions 164
  photoejection of 108
  scavenger 106
  spectroscopy 277
    for chemical analysis (ESCA) . 70
  spin resonance (ESR) 174
  spectrum
    of CO_2 adduct of DMPO 181f
    O_2 adduct of DMPO 177f
    of OH adduct of DMPO . 176f
  splittings, β-hydrogen 177
  splittings, nitrogen 177

INDEX

Electron (*continued*)
  stimulated desorption 38–39
  thermalization times, p-GaP ..164–165
  transfer 99, 146, 233
    agents, photochemical 141
    band bending to prevent back 274
    barrier to 274
    inner-sphere 286
    mediated 285–288
    oxidation of solution species by 285
    photoassisted 109
    processes 59–65, 123
      detection of 178
    rates, Marcus–Weller theory
      of 140
    selective 69
    in n-type semiconductor 259
  tunneling 39, 41, 42, 224
  thickness 219
Electronic charge 107
Electronic states, intraband
  gap 189–191, 194, 197
Electrophotographic imaging 179
Electroreduction of molecular
  oxygen 131
Electrostatic interaction,
  strong 226, 229f
Electrostatic interaction,
  weak 226, 227f
Elimination processes, β- and
  reductive- 24
ELS (*see* Electron loss spectroscopy)
Emission
  for CdS, difference between in-
    and out-of-circuit 200, 202
  from CdS, effect of temperature
    on 207
  curves for doped CdS, current–
    voltage- 202f, 203f, 205f
  from doped CdS, effect of
    band bending on 198
  from doped CdS, effect of
    electrolytes on 197–198
  efficiency 207
  intensity, effect of photocurrent
    on 203–204
  properties of doped CdS, current–voltage- 200–206
  quantum yield 208
  spectra 93f
    of assemblies 121f
    of doped CdS .196f, 197f, 198f, 208f
    of fluoranthene 117, 118f
    of polystyrene immobilized
      sensitizer 7–9
    of silica-functionalized
      sensitizer 7–9
Energetics for derivatized electrode,
  interfacial 272–274
Energetics of photoenhanced
  reduction on p-type semiconductors 167–168

Energy
  conversion characteristics for
    doped CdS 207t
  conversion efficiency, optical-
    to-electrical 290–292
  diagram of heterojunction model 158f
  efficiency 30, 31f
  electron 124
  interaction 151
  level diagram
    for electrochemical photo-
      voltaic cells 156f
    for p-GaP photocathode 164
    for photoelectrolytic and
      photocatalytic cells 157f
  migration 92
  of photoionization 102
  photon 116–117, 123
  storage, heterogeneous photo-
    sensitizers for 1–12
  storage system, solar 1–12
  transfer distance 91
    dependence of 48
  utilization of incident 116
Environments, synthetic reaction 113–136
Enzyme systems models 48
Epitaxial growth 223
Equilibrium current-potential
  curves 289–292
Equilibrium solvation structures .. 163
ESCA (Electron spectroscopy for
  chemical analysis) 70
ESR (Electron spin resonance) ... 174
Ethane, preparation of 1-phenyl-
  phosphino-2-diphenylphos-
  phino 16
Ether (CE), 18-crown-6- 176
tris(ethylenediamine)cobalt(III) . 145
Ethylenediamineteraacetic acid
  (EDTA), photooxidation of . 179
Ethyl octadecanoate, hydrolysis of . 83
Exciplex systems 98
Excitation
  $(a_{1u})^1(e_g)^1$ 149
  beam, chopping of 283
  beam, penetration depth of 200
  energy 60, 100
  intensity on photocurrent, effect
    of 202
  of photoelectrodes, laser 191
  spectra of assemblies 120f, 121f
  spectrum for fluoranthene 119f
Excited state(s)
  deactivation
    mechanisms for 188–189
    rate constants 200
  routes, competition
    between 200–209
  of semiconductor electrodes .. 188f
  degenerate 148–149
  luminescent characterization of . 189
  quenching 47–65
  reactions 92

Exciton
 collection 116–123
 distance 123
 diffusion lengths 120, 220
 formation 189
 trapping 120
Excitonic binding energy 189–191
Extinction coefficient 3–5
 of metals 28–29
Extrusion of CO 19–20

F

Faradaic activity 132
Fatty acids, saturated 49, 53
FDMS (see Field desorption mass spectra)
Fermi level 42, 158f, 161f, 186, 272
 electrolyte redox potential and electrolyte 158
 semiconductor 216
Ferricyanide/ferricyanide couple . 145
Ferrocene(s)
 derivatization with 272
 oxidation of surface-attached ... 279
 units, oligomers of electroactive . 277
Ferricenium centers, hole transfer to form 282
Field desorption mass spectra (FDMS) 89, 93
 of perchlorate and chloride salts . 90f
Film shrinkage, first-order 82
Filtering effect, optical 130
Flash photolysis of TMPD 102
Flat-band potential 272
 of phthaloxycyanine electrodes 132–134
Flocculation in pigment dispersions, prevention of 179
Fluoranthene 115
 emission spectra of 117, 118f
Fluorescence
 activation spectrum 101
 quenching 55–56, 64, 120–123
  effect of selectively placed Au on 121–122
 yield 99
Fluorescent dyes 61
Fluoride-sensitive electrode 99
Fluorimeter, monolayer 91
Fluorometry 115
Forbidden dipole terms 150
Förster theory calculations 91
Förster's transfer 120
Fourier transform IR (FTIR) 89f
 spectroscopy 71
Frank–Condon transition 40
Frequency factor of the quantized states 161
FTIR (see Fourier transform IR)

G

Gallium arsenide
 cell, solar energy conversion efficiency for n-type 269–270, 271
 current density–voltage curve for 230f

Gallium arsenide (continued)
 electrode 216
 etched 219f, 224
  current–voltage curves for ... 228f
  electron micrograph of 218f
 photocorrosion of 216
 Ru(II) chemisorption 227
 surface states 224
Galvanic
 cell, photo-, diffusion-controlled 139–141
 -devices, photo- 85–87
 photo-, open circuit voltage ..147–148
 systems, photo-, potential 97–110
p-Gallium phosphide
 carrier density of 164
 cathodes, photoenhanced reduction of N_2 on 165–168
 conduction band of 164
 electrodes, photoenhanced reduction on 167f
 electron diffusion times 164
 electron thermalization times .164–165
 photocathodes 164–165
 valence band bending of 164
Gas
 –assembly–interface 56
 chromatography 16, 248
 interface, photochemistry at the solid– 233–252
 -phase spectra 148–149
 –water interface 70
Generated, photo-, hole in n-type semiconductor electrodes 161f
Germanium
 attenuated total reflectance (ATR) plates 71–73, 89
 derivatized n-type 285
 voltammograms for n-type 284f
Glass, sensitizers grafted to silica or 6–7
Glass transition temperature 50, 51
Glutaraldehyde 74
Gold
 contact 124
 derivatized 275–277
 on fluorescence quenching, effect of selectivity placed 121–122
Grain boundaries in polycrystalline electrodes 193–194, 201
Grafting of chromophore .... - 3
Grafting monolayers 77f
Ground states, spin degenerate ... 152

H

Heat capacity 40
Heating rate on desorption, effect of 32f, 34f
Heats of adsorption of CO on Pt .. 247
Heptylviologen 145
Heterogeneous catalysis 28–43
Heterojunction model of semiconductor–electrolyte junctions 158–160

INDEX

High performance liquid chromatography (HPLC) 79
 analysis 81-82, 92
H_2O (see Water)
Hole(s)
 capture processes, competitive .. 278
 diffusion times, n-TiO_2 162
 –electron pair 141
  band bending to separate 274
  electron– 178, 200, 233
   recombination of 188, 200
   recombination inhibition ... 186
   separation of 271
 generation, stability of silane
  surface layer toward 256
 injection, hot 164
 limited current 286
 migration 216
  barrier to 220f
 mobility 162
 thermalization times, n-TiO_2 ... 163
 transfer 285
  to form ferricenium centers .. 282
 tunneling times, n-TiO_2 161–162
 in valence band, photogenerated 282
Homogeneous catalysis 13
Homogeneous solution effect on
 photoionization 97–110
Hot
 carrier injection 58
  across semiconductor-electrolyte interfaces 160–165
 electron injection 164–165
 hole injection 164
HPLC (High performance liquid
 chromatography) 79
Hückel calculations, extended 149
Hydrogen
 abstraction 2, 50, 51
 -bonding solvents 52
 peroxide (H_2O_2), application of
  spin trapping to photosynthesis of 180–183
 phthalocyanine electrode, photoelectric effect at 135f
 phthalocyanine electrodes, differential capacitance of 134f
γ hydrogen 50
β-hydrogen ESR splittings 177
Hydrogenation of alkenes 19
Hydrolysis kinetics 79
Hydrophobic compounds at interfaces, reactions of surfactant
 and 47–65
Hydrophobic compounds in organizates, reactions of surfactant
 and 47–65
Hydrosilation of alkenes 19
Hydroxyl radical(s) 174
 adduct of DMPO 175–176
8-Hydroxyquinoline 55
Hysteresis
 compression–expansion 88
 in cyclic scans 225
 loop 88

I

Ideal cell 142
Indium
 contact 124
 –H_2Pc–Au cells 124
 –$ZnPc$–Au cells 124–131
Induced, redox processes, photo-,
 interfacial 59–65
Infrared (IR)
 assignments for iron carbonyl–
  phosphine complexes 18t
 filter 239, 248
 spectroscopy 37, 70
Insulatory, electron excitation of .. 39
Intensity dependence of cation
 yield 104f
Intensity in sodium lauryl sulfate
 solution 105f
Interaction energy 151
Interface(s)
 air–water 47–48, 53, 63, 70, 87
 charge
  separation at 124
  transfer at metal–semiconductor 235
  transfer semiconductor–
   electrolyte 155–169
 depletion region at electrode-
  solution 133–134
 depletion region at n-type semiconductor–electrode 187f
 gas–assembly 56
 gas–water 70
 photochemistry at the solid–
  gas 233–252
 reactions of surfactant and hydrophobic compounds at 47–65
 solute penetration at assembly-
  solution 55–59
 solution–assembly 56
 states, semiconductor 281
Interfacial potential well for n-type
 semiconductors 159f
Intermolecular photoreactions 49–52
Intraband relaxation 160
Intramolecular photoreactions 49–52
Ion
 bombardment 238–240
  cleaning by ion 247
  in situ surface cleaning by ... 236
 desorption 39
 exchange between semiconductor
  surface and solution 218–223
 on surface states, effect of
  chemisorbed 225–227
Ionic radius 107
Ionization, photo-
 in alcohols 101–102
 in alkanes 100–101
 in anionic micelles, quantum
  yields for 106t
 of AP, laser 102–104
 condensed phase effect on 97–110
 energy of 102

Ionization, photo- (*continued*)
 homogeneous solution effect
  on 97–110
 micellar systems effect on 97–110
 onset of 98
 onset in micelles 104–106
 of solutes in nonpolar liquids .. 100f
Ionization, potential 97–110
 of anthracene 109
 in the gas phase 107
 of pyrene 109
 in solution 107
 of TMPD 101
Iridium complex sensitizer 11
Iron
 carbonyl
  catalyzed alkene reactions ... 13–25
  complexes, photoprocesses in . 19–20
  –phosphine complexes, IR
   assignments for 18t
  complexes, cyclic voltammograms 133f
  coordination sphere 19–24
Isobacteriochlorin 62–63
Isomerization
 of alkenes 19
 π-allyl hydride intermediate for . 23, 24
 1-pentene 20–23
 quantum yield 21–23
 trans-cis 54
Isooctane 100
Isosbestic points 53

K

Keithley picoammeter 98
Kinetics
 controlled cell reaction 216–218
 first-order 81
 hydrolysis 79
 reaction 81

L

Langmuir–Blodgett technique 56, 77
Laser
 Ar ion 191
 excitation 106
  of photoelectrodes 191
 photoionization of AP 102–104
 photolysis 99
  AP after 103f
 flash 100
 of TMPD 102
Lattice
 heating 28–36
 modification, conductivity
  dependence on 152
 vibrations 188
Least-squares method of resolution 117
LEED (*see* Low energy electron
 diffraction)

Ligand(s)
 displacement 262
 dissociation 14
 exchange with anchored
  metalloporphyrins 56
 phosphine 17–18
Light
 intensity on electrode reactions,
  effects of 278f
 intensity vs. open-circuit
  photovoltage in sandwich
  cells 131f
 reabsorption and reemission of
  emitted 201
Lipid bilayers 47
Liposomes 47
Liquid
 chromatography (HPLC), high
  performance 54
 junction chemistry, semiconductor 215–230
 junction solar cell 155–156
  shunting of by a surface state 225f
Low energy electron diffraction
 (LEED) 236
 optics 238
 pattern 240
Luminescence 91
 effect of electrode potential on . 204
 efficiency, determination of 207
 quantum yield 91
Luminescent
 characterization of excited states 189
 doped electrode 190f
 properties of semiconductor
  photoelectrodes 185–211

M

Madelung potential 41
Magnetic circular dichroism
 (MCD) 140
Magnesium electrode, current–
 voltage curves for 145f
 Q band region of α-NiPc 149f
 spectra 148–149, 152
  of phthalocyanines 148–152
  of β-ZnPc 151f
 spectrometer 144f
Marcus–Weller theory of electron
 transfer reaction rates 140
Mass spectra, field desorption
 (FDMS) 89
Mass spectroscopy, vapor-phase
 chormatography– 74
Matrices, functionalized insoluble . 1
Matrix isolation 14
MCD (*see* Magnetic circular
 dichroism)
Mechanistic consequences of short-
 lived radicals at surfaces .. 173–183
Mercury 248
 lamp 239
 –Xenon arc lamp 143
Merrifield polymer 2

Metal(s)
  diimine complexes 141
  electron excitation of 39
  extinction coefficients of 28–29
  ions, aquo complexes of 58
  –metal bond cleavage 14
  phthalocyanine films 139–154
    photoelectrochemistry of ..139–154
    spectroscopy of 139–154
  –semiconductor interfaces, charge
    transfer at 235
  work function of ...124–125, 141, 158
Metallation, rate of 58–59
Metallic films, deposition of ...114–115
Metalloporphyrins 56, 61–64
  ligand exchange with anchored . 56
Methane production from CO_2
  and H_2O 236
  over $SrTiO_3$-Pt 247–249
Methane, thermal generation of ... 249
Methanol, radiolysis of 102
p-Methylacetophenone 50–51
3-Methyl-2-benzothiazolinone
  hydrazone (MBTH) 73–74
  absorption spectra of stear-
    aldehyde– 75f
  hydrochloride 71
Methylcyclohexane 52
Methylene blue 71
Methylene blue adsorption,
  orientation of 85–87
MGR model 40–43
Micelles (see also Organizates) 47–
                                  48, 271
  cationic 98, 109
  photoionization onset in 104–106
  quantum yields for photoioniza-
    tion in anionic 106t
  SDS 52
  quantum yield for 52
  sodium lauryl sulfate 107
Micellar systems effect on photoionization 97–110
Microenvironment(s) 51
  degree of congestion 56
  ketone 50
Microscope, polarizing 53
Molecular
  absorption, onset of 108
  films, deposition of 114–115
  models, Corey–Pauling–Koltun
    (CPK) 85–86
  orbital calculation 150
Monochromator 98
Monolayer(s)
  dye films, optical densities of ... 87
  film(s)
    characterization and reactions
      of surfactant 69–94
    operations 77f, 80f
    techniques 71–94
  fluorimeter 91
  quartz-supported 93
  trough, multicompartment 71–73

Multilayer(s)
  assemblies 47–55
  defect channels in 58–59
  fluorescence quenching of
    cyanine dye in 64
  reactions in 55–59
  degree of order in 51
  restriction of motion in 51

N

Nickel, CO desorption 28–36
Nickel phthalocyanine (NiPc)
  electrodes, current–voltage
    curves for 146f
α-Nickel phthalocyanine (α-NiPc)
  absorption Q band region of ... 149f
  MCD Q band region of 149f
  Q band 149
Nitroaromatics on silanized polycrystalline TiO_2, voltammograms of surfaced-attached .. 260f
Nitroaromatics, surface- attached 259–261
Nitrogen (N_2)
  electrolytic fixation of 166
  ESR splittings 177
  on p-GaP cathodes, photoen-
    hanced reduction of 165–168
m-Nitrotoluene, solution formal
  potential of 260
Norbornadiene
  conversion to quadricyclene 1–12
  triplet energy of 3

O

n-Octadecanol 76, 79
Octanophenone 52
Octupole transition rule 152–153
Ohmic contact 125
Oil diffusion pump 143
Oligomers of electroactive ferrocene units 277
Open-circuit photovoltage 130
  limiting 216
  in sandwich cells, light
    intensity vs. 131f
Open circuit voltage, photogalvanic 147–148
Optical
  conductivity 40
  densities of monolayer dye films . 87
  density of polycrystalline doped
    CdS 195f
  energy conversion 206
  efficiency 208
  sustained 189
  filtering effect 130
  properties of doped CdS
    electrodes 194
  spectroscopy 70
Optics, front-face 117
Optics, LEED 238

Optimum band gaps for one-step
 solar energy conversion 216
Organic semiconductors 141
Organizates (see also Micelles)
 probe molecules in 48–55
 properties of 48–55
 reactions of surfactant and
  hydrophobic compounds in 47–65
Organometallic complexes, transition metal 13
Organometallic photocatalysis 13–25
Orientation, chromaphore 86
Orthogonal cross-sectional areas ... 86
Oscillator, quartz crystal 115
Output stability of CdSe cell,
 short-circuit 222f
Oxidation
 of CO, Pd-catalyzed 36–37
 current, dark 281
 mediated 286
 photo- 56
 of solution species by electron
  transfer 285
 of surface-attached ferrocene ... 279
Oxide
 growth, kinetic inhibition of 282
 growth, peak shifting from 283
 surfaces, photodesorption from . 246
Oxygen (O_2)
 adsorbed on $SrTiO_3$, UPS
  spectra for 245f
 chemisorption on Pt polycrystalline foil surfaces 247
 chemisorption on $SrTiO_3$(III)
  crystal face 240–247
 chemisorption on $SrTiO_3$, work
  function 244
 dissociation 37
 electroduction of molecular 131
 photoreduction, rate of 182
 on Pt surfaces, chemisorption of 247
 on $SrTiO_3$ adsorption characteristics of 236

P

Palladium-catalyzed oxidation of
 carbon monoxide 36–37
Palladium porphyrin substrates ... 61
Paraffins, packing of crystalline
 linear 51
Paraquat 60
Peak
 -to-peak ratios of $SrTiO_3$, AES .. 240
 positions for derivatized electrodes, anodic 275t
 shifting in derivatized Ge 285
 shifting from oxide growth 283
 widths at half height, derivatized Si 281
1-Pentene
 isomerization 20, 22t, 21–23
 reaction, quantum yield for 21
 with silicon hydrides, reaction
  of 23–25

Penetration depth of excitation
 beam 200
Perchlorate and chloride salts,
 FDMS of 90f
Perovskite structure 240
Perylene 104, 115
 diffusion of 120
 excitation of 117
1,10-Phenanthroline 132
Phenothiazine 102
 cation polarization energy for .. 108
 di-cation 106
1-Phenylphosphino-2-diphenyl-
 phosphino ethane, preparation of 16
Phonon–carrier collisions .. 160, 163–164
Phonon–electron collisions 164
Phosphorescence spectra of silica-
 functionalized sensitizer 7–9
Photolysis 92
 in benzoquinone 106–107
 laser 99
 AP after 103f
 flash 100
 of TMPD 102
 steady state 100
Photon(s)
 absorption 41
 -assisted reaction of reduced
  oxide–metal contact system 249
 energy(ies) 29, 38,
                                116–117, 123
 flux 239
 -induced desorption 38–39
 low energy 200
 surface interaction 39–40
Phthalocyanine(s) (Pc) 115, 175
 absorption spectra of 148–152
 crystal modification 147–148
 deposition of 143
 determination 148–152
 electrodes, photoelectrochemistry at 131–135
 films 123–131
 metal 139–154
 MCD spectra of 148–152
 spin trapping on aqueous
  dispersions of 177
 vibronic assignment of 153
Physics of photoelectrochemistry . 186
Picoammeter, Keithley 98
Pigment dispersions, prevention of
 flocculation in 179
Pigments, ultrasonic dispersion of . 175
Piperidine 56
Platinum (Pt)
 CH_4 production from H_2O, and
  CO_2 over $SrTiO_3$– 247–249
 chemisorption of H_2O, O_2, CO,
  and CO_2 from 247
 electrodes, derivatized 275–277
 foil conterelectrode 201
 heats of adsorption of CO on ... 247

Platinum (Pt) (*continued*)
  polycrystalline foil surfaces, O_2
    chemisorption on 247
  –PtO modified electrodes 253–267
  –PtO–py, XPES spectra of 264f
  –PtO surface, silanized 256–258
  sulfur and carbon removal from . 239
  –SrTiO$_3$, adsorption and reactions
    of CO_2 and H_2O on 233–252
  –SrTiO$_3$, foil sample 239f
Point dipole model 150
Polyacrylophenone 2
Polarization energy(ies),
  cation 101t, 107–109
Polycondensation chemistry 74–79
Polycrystalline doped CdS, optical
  density of 195f
Polycrystalline electrodes, grain
  boundaries in 193–194, 201
Polymer
  -anchored catalysts, photoactiva-
    tion of 13–25
  anchoring system(s) 14–15, 19
  catalyst systems 15t
  swelling 23
Polymeric films 115
Polypeptides 78
Polystyrene 115
  adsorptive affinity of 10, 11f
  immobilized sensitizer 6–12
Polysulfide electrolyte 193
Porphine, tetra (octadecyl)ester
  of $\alpha,\beta,\gamma,\delta$-tetra(p-carboxy-
    phenyl) 62
Porphyrin(s)
  anchored 56
  rate constants for metallation of . 59
  substrates, Pd 61
Potassium Chloride (KCl) as supporting electrolyte 142–144
Potential
  barrier, transmission coefficient
    for 162
  curves, equilibrium current- . 289–292
  energy, band-bending 160, 163
  energy curve 42
  properties for derivatized Si,
    dark current- 281
  well dimensions for n-TiO$_2$ 162
  well for n-type semiconductors,
    interfacial 159f
Preferential binding of reactant
  molecules 69
Pressure–area curve(s) 50, 85f, 93f
Pressure–area isotherms 50, 71, 76
Probe catalytic chemistry 15
Probe molecules in organizates ... 48–55
Produced current, photo- 98
Production of cations, photo- 106
Pt (*see* Platinum)
Pulse radiolysis of AP 102
Pulse radiolysis of p-xylene 102
Pump, photo- 179

Pyrene 98, 101, 106
  cation polarization energy for .. 108
  ionization potential of 109
Pyridine 56

Q

Q band 148–152
Quadricyclene
  formation, quantum efficiency for 5–7
  norbornadiene conversion to ... 1–12
  –norbornadiene interconversion . 2–3
Quantized states, frequency factor
  of the 161
Quantum
  effect, photodesorption via 28–36
  efficiency(ies) (*see also* Current
    efficiency) 50
    for electron flow in external
      circuit 206
    at high intensities, electrode ... 290
    for quadricyclene formation . 5–7
    of sandwich cells 130
    mechanical tunneling 159–165
  yield(s) 5–7
    for electron flow 274
    isomerization 21–23
    luminescence 91
    for 1-pentene reaction 21
    for photoionization in anionic
      micelles 106t
    for photolyses 92
    for SDS micelles 52
Quartz crystal oscillator 115
Quartz-supported monolayers 93
Quenchers, diffusable solute
Quenching
  acid 81
  concentration 92
  excited-state 47–65
  fluorescence 55–56, 64, 120–123
  effect of selectively placed
    Au on 121–122
  by impurities 6
  interfacial 61
  by reduction 77
Quinone/hydroquinone couple . 145–148

R

Radical(s)
  addition reaction, short-lived
    radical detection by 174
  adduct
    of DMPO, hydroxyl 175–176
    of DMPO, superoxide 176–177
    formation, time dependence of 181f
  hydroxyl 174
  intermediates in photo-
    chemistry 173–183
  superoxide 174
  at surfaces, mechanistic conse-
    quences of short-lived .. 173–183
  at surfaces, spin trapping of
    short-lived 173–183

Radiolysis
  of AP, pulse 102
  of aqueous SF_6 100
  of methanol 102
  of p-xylene, pulse 102
Rate constants 99
  first-order 81
  for metallation of porphyrins ... 59
Rate enhancement 83
Reactant molecules, preferential
  binding of 69
Reactants in surface studies,
  advantages of gaseous 235
Reaction
  kinetics 81
  path of H_2O_2 photosynthesis ... 180
  photo-, of functionalized sur-
    factant molecules 48–55
Reactivity 47–65
Rectification of photoelectrodes . 140–153
Redox
  carrier pairs, photo- 123–124
  chemistry 60
  couple(s) 139–140, 155–156, 216, 272
    derivatization with 271–272
    voltammetric behavior with .. 224
  potential 186, 216
    electrolyte, and electrolyte
      Fermi level 158
  processes, interfacial photo-
    induced 59–65
  reactions, electrolyte 188
Red shifting in emission and
  absorption spectra 149–195
Reducing agent 166–167
Reduction
  on p-GaP electrodes, photo-
    enhanced 167f
  of N_2 on p-GaP cathodes,
    photoenhanced 165–168
  photo-, rate of oxygen 182
  of titanium tetraisopropoxide,
    stepwise 166
  on p-type semiconductors, ener-
    getics of photoenhanced 167–168
Reductive photoaddition 63
Refractive index 107
Reineckate chemical actionometry . 143
Relaxation
  dipolar 163
  intraband 160
  time(s) 160
  in electrolyte, n-TiO_2 163–164
Resin
  catalyst 17
  phosphinated 14, 16–17
  styrene–divinylbenzene 14–17
Resistive heating 30
Resonance tunneling calculations . 162
Resonance tunneling effects 160
Reverse
  charge transfer 163
  saturation current density ...127–128
  tunneling 163–164

Reversible voltammetry 132
Ruthenium (Ru)
  (II)$tris(2,2'$-bipyridine),
    analogs of 79
  (II)$^{2+}tris(2,2'$-bipyridine) 60
  (II) chemisorption on GaAs ... 227
  complexes
    (II)$tris(2,2'$-bpy) photo-
      physics of 94
    $E^{o'}$ values and reduction peak
      potentials for surface-
      attached 266t–267t
    stability of 266
    surface-attached 262–267
    voltammograms of 258f
    of surface-bound 265f
  surfactants 88
  by XPES, detection of 263

S

Saltiel plot of stilbene isomerization 10f
Sandwich cells 124–131
  current–voltage characteristics
    of 126–127f
Saturated fatty acids 49
Scattering time 163
Schottky barrier 235, 274
  cells, photovoltaic activity
    from 123–131
  solar cell 281
  theory 141
Schottky junction 124–131
  cells 224–225
Schrodinger equation 159
SDS (see Sodium dodecyl sulfate)
Selectivity, electrochemical 85
Selectivity transport ability 74
Semiconducting electrode(s) ..139–154
  photoactive 155
Semiconducting oxides, kinetically
  inert n-type 270
Semiconductor(s)
  cells, regenerative 140
  devices 124
  electrode(s)
    dark 187f
    photogenerated hole in
      n-type 161f
    p-type 132
  –electrolyte interfaces, charge
    transfer 155–169
  energetics of photoenhanced
    reduction on p-type 167–168
  induced bending in 164
  interfaces, charge transfer at
    metal- 235
  interface states 281
  –liquid junction chemistry ...215–230
  organic 141
  photoelectrodes, chemically
    derivatized 269–292
  photoelectrodes, luminescent
    properties of 185–211

INDEX

Semiconductor(s) (*continued*)
  photovoltaic cell 114
  surface and solution, ion
    exchange between 218–223
  n-type 41, 141–142, 186
    –electrolyte interface, depletion region at 187f
    electron transfer in 259
    interfacial potential well for .. 159f
    photoanodic decomposition of 270
    a technique for the stabilization of 269–292
  p-type 124, 141–142
  two-stimuli response of 285
Sensitization indices of heterogeneous photosensitizers and models 8t
Sensitization of stilbene isomerization 9–10
Sensitizer(s)
  dimethylaminobenzophenone
    as a 3–5
  grafted to silica or glass 6–7
  iridium complex 11
  photo-, heterogeneous, for
    energy storage 1–12
  polystyrene immobilized 6–12
  self-quenching 1
  silica-functionalized 5–12
  singlet absorption band of 3
  structure of 4f
Sequential hydrolyses 81
Sequential chemical reactions 73–74
Sequestering reagent 20
Shelf life of derivatized
  electrodes 283
Shockley equation, modified ... 126–128
Short-circuit photocurrent density . 130
Short-lived radical detection by
  radical addition reaction 174
Si (*see* Silicon)
Silane
  3-(2-aminoethylamion)propyl-
    tridilane, 3-(2-methoxy) .. 254
  coverage of oxides, determination of 254–256
  multilayers, stability of 261
Silanization reaction 254
Silanization, XPES spectra of polycrystalline TiO_2 electrode
  before and after 255f
Silica-functionalized sensitizers ... 5–12
Silica, sensitizers grafted to 6–7
Silicon (Si)
  dark current-potential properties
    for derivatized 281
  hydrides, reaction of 1-pentene
    with 23–25
  peak widths at half height,
    derivatized 281
  solar cell 124
  –TiO_3, UPS difference spectra for
    CO_2 and CO adsorbed on
    stoichiometric 246f

Silicon (Si) (*continued*)
  n-type
    derivatized 277–288
    in cell configuration 289–292
    with chopped excitation
      beam, voltammograms
      for 283f
    photoelectrodes, current–
      voltage curves for 291f
    photoelectrodes, output
      voltage for 279
    voltammograms for ... 287f, 288f
  electrode, cyclic voltammetry
    of 277–280
  electrode, time dependency of
    photocurrent from 290f
Singlet absorption band of
  sensitizers 3
Sodium
  borohydride 71
  dodecyl sulfate (SDS) 50
    micelles 52
    quantum yield for 52
  lauryl sulfate 98
    micelles 107
  solution, intensity in 105f
Solar cell(s)
  liquid junction 155–156
    shunting of by surface state . 225
  photoelectrochemical 215–230
  Schottky barrier 281
  silicon 124
Solar energy
  conversion 155
  efficiency for n-type GaAs
    cell 269–270, 271
  optimum band gaps for one-
    step 216
  converters 174
  storage system 1–12
Solid–gas interface, photochemistry
  at the 233–252
Solid-state spectra 148–149
Solute penetration at assembly–
  solution interface 55–59
Solution–assembly interface 56
Solution spectra 148–149
Solvation structures, equilibrium .. 163
Solvent
  effects on luminescence 91
  effect on photoionization 108
  interaction(s) 20
  cation– 108
Space–charge components 133
Space–charge region 125, 179
Spectral distribution of luminescence 91f
Spectrophotofluorometer 115, 191
  emission 201
Spectrophotometer 144
Spectroscopic properties of heterogeneous photosensitizers and
  models 8t
Spectroscopy, IR 37

Spectroscopy of metal phthalo-
    cyanine films 139–154
Spectroflourimeter 99
Spin
    coating, deposition by 115
    coating to form multilayers 261
    degenerate ground states 152
    trapping
        application to photosynthesis
            of H_2O_2 180–183
        on aqueous dispersions of CdS 177
        on aqueous dispersions of
            phthalocyanine 177
        to dispersions, application
            of 177–179
        effect of surfactants on 179
        mechanism 178f
        in oxygenated nonaqueous
            dispersions 178
        reactions 174–175
    of short-lived radicals at
        surfaces 173–183
$SrTiO_3$ (see Stronium titanium
    trioxide)
Stability
    to adsorbed impurities, elec-
        trode 224–230
    of doped CdS electrodes 191–194
    of electroactive surfaces 275–277
    of ruthenium complexes 266
Stabilization of n-type semiconduc-
    tors, technique for 269–292
Stainless steel, CO desorption from 33–34
Standing wave 39
Stationary, photo-state composition
    of stilbene isomerization 10f
Steady-state photolysis 100
Stearaldehyde 71
    concentration, calibration curve
    –MBTH, absorption spectra of .. 75f
        for 78
    reduction of 73
    surface viscosity of 76f
Stearyl alcohol 71
Stern layer 52
Stern–Vohner constant 61
Stereoselectivity 54
Stilbene isomerization, sensitiza-
    tion of 9–10
Strontium titanium trioxide
    ($SrTiO_3$)
    adsorption characteristics of CO_2,
        H_2O, CO, and O_2 on 236
    AES peak-to-peak ratios of 240
    CO_2 chemisorption 244
    (111) crystal face, chemisorption
        of H_2O, O_2, CO, and
        CO_2 on 240–247
    crystal surface, properties of ... 240
    –Pt, CH_4 production from H_2O
        and CO_2 over 247–249
    –Pt foil sample............... 239f

Strontium titanium trioxide ($SrTiO_3$)
    (continued)
reduced
    AES spectra of stoichiometric
        and 241f
    by Ar^+ bombardment, UPS
        difference spectra for
        H_2O adsorption on .242–244
    ELS spectra for stoichio-
        metric and 242f
    UPS spectra of stoichio-
        metric and 241f
    single-crystal samples, prepara-
        tion of 238–239
    UPS spectra for H_2O adsorbed on 243f
    UPS spectra for O_2 adsorbed on . 245f
    work function 244
        effect of Ti^{3+} on 242
Styrene–divinylbenzene resin 14–17
Sublimation, deposition by 143
Sulfur and carbon removal from
    Pt samples by heating in O_2 . 239
Superoxide radical(s) 174
    adduct of DMPO 176–177
    –H_2O radical equilibrium 176
Supporting electrolyte, KCl as .142–144
Surface
    acidity probe 83–84
    active network, two-dimensional 74
    analysis chamber, UHV 237f
    –area curves 87, 88f
    -attached
        ferrocene, oxidation of 279
        nitroaromatics 259–261
        oxidant, cathodic return peak
            for 286
        ruthenium complexes 262–267
        $E^{o'}$ values and reduction peak
            potentials for 266t–267t
    -bound ruthenium complexes,
        voltammograms of 265f
    characterization 114
    charge density 81
    cleaning by ion bombardments,
        in situ 236
    composite alteration from elec-
        tron bombardment 240
    composition determination 239
    durability of derivatized 282
    incorporation of impurities,
        beneficial 224
    mechanistic consequences of
        short-lived radicals at ...173–183
    modification, electrode 215
    modified electrodes, tailor-
        made 253–267
    potential 41
    properties of $SrTiO_3$ crystal 240
    quality, electrode 192–193
    reactions 233–252
        methods for following 236

Surface (*continued*)
spin trapping of short-lived
radicals at 173–183
stability of electroactive 275–277
state(s)
effect of chemisorbed ions
on 225–227
electrode 224–230
shunting of liquid-junction
solar cell by 225f
studies, advantages of gaseous
reactants in 235
traps 200
viscosity 71
of stearaldehyde 76f
Surfactant
cationic 87
ester hydrolysis 79–83
molecules
functionalized 47–65
neutral 92–94
photoreactions of functionalized 48–55
products, nonvolatile 51
reactions of
and hydrophobic compounds
at interfaces 47–65
and hydrophobic compounds
in organizates 47–65
monolayer films, characterization and 69–94
ruthenium 88
on spin trapping, effect of 179
tailored 70
Symmetry, e_g 152–153
Symmetry sites in unit cell 150
Synchrotrom radiation 38–39, 42
Synthesis(ses)
photo-, of H_2O_2, application of
spin trapping to 180–183
photo-, yielding small organic
molecules 234
of surface-attached Ru complexes 262t
Synthetic reaction environments 113–136

T

Teflon cell 143
Tellurium-doped CdS electrode 189–211
Temperature
dependence of CO oxidation ...36–37
on emission from CdS, effect of . 207
gradient 29
Tetracene 101
Tetrahydrofuran (THF) 16, 79, 83
Tetramethylbenzidene (TMB) .101–109
cation
polarization energy for 108
N,N'-Tetramethylparaphenylenediamine (TMPD)
cation 102
polarization energy for 108

N,N'-Tetramethylparaphenylenediamine (TMPD) (*continued*)
flash photolysis of 102
ionization potentials of 101
laser photolysis of 102
Tetramethylsilane (TMS) 100–101
Thermal
conduction 40
desorption 40
generation of CH_4 249
velocity, carrier 162
Thermalization 160
time(s) 160
p-GaP electron 164–165
n-TiO$_2$ hole 163
Thermalized injection 168
Thermocouple, iron-constantan ... 30
Thermodynamic potentials for
excited state deactivation
reactions 189
Thermoelectric cold stage 115
THF (Tetrahydrofuran) 16, 79, 83
Thiazine dyes, dimerization of ... 87
Thiazine, photo- 104
Thickness monitor 115
Thin film systems 113–136
antenna effect in 117
Thionine, adsorption of 85
Time
dependence of CH_4 production
over metal sandwich 248f
dependence of photocurrent from
n-type Si electrode 290f
dependence of radical adduct
formation 181f
-dependent relaxation 76
-of-flight methods for desorption detection 38
Tin dioxide (SnO_2)-coated glass .. 143
n-Tin dioxide (n-SnO$_2$)
electrodes, current–voltage
curves for 144–148
Titanium dioxide (TiO_2)
adsorption characteristics of CO_2,
H_2O, CO, and O_2 on 236
electrodes, electrochemical behavior of 259
electrode before and after salanization, XPES spectra of
polycrystalline 255f
modified electrodes 253–267
surface reduced by Ar$^+$
bombardment, UPS
difference spectra for
H_2O adsorption on 242–244
valance band of 161
voltammograms of surfaceattached nitroaromatics on
silanized polycrystalline ... 260f
XPES spectra of polycrystalline,
silanized 257f
–py, XPES spectra of 264f

n-Titanium dioxide
hole
 diffusion times 162
 thermalization times 163
 tunneling times 161–162
 potential well dimensions for ... 162
 relaxation times in electrolyte 163–164
[Ti^{3+}] on $SrTiO_3$ work function,
 effect of 242
Titanium tetraisopropoxide and
 $AlCl_3$ dissolved in 1,2-di-
 methoxyethane 165
Titanium tetraisopropoxide, stepwise reduction of 166
TMB (see Tetramethylbenzidene)
TMPD (see N,N'-Tetramethylpara-
 phenylenediamine)
TMS (Tetramethylsilane) 100–101
Topological control 54
Transition(s)
 dipole 152
 moments 150
 in-plane 151
 long axis 148
 metal(s)
  organometallic complexes ... 13
  reactions over 27–43
  surfaces, CO oxidation to
   CO_2 over 246
 out-of-plane 152–153
 rule, octupole 152–153
 short axis 148–149
 vibronic 152
0–0 Transition 148, 152–153
Transmission coefficient for
 potential barrier 162
Triethylamine 61
Triphenylphosphine 22
Triplet energy
 of dimethylaminobenzophenone . 3
 of norbornadiene 3
 of silica-functionalized sensitizer 7
 of stilbene 9
Tungsten, CO desorption from ... 33
Tunneling
 electron 224
 probabilities 160
 quantum mechanical 159–165
 resonance, calculations 162
 resonance, effects 160
 reverse 163–164
 time(s) 160
  quantum mechanical 162–163
  n-TiO_2 hole 161–162
 thickness, electron 219
n-Type semiconductor(s) ..141–142, 186
 electrodes, photogenerated
  hole in 161f
 interfacial potential well for ... 159f
p-Type semiconductor(s) ..124, 141–142
 electrodes 132
 energetics of photoenhanced
  reduction on 167–168

U

Ultrahigh vacuum (UHV)
 chamber 236
 photoelectron spectroscopy
  (UPS) 236, 238, 240
 difference spectra
  for CO_2 and CO adsorbed
   on stoichiometric
   $SiTiO_3$ 246f
  for H_2O adsorption on
   $SrTiO_3$ reduced by
   Ar^+ bombardment ..242–244
  for H_2O adsorption on
   TiO_2 surface reduced
   by Ar^+ bombardment ...242–244
 spectra
  for H_2O absorbed on
   $SrTiO_3$ 243f
  for O_2 adsorbed on $SrTiO_3$ 245f
  of stoichiometric and
   reduced $SrTiO_3$ 241f
 surface analysis chamber 237f
Ultrasonic dispersion of pigments . 175
Unit cell, symmetry sites in 150
Unsaturation, coordinative 13–25
UPS (see Ultrahigh vacuum
 photoelectron spectroscopy)

V

Vacuum (UHV) chamber,
 ultrahigh 236
Valence band
 of CdS 219
 bending of p-GaP 164
 photogenerated holes in 282
 structure 240
 of TiO_2 161
Valerophenone 52
van der Waals forces 124
van der Waals–London interactions 87
Vapor-phase chromatography
 (VPC) 71, 73–74, 92
 analysis 78f, 79
 –mass spectroscopy 74
Vesicles 47
Vibration, e_g 153
Vibronic
 assignment of phthyalocyanines . 153
 selection rule 152
 series 148
 transitions 152
Viologen, methyl 178
Viscosity, surface 71
Voltage, current–
 characteristics of sandwich
  cells 126f, 127f
 curves 143
 for CdSe 222f
 for $CuInS_2$ cell 223f
 for derivatized n-Si photoelectrode 291f
 for etched GaAs 228f

Voltage current (*continued*)
  curves (*continued*)
    for idealized cell 226f
    for n-SnO$_2$ electrodes 144–148
    density curve for GaAs 230f
    —emission curves for doped
      CdS 202f, 203f, 205f
    —emission properties of doped
      CdS 200–206
Voltage, photo-
  open-circuit 130
  limiting 216
  in sandwich cells, light
    intensity vs. 131f
Voltaic, photo-
  activity from Schottky barrier
    cells 123–131
  cell, semiconductor 114
  cells, electrochemical 155–156
  devices 139–141
Voltammetric behavior with
  redox couples 224
Voltammetry
  cyclic
    of derivatized electrodes in
      nonaqueous electrolyte 275–277
    of n-type Si electrode 277–280
  reversible 132
Voltammograms 132
  complex(es)
    of a chromium 258f
    of iron, cyclic 133f
    of ruthenium 258f
    of surface-bound ruthenium . 265f
  for derivatized n-type Si ... 287f, 288f
  for derivatized n-type Si with
    chopped excitation beam .. 283f
  of surface-attached nitroaro-
    matics on silanized poly-
    crystalline TiO$_2$ 260f
  for n-type Ge 284f
VPC (*see* Vapor-phase chromatography)

W

Water (H$_2$O)
  adsorbed on SrTiO$_3$, UPS
    spectra for 243f
  adsorption 242
  on SrTiO$_3$ reduced by Ar$^+$
    bombardment, UPS dif-
    ference spectra for ...242–244
  on TiO$_2$ surface reduced by
    Ar$^+$ bombardment, UPS
    difference spectra for .242–244

Water (H$_2$O) (*continued*)
  band gap of 158
  CH$_4$ production from CO$_2$ and . 236
  CO, CO$_2$, and O$_2$ on SrTiO$_3$,
    adsorption characteristics .. 236
  and CO$_2$ over SrTiO$_3$–Pt, CH$_4$
    production from 247–249
  dissociation of 234
  in an electrochemical cell,
    dissociation of 235
  electron affinity of 158
  O$_2$, CO, and CO$_2$ on
    SrTiO$_3$(111) crystal face,
    chemisorption of 240–247
  on Pt–SrTiO$_3$, adsorption and
    reactions of CO$_2$ and ... 233–252
  on Pt surfaces, chemisorption of 247
Wave function 159–160
Wavelength dependence of emission and photocurrent from
  doped CdS 199t
Wavelength on desorption, effect
  of 33f, 35f
Wilhelmy balance 73
Work function(s)
  effect of [Ti^{3+}] on SrTiO$_3$ 242
  metal 124–125, 141, 158
  SrTiO$_3$ 244

X

X-Ray photoelectron spectroscopy
  (XPES) 254, 259, 261, 266
  amidization to study, use of 256
  detection of ruthenium by 263
  spectra
    of polycrystalline, silanized
      TiO$_2$ 257f
    of polycrystalline TiO$_2$ elec-
      trode before and after
      after silanization 255f
    of Pt–PtO–py 264f
    of TiO$_2$–py 264f
p-Xylene, pulse radiolysis of 102
p-Xylene triplet state 102

Z

Zinc phthalocyanines (ZnPc) ..124–131
  electrode, current-voltage curves
    for 145f, 147f
  MCD spectra of 151f

Jacket design by Carol Conway.
Editing and production by Robin Allison.

The book was composed by Service Composition Co., Baltimore, MD, printed and bound by The Maple Press Co., York, PA.